DATE OF RETURN

POWER TRANSMISSION AND VIBRATION CONSIDERATIONS IN DESIGN

DESIGN TECHNOLOGY SERIES

Power Transmission and Vibration Considerations in Design

Terance V. Duggan

A.R.T.C.S. (Hons.), C.Eng., M.I.Mech.E., M.I.E.D., F.R.S.A.

Senior Lecturer in Mechanical Engineering
(Group Leader – Fatigue Research Group,
Portsmouth Polytechnic

LONDON
ILIFFE BOOKS

THE BUTTERWORTH GROUP

ENGLAND

Butterworth & Co (Publishers) Ltd
London: 88 Kingsway, WC2B 6AB

AUSTRALIA

Butterworth & Co (Australia) Ltd
Sydney: 586 Pacific Highway Chatswood, NSW 2067
Melbourne: 343 Little Collins Street, 3000
Brisbane: 240 Queen Street, 4000

CANADA

Butterworth & Co (Canada) Ltd
Toronto: 14 Curity Avenue, 374

NEW ZEALAND

Butterworth & Co (New Zealand) Ltd
Wellington: 26-28 Waring Taylor Street, 1
Auckland: 35 High Street, 1

SOUTH AFRICA

Butterworth & Co (South Africa) (Pty.) Ltd
Durban: 152-154 Gale Street

First published in 1971 by
Iliffe Books, an imprint of
the Butterworth Group

ISBN 0 592 00050 8 Standard
0 592 00051 6 Limp

Printed in Hungary

Contents

Preface 7

Nomenclature *11*

1 Introduction to Power Transmission Methods 1

2 Vibration Considerations in Design 10

3 Design of Shafts and Couplings 52

4 Bearings, Wear and Lubrication Considerations in Design 93

5 Flexible Power Transmission Elements 180

6 Design of Gears 204

7 Brakes and Clutches 287

Appendix A Keys, Splines and Serrations 316

Appendix B Material Properties and Design Data Sheets for Gears 339

Subject Index 357

Preface

This second book in the series Design Technology is complementary to my first book entitled *Applied Engineering Design and Analysis*, and many of the objectives are naturally the same. It is intended primarily for students of engineering, and should, ideally, be read in conjunction or following *Applied Engineering Design and Analysis*. However, the present book has been written so that it may, if desired, be read as a separate text, and is quite complete in itself. I hope that as well as satisfying parts of the syllabus in Engineering Design for a variety of courses such as degree, diploma and H.N.C., the book will be found useful by many practising engineers. My original intention was to write a single book which I had hoped would be suitable for the design contents of most undergraduate courses, but it soon became evident that any such attempt could only treat the subject in a somewhat superficial manner, without producing an extremely lengthy text. Further, the content in the design syllabus of one university or college is likely to differ considerably from that of another, since the subject of engineering design it by no means as well established as most other subjects of mechanical engineering. However, it is hoped that the volumes in this series will meet the needs of the many and varied design syllabuses in undergraduate courses, and that they will, to some extent, provide something of a basis for future undergraduate courses in the subject.

It is assumed that the reader has an elementary knowledge of applied mechanics and mathematics, and that the subject of engineering design will be studied simultaneously with other subjects of engineering.

In the present text, as the title suggests, I have limited the discussions to power transmission and vibration consideration, the object being to treat these topics in some depth, whilst at the same time maintaining the emphasis on a methodical design procedure. Particular attention has been given to the conversion of a physical system into a mathematical model, and the importance of design decisions and the assumptions necessary in analysis have once again been stressed.

Following the first chapter, in which a general introduction to power transmission methods is given, I have attempted to discuss in detail vibration considerations; the design of shafts and couplings; bearings, wear and lubrication; flexible power transmission elements; the design of gears; and finally clutches and brakes. These topics may be quite properly considered the 'bread and butter' of machine design.

The essential purpose of this book, like the first in the series, is a sincere attempt to establish a respectable position for engineering design. In attempting this, I am even more aware than ever of its limitations, and I can only hope that, by the time the series is complete, a coherent philosophy will have emerged which, through the consideration of real engineering (as distinct from a consideration of design philosophy itself) contributes towards the acceptance of Engineering Design as a discipline in its own right.

This book is not intended, nor presumes, to supersede the various excellent texts dealing with particular aspects of power transmission elements, lubrication and bearings, and vibration problems. However, I believe that each chapter is sufficiently comprehensive to enable the material to be useful not only from a students viewpoint in learning something about design, but also from the practising engineer's viewpoint in application to real design problems. To enable any particular aspect to be studied in greater depth, each chapter has a carefully selected and comprehensive bibliography.

A dual system of units has been used throughout the text, namely the S.I. metric system and the traditional English system. Further, although reference is made in particular to methods used in the United Kingdom, the subject matter has been presented in such a way that complete generality has been maintained, and its use in other countries is in no way prohibited.

I should like to express my gratitude to all those who have helped to make this book possible. I am once again indebted to the authors of standard texts, references and special papers, too numerous to mention individually, but who throughout the years have made available information and data so valuable to the designer, and which is frequently taken for granted. In producing this book I have received considerable co-operation from various bodies, particularly with regard to the provision of photographic and other art material, and for permission to reproduce such material and other extracts in the text of this book, I am sincerely grateful. Accordingly, my thanks go to the following:

Crofts (Engineers) Ltd., Bradford, England,
National Aeronautics and Space Administration, Washington, D.C., U.S.A.,
British Standards Institution, London, England,
Messrs. A.A. Raimondi and J. Boyd of the Westinghouse Electric Corporation,
Research and Development Centre, Pittsburgh, U.S.A.,
The Skefko Ball Bearing Company, Ltd., Luton, Beds., England,
J. H. Fenner & Co. Ltd., Hull, England,
R.W. Transmissions Ltd., Hebburn, Co. Durham, England,
David Brown Gear Industries Ltd., Huddersfield, England,
Ferodo Ltd., Stockport, England.

I must also express my gratitude to my Publishers, for their co-operation during the preparation of this book, particularly for their patience and understanding concerning the various delays which have taken place.

Finally, I am grateful to my wife Joy, for her patience and understanding during the preparation of this work, without which I doubt whether the book would ever have been completed.

Terance V. Duggan

Nomenclature

A area
constant
A_s exterior surface area dissipating heat
a amplitude of vibration
centre distance
material constant
a_r modified or extended centre distance
B constant
b linear dimension
width
b_{e2} effective facewidth of wormwheel
b_n face width in normal plane
b_{r2} length of root of wormwheel teeth
C centre distance
constant
diametral clearance
dynamic capacity of rolling contact bearings
thermal capacity per unit mass (specific heat capacity)
$[C]$ damping coefficient (matrix notation)
C_f face width factor
C_{fn} face width factor in normal plane
C_L loading factor
C_s size factor
C_v velocity factor (spur gears)
c bottom tooth clearance
damping coefficient
linear dimension
material and geometric constant
radial clearance in journal bearing
c_c critical damping coefficient
D diameter
linear dimension
DE dissipative energy
d diameter
linear dimension
d_{a1} tip diameter of pinion
d_{a2} tip diameter of gear
d_{b1} base circle diameter of pinion
d_{b2} base circle diameter of gear
d'_1 pitch diameter of pinion
d'_2 pitch diameter of gear
d'_{1n} diameter of pinion referred to normal plane
d'_{2n} diameter of gear referred to normal plane
d_{vm} virtual mean diameter
E modulus of elasticity
E_p potential energy
E_r rotational kinetic energy
E_t translational kinetic energy
e journal displacement
eccentricity
F applied force
F_a accelerating force
axial bearing load
F_b allowable tangential tooth load per unit face width based on bending
F_c allowable tangential tooth load

	per unit face width based on wear
F_d	dynamic tooth load
F_i	initial pre-load
F_m	equivalent mean load acting on rolling contact bearing
F_n	normal force
F_o	applied force
F_r	radial bearing load
	radial or separating force
F_s	separating force
F_t	force transmitted to foundation
	tangential tooth force
$F(t)$	periodic disturbing force
F_w	allowable tangential tooth load per unit face width based on wormwheel wear
F_x	belt or chain force factor
F_2	force required to deform tooth amount of effective error
f	coefficient of friction
	frequency of vibration, vib/s (Hz)
f'	virtual coefficient of friction
G	modulus of rigidity or shear modulus
g	gravitational constant
H	heat generated or heat dissipated
h	film thickness
	linear dimension
$\bar{h}$	film thickness corresponding to maximum value of pressure
h_{a1}	addendum of pinion
h_{a2}	addendum of grear
h_{f1}	dedendum of pinion
h_{f2}	dedendum of gear
h_0	minimum oil film thickness
hp	horse power
I	mass moment of inertia
	second moment of area
I_m	equivalent second moment of area
J	Joule's equivalent
	polar mass moment of inertia
	polar second moment of area
K	coefficient of heat transfer
	material and form factor
	torsional stiffness
KE	kinetic energy
K_a	axial load acting on bearing
K_f	fatigue strength reduction factor
K_s	surface finish factor
K_t	theorectical stress concentration factor
$[K]$	stiffness coefficient (matrix notation)
K_1	geometrical constant
K_2	geometrical constant
k	constant in McKee equation
	face width factor
	radius of gyration
	spring stiffness
k_1	correction factor for pinion
k_2	correction factor for gear
L	bearing life
	linear dimension
	total belt length
l	axial length of bearing
	linear dimension
M	bending moment
	mass
MF	magnification factor
$[M]$	inertia coefficient (matrix notation)
M_t	twisting moment or torque
m	mass
	mass of belt per unit length
	module
N	frequency of vibration, vib/min
	normal force
	speed of rotation, rev/min
N_c	critical or whirling speed, rev/min
N_L	Ocvirk load number
n	constant
	integer or number
	number of braking or clutching operations
	speed of rotation, rev/s
P	axial force
	applied external force
	belt or chain pull
	diametral pitch
	equivalent radial load acting on rolling contant bearing
	radial force on journal bearing
PE	potential energy
P_n	normal diametral pitch
p	angular frequency, rad/s
	circular pitch
	exponent in rolling contact bearing life equation
	pitch of chain sprocket teeth
	pressure intensity
p_b	base pitch
p_n	normal circular pitch
p_x	axial pitch of helical gear
p_z	lead of worm thread

p_{max} maximum value of pressure intensity
Q shear force
interference factor for bevel gears
ratio factor
volume of fluid flowing
Q_i generalised external force
Q_s quantity of oil lost by side leakage
q diameter factor or diameter quotient
notch sensitivity index
q_i generalised co-ordinate
$\dot{q}_i$ dq_i/dt = generalised velocity
$\ddot{q}_i$ d^2q_i/dt^2 = generalised acceleration
R cone distance
reaction to external forces
R_m mean radius
RF reserve factor (or factor of safety)
r radius
r_m pitch radius of tooth at mid-point
S Sommerfeld number
S_a intrinsic fatigue strength
S_a' corrected fatigue strength
S_b bending strength factor for gear material
S_c surface fatigue strength for gear material
S_d design strength
S_e intrinsic fatigue strength at endurance limit
S_e' component fatigue strength at endurance limit
S_p yield or proof strength
S_{su} ultimate shear strength
S_u ultimate tensile strength
S_w working strength
s gear tooth thickness
T average belt tension
number of teeth in chain sprocket
thickness
T_1 tension on tight side of belt or band brake
T_2 tension on slack side of belt or band brake
T_c centrifugal tension in belts
T_F transmissibility
t linear dimension
thickness of belt
thickness of key
time
U velocity
u gear ratio
velocity
V rotational factor for rolling contact bearings
velocity of belt or chain
v velocity
W power (watts)
width
w width
X component of force in x direction
radial load factor for rolling contact bearings
X_b speed factor for strength
X_c speed factor for wear
x linear dimension
linear displacement
$\dot{x}$ dx/dt = linear velocity
$\ddot{x}$ d^2x/dt^2 = linear acceleration
Y component of force in y direction
thrust load factor for rolling contact bearings
strength factor for gears
Y_1 strength factor for helical gears
y linear dimension
linear displacement of deflection
Lewis form factor
y_n Lewis form factor for formative gearset
y_o amplitude of vibration
y_{st} static deflection or displacement
Z dynamic or absolute viscosity of lubricant
section modulus in bending
zone factor (gearing)
Z_t section modulus in torsion
z linear or angular displacement
z_1 number of teeth in pinion
z_2 number of teeth in gear
z_{1n} number of teeth in pinion referred to formative gearset
z_{2n} number of teeth in gear referred to formative gearset
α angle of lag
half V-belt pulley groove angle
pressure angle
α_{ij} flexibility influence coefficient
α_n normal tooth pressure angle in axial plane
α_r running pressure angle for gears
α_x worm thread pressure angle
β angular measurement
constant

helix angle
spiral angle
β' spiral angle at the pitch circle
γ lead angle of worm thread
mass density
γ_M slope due to application of unit moment
γ_Q slope due to application of unit force
δ radial wear of friction material
static deflection
δ'_1 pitch angle of bevel gear pinion
δ'_2 pitch angle of bevel gear wheel
ε eccentricity ratio
ε_β face contact or overlap ratio
ε_γ profile contact ratio
ζ damping ratio
η angular damping coefficient
θ angle of lap
angular displacement
slope
temperature
$\Delta\theta$ change in temperature
$\dot{\theta}$ $d\theta/dt$ = angular velocity
$\ddot{\theta}$ $d^2\theta/dt^2$ = angular acceleration
λ constant
μ dynamic or absolute viscosity of a fluid
ν kinematic viscosity of a fluid
Σ shaft angle
summation
σ direct or normal stress
σ_c direct or normal compressive stress
σ_m mean normal stress
σ_s equivalent static normal stress
σ_{alt} alternating normal stress
τ shear or torsional shear stress
periodic time
τ_{alt} alternating shear stress
τ_m mean shear stress
τ_s equivalent static shear stress
φ constant
function
position of minimum oil film thickness
ψ phase angle
Ω_b spiral angle factor
Ω_c overlap ratio factor
ω angular velocity, rad/s

1.4. FLEXIBLE POWER TRANSMISSION ELEMENTS

Flexible power transmission methods are generally considered to mean belt, rope and chain drives. Belt and chain drives have some distinct advantages over geared drives, in that they are ideally suitable for relatively long centre distances, where a positive and definite velocity ratio is not too important, and they have the ability to absorb shock and vibration. However, probably the major avantage of these flexible elements is the considerable saving in cost associated with them. Unlike gears which are generally specifically designed (although standard gears are available from various manufacturers'), belt and chain drives are readily available as standard commercial units.

It is useful to consider gears as a positive and rigid method of transmitting power; belts (which may be subject to slip, particularly flat belts at high speed) as a very flexible and usually not very positive method of transmitting power; and chains as intermediate between the two, since they are capable of providing a more compact and positive drive than belts, but they require more accurate alignment of the shafts and need some form of lubrication.

The main application of wire rope is in the field of haulage and hoisting, where the high strength to weight ratio makes it generally more suitable than hemp or the common link chains. It is not now normally used for conventional power transmission as understood here.

One final element which should be mentioned in connection with flexible elements is the common flexible shaft. This overcomes the disadvantage associated with a solid shaft in that it is capable of transmitting power round a corner, without the necessity for costly parts, such as gears, belts or chains. Such elements are limited to low power requirements, and are commercially available as standard parts; selection should be made from manufacturers' catalogues.

1.5. LUBRICATION AND BEARINGS

Any power transmission shaft must be supported in such a manner that relative motion between the shaft and the supports in the direction of the applied load (force) is prevented, whilst being free to rotate with the minimum amount of friction. This is achieved by the use of bearings, which may be either so-called *plain*, *sliding* or *sleeve*, or may be by bearings which have *rolling contact* elements between the surfaces; these latter type of bearings are sometimes referred to as *anti-friction* and include the common ball and roller bearings. With sliding bearings, the designer is frequently called upon to design them from first principles (although more and more standard types are becoming available), whilst for *rolling contact bearings*, the selection is made from manufacturers' catalogues, the designer being required to make the selection based upon the requirements of the nature and type of load, speed and life.

Whenever relative motion between two parts occurs, the friction created generates heat and wear of the parts. If metal to metal contact takes place, then *dry* or *Coulomb friction* results, and if the frictional forces developed are sufficiently high, breakdown of the surfaces or seizure may occur. Common examples of machine elements in which relative motion between moving parts takes place are to be found in cam mechanisms, between the cam follower and the cam profile, between mating gear teeth, and between the elements of the bearings mentioned above. In order to reduce the amount of friction between moving parts, with a consequent reduction in the heat generated and wear of the parts, the *dry friction* is replaced by the molecular friction which occurs in a *lubricant*. A lubricant may be defined as any substance which, when inserted between the moving surfaces, accomplishes the purpose of reducing friction[4].

1.6. FLYWHEELS

When energy is being transmitted from one machine or machine element to another, sometimes the flow of energy is not constant, either because the power delivery source varies, or because the power demand from the delivery source is not constant, or possibly a combination of the two. In such instances, it may be desirable to keep the energy fluctuation to a minimum, and one way of achieving this is by the inclusion of a flywheel in the system. A flywheel is a machine element capable of storing and delivering kinetic energy, depending upon the supply and demand requirements, and such an element is designed depending upon the co-called coefficient of speed fluctuation which is permited. In effect, it may be considered as a mechanical filter, which serves to absorb energy when there is a surplus, and to supply energy (from the kinetic energy stored) when the need is greatest.

1.7. MECHANISMS

In addition to the above mentioned machine elements concerned primarily with the transmission of power, the design engineer must often concern himself with other mechanisms and devices for converting one type of motion into another. From a purely academic point of view, such mechanisms are usually dealt with under the subject Mechanics or Theory of Machines, in which various linkages are discussed analytically, and the motion of such mechanisms is studied. From these studies, the displacements, velocities and accelerations can be determined. If the various members in the mechanism have both mass and geometry, the resulting inertia forces and consequential stresses and deformations (elastic) can also be established, and the analysis yields the suitability or otherwise of the arrangement. However, this is a problem in analysis which, although a major part of the design process[1], is not usually the way in which a problem presents itself in a design situation. Rather, it is the reverse process which confronts the designer, i.e. the synthesis of mechanisms, in which a suitable arrangement and geometry of parts (not to mention material selection and production methods) to satisfy a particular type of motion, must be established;

as already mentioned the analysis, although a very important aspect, is nevertheless not the whole problem. The synthesis of mechanisms is a most involved subject, and that this is so can be seen from consideration of the design of a cam mechanism, as illustrated already by the author in a previous book[1]. The power screw, is another common mechanism and is a device used for converting rotary motion into linear motion. For a more involved and complete study of the synthesis of mechanisms, the reader is referred to Ruzinov[5] and Hall[6].

1.8 CONCLUDING REMARKS

In this somewhat brief introductory chapter, an attempt has been made to familiarise the reader with the more important methods of power transmission, but this has been attempted in only a superficial manner. Rather, the intention has been to make the nature of the design process clear, and to point out the various advantages and disadvantages associated with the various elements mentioned. A rather interesting and unusual approach to machines, energy and power is given by Dr. Duncan Morrison[7], and it is suggested that the reader refer to this book.

In the following chapters, the elements discussed briefly in this chapter will be dealt with in some detail, so that at the completion of the book, it is hoped that the reader will be in a position to apply a methodical approach to the design of systems for the transmission of power.

REFERENCES

1. DUGGAN, T. V., *Applied Engineering Design and Analysis, Design Technology Series*, Iliffe Books Ltd., London (1970)
2. DIXON, J. R., *Design Engineering: Inventiveness, Analysis and Decision Making*, McGraw-Hill Book Company, New York (1966)
3. TUPLIN, W. A., *Gear Design*, The Machinery Publishing Co. Ltd., London (1962)
4. SHIGLEY, J. E., *Machine Design*, McGraw-Hill Book Company, New York (1956)

5. RUZINOV, L. D., (edited by R. L. Napier), *Design of Mechanisms by Geometric Transformation*, Iliffe Books Ltd., London (1968)
6. HALL, A. S., *Kinematics and Linkage Design*, Prentice-Hall Inc., Englewood Cliffs, N. J. (1961)
7. MORRISON, D., *Engineering Design: the Choice of Favourable Systems*, McGraw-Hill Publishing Co. Ltd., London (1968)

Chapter 2

Vibration Considerations in Design

Perhaps one of the most important concepts with which the design engineer must be familiar is that relating to vibrations in machines and structures. In most instances, vibratory motion is an undesirable feature, although sometimes it may be a deliberately introduced design feature, as for example a fatigue testing machine which operates on the resonance principle. The subject of vibrations represents a good example of how the design engineer must formulate a mathematical model from a physical concept, in such a way that the mathematical formulation represents the physical system in a realistic manner, and at the same time permits of an analysis in a reasonable period of time. Any elastic system having the elements of mass and elasticity and subjected to the application of a disturbing force is capable of vibrating, and the engineer must so design the system that such vibrations which do occur are not damaging; in the main, the engineer must design against vibration.

In a single chapter in a book dealing with design technology, it is not possible to cover the whole field of vibration; indeed, it would be foolish to try. Nevertheless, the designer will be called upon to utilise some vibration concepts in many design situations, and this chapter will therefore review some of the more important methods of analysis.

2.1. IDEAL MATHEMATICAL MODEL

The elements in the mathematical model representing the actual physical machine or structure can be divided into (a) mass, which is assumed perfectly rigid, (b) elasticity, which is assumed massless and non-dissipative and (c) damping or dissipation, assumed massless and inelastic. These elements of the mathematical model are described by Crede[1] and should already be familiar to the reader.

When an elastic body is displaced from its position of equilibrium by the application of an external force, work is done in causing the displacement, and this work is stored in the body in the form of strain energy. Provided that yielding of the material does not occur, almost all of the work is recoverable, and when the external force is removed, the energy in the system is dissipated in an attempt to restore the body to its equilibrium position; mechanical vibrations consequently ensue. If, on removal of the external force, the only forces acting on the body are the internal elastic forces, the vibrations are termed *free vibrations*. It is an observed fact, however, that the vibration amplitude gradually decreases with time until eventually the system comes to rest again in its equilibrium position. The reason for this is that the energy in the system is used in overcoming internal and external resistance to the motion. Where such resistances damp the vibrations, the system is said to execute *damped vibrations*. When a system is repeatedly subjected to an applied force, i.e. a periodic force, the ensuing vibrations are termed *forced vibrations*.

The simplest ideal mathematical model, representing the elements of mass, elasticity and damping, is indicated in Fig. 2.1.
In this model, a mass m is subjected to a displacement y by a periodic disturbing force $F(t)$. The elasticity of the system is represented by the spring having a stiffness k, and the damping by the dashpot (indicating viscous damping) having a damping coefficient c. Viscous damping permits of mathematical analysis, since in this instance it may be assumed that the damping is directly proportional

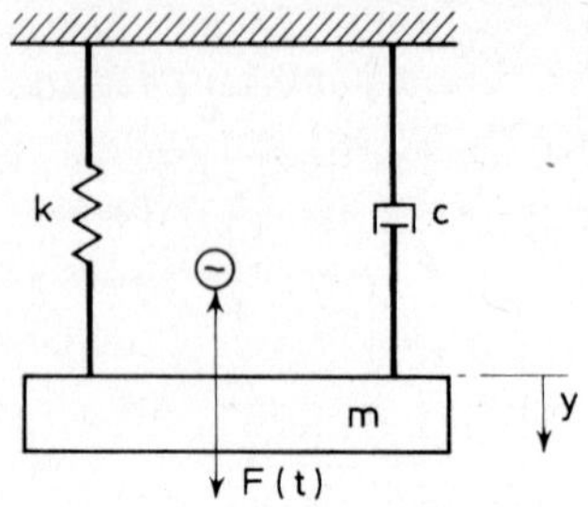

Fig. 2.1. Ideal mathematical model

to the velocity. Consequently, when the damping is other than viscous, it is usual to represent the total damping in the system by the so-called *equivalent viscous damping*.

2.2. NATURAL FREQUENCY OF A SYSTEM

When the vibration of a system continues after the removal of the disturbing force, such a vibration is termed a *natural vibration*, and the number of vibrations or cycles which occur in unit time is termed the *natural frequency;* the time for a single cycle is termed the *period* of vibration, and this is the reciprocal of the natural frequency. If a system is caused to continue vibrating due to a periodic disturbing force, such vibrations are termed *forced vibrations*. When the forcing frequency is equal to the natural frequency of the system, a condition of resonance occurs, and if the damping in the system is small, the amplitude of vibration tends to infinity and the system is in danger of destroying itself. Consequently, the design engineer must usually ensure that a condition of resonance does not occur, although sometimes it is used as a specific design factor, in which case sufficient damping must be provided to limit the amplitude to an acceptable value.

The basic equations for free vibrations are established in the theory for simple harmonic motion[2], and expressed mathematically

$$\frac{d^2z}{dt^2}+p^2z = 0 \tag{2.1}$$

where z is a linear or angular displacement, and p is a constant for any particular system, and is termed the *angular frequency*. This term is used because simple harmonic motion can be represented by the projection from the end of a rotating vector having a length equal to the amplitude of vibration (or oscillation), and having an angular velocity p rad/s. Solving equation 2.1 by the usual methods of linear differential equations, it is found[2] that

$$z = a \sin(pt+\psi) \tag{2.2}$$

where a and ψ are arbitrary constants representing the amplitude of vibration and the phase angle of motion respectively. The *period* of vibration is given by

$$\tau = \frac{2\pi}{p} \text{ s} \tag{2.3}$$

and the *frequency* of vibration is

$$f = \frac{1}{\tau} = \frac{p}{2\pi} \text{ vib/s (Hz)} \tag{2.4}$$

The various types of free vibrations commonly encountered have been discussed by the author elsewhere[2].

2.3. DEGREES OF FREEDOM

When a system is restrained so that it can move in only one direction, its position at any instant is completely specified by one parameter, measured from the position of equilibrium. Such a system is said to possess *one degree of freedom*, and only one mode of vibration is possible.

If a system is restrained in such a manner as to allow displacement in two directions, so that its position and motion are completely specified by two independent quantities, such a system will have two degrees of freedom, and two distinct modes of vibration are possible.

In general, the number of degrees of freedom of a system is specified by the number of independent quantities required to define its position and motion. For systems possessing more than one degree of freedom, a solution is usually obtained by establishing the differential equation of motion for each mode and solving simultaneously. There will, of course, be as many modes of vibration as there are degrees of freedom.

2.4. FORCED-DAMPED VIBRATION FOR SYSTEM HAVING ONE DEGREE OF FREEDOM

The differential equation of motion for the idealised system shown in Fig. 2.1 (assuming that the mass m is restrained to move only in the vertical direction, as indicated) is obtained by the application of Newton's second law. Thus,

Accelerating Force + Damping Force + Restoring Force
= Periodic Disturbing Force

Stated mathematically

$$m\frac{d^2y}{dt^2}+c\frac{dy}{dt}+ky = F(t) \tag{2.5}$$

If the periodic disturbing force is given by $F(t) = F_o \cos \omega t$, then equation (2.5) may be expressed in the more usual form[3]

$$\frac{d^2y}{dt^2}+2n\frac{dy}{dt}+p^2y = \frac{F_o}{m}\cos \omega t \tag{2.6}$$

where

$$p^2 = \frac{k}{m} \tag{2.6a}$$

$$2n = \frac{c}{m} \tag{2.6b}$$

p being the undamped natural angular frequency and ω the angular forcing frequency.

The complete solution to equation 2.6 consists of two parts, the complementary function and the particular integral. The *complementary function* may have three different solutions depending upon whether $n > p$, $n = p$ or $n < p$. When $n > p$, the damping is said to be *heavy;* when $n = p$ the damping is said to be *critical;* when $n < p$ (the most practical case) the damping is termed *light*. Thus, the solutions for these three cases can be shown[3] to be given by the following complementary functions:

For *heavy* damping:

$$y = (A_1 \exp(\lambda_1 t) + A_2 \exp(\lambda_2 t) \tag{2.7}$$

where $\lambda_1 = -n + \sqrt{(n^2 - p^2)}$

and $\lambda_2 = -n - \sqrt{(n^2 - p^2)}$

A_1 and A_2 are constants of integration which may be determined from the boundary conditions.

For *critical* damping:

$$y = (A_1 t + A_2) \exp(-nt) \tag{2.8}$$

where A_1 and A_2 are again constants of integration which may be obtained from the boundary conditions.

For *light* damping:

$$\begin{aligned} y &= \exp(-nt)(A_1 \cos \beta t + A_2 \sin \beta t) \\ &= C \exp(-nt) \sin(\beta t + \varphi) \end{aligned} \tag{2.9}$$

where $\beta = \sqrt{(p^2 - n^2)}$

and C and φ are constants to be determined from the boundary conditions.

The *particular integral* for equation 2.6 can be shown[3] to be given by the equation

$$y = \frac{F_0}{m\sqrt{[(p^2 - \omega^2)^2 + 4n^2\omega^2]}} \cos(\omega t - \alpha) \tag{2.10}$$

where $$\alpha = \tan^{-1} \frac{2n\omega}{p^2 - \omega^2} \tag{2.11}$$

and is the angle of lag, i.e. the forced vibration lags behind the periodic force by the angle α.

An alternative, and somewhat more modern, method of representing equation (2.10), is in terms of the *damping ratio* or *fraction of critical damping*, designated by the Greek symbol ζ, and defined as the ratio of the actual damping in the system to the critical damping[4]. Thus,

$$\zeta = \frac{c}{c_c}$$

where c = damping in the system, and
c_c = critical damping of the system
$= 2\sqrt{(k\,m)}$

Re-arranging equation 2.10 and writing it in terms of ζ

$$y = \frac{F_0/k}{\left\{\left[1-\left(\frac{\omega}{p}\right)^2\right]^2+\left[2\zeta\frac{\omega}{p}\right]^2\right\}^{1/2}}\cos(\omega t-\alpha) \qquad (2.12)$$

and

$$\alpha = \tan^{-1}\frac{2\zeta\left(\frac{\omega}{p}\right)^2}{1-\left(\frac{\omega}{p}\right)^2} \qquad (2.13)$$

The *complete solution* to equation 2.6 consists of the complementary function, expressed by one of equations 2.7, 2.8 or 2.9, and the particular integral given by equation 2.10 or 2.12. It is very important to note that during the *transient* condition, the complementary function must be included in the solution. However, the complementary function, which represents the damped vibrations of the body, is soon destroyed by friction and the *steady state* vibration which remains is specified by equation 2.12. The *amplitude* of vibration, i.e. the maximum value of y, say y_0, clearly occurs when $\cos(\omega t-\alpha) = 1$, i.e.

$$y_0 = \frac{F_0/k}{\left\{\left[1-\left(\frac{\omega}{p}\right)^2\right]^2+\left[2\zeta\frac{\omega}{p}\right]^2\right\}^{1/2}} \qquad (2.14)$$

Now since F_0/k is the static deflection of an elastic body subjected to an applied force F_0, it is clearly seen from the above relationship that the amplitude of vibration may be obtained by multiplying the static deflection by a factor, termed the *magnification factor* (abbreviated *MF*), such that

$$y_0 = y_{st} MF$$

where
$$y_{st} = \frac{F_0}{k}$$

and
$$MF = \frac{1}{\left\{\left[1-\left(\frac{\omega}{p}\right)^2\right]^2 + \left[2\zeta\frac{\omega}{p}\right]^2\right\}^{1/2}} \quad (2.15)$$

When the forcing frequency ω is equal to the natural frequency of the system p, a condition of *resonance* occurs, and this is usually a

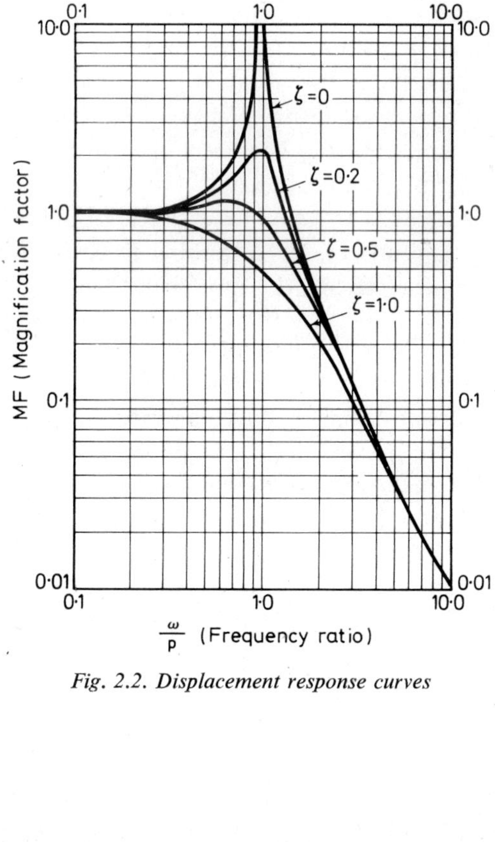

Fig. 2.2. Displacement response curves

condition to be avoided. The value of MF depends upon the ratio of (ω/p) and also upon the damping ratio ζ. In the majority of practical cases, the damping is generally small and may usually be neglected without significant error. For this instance the magnification factor becomes

$$MF = \frac{1}{|1-(\omega/p)^2|} \tag{2.16}$$

If the value for the magnification factor corresponding to different values for the ratio (ω/p) is plotted, for different values of ζ, the curves so produced are termed *displacement response* curves; such curves are shown in Fig. 2.2.

2.5. FORCE TRANSMISSIBILITY AND VIBRATION ISOLATION

It is frequently desirable to reduce the amount of force transmitted to the foundation, due to the vibration of machinery, and this may be achieved by mounting the machines on *vibration isolators*, in the form of springs and dampers. The resulting force transmitted to the foundation is consequently the sum of the spring force and the force due to damping. The ratio of the amplitude of the transmitted force to the amplitude of the disturbing force is defined as the transmissibility, designated T_F. Thus, stated mathematically,

$$T_F = \frac{F_t}{F_0}$$

where F_t = force transmitted to the foundation

$$= ky + c\frac{dy}{dt}$$

It can be shown[1] that

$$T_F = \left(\frac{1+\left(2\zeta\frac{\omega}{p}\right)^2}{\left(1-\frac{\omega^2}{p^2}\right)^2+\left(2\zeta\frac{\omega}{p}\right)^2}\right)^{1/2} \tag{2.17}$$

The value for T_F corresponding to different values for the ratio (ω/p) is plotted for different values of ζ in Fig. 2.3. Thus, the actual force transmitted to the foundation is obtained by multiplying the disturbing force, F_0, by the transmissibility T_F.

Although the engineer should attempt to design machinery so that vibration is minimised, nevertheless it will be impractical in

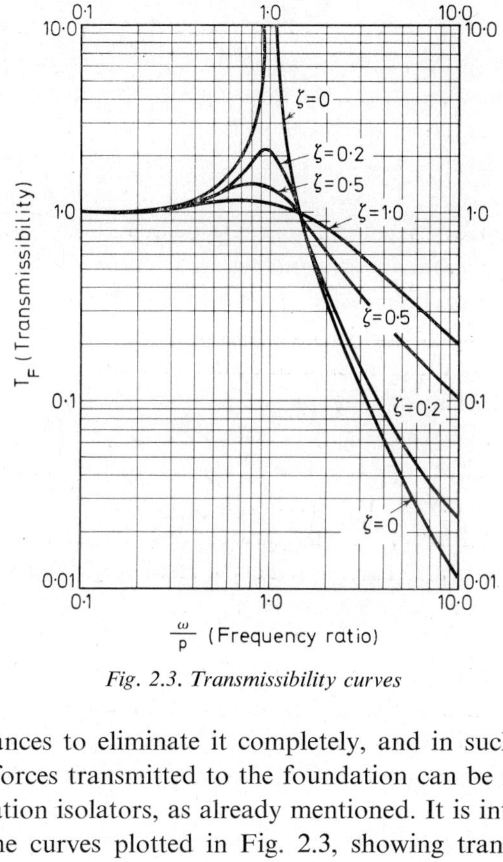

Fig. 2.3. Transmissibility curves

many instances to eliminate it completely, and in such instances vibratory forces transmitted to the foundation can be reduced by using vibration isolators, as already mentioned. It is interesting to examine the curves plotted in Fig. 2.3, showing transmissibility verses the ratio (ω/p). When the ratio of (ω/p) is equal to $\sqrt{2}$, it may be observed that the value for T_F is equal to unity.

Further, some important and useful factors which are necessary when designing a spring suspension system to isolate a machine from its foundation, are also indicated in Fig. 2.3. In the first instance, it is seen that the values for T_F are in all cases greater than unity for values of (ω/p) less than $\sqrt{2}$, but that above this value, the transmissibility decreases. Note also that an increase in damping has the effect of reducing T_F for ratios of (ω/p) up to the value $\sqrt{2}$, but that above this value the reverse is true, i.e. a reduction in damping has the effect of reducing T_F. The obvious conclusion to be made from these observations is that suspension springs should be designed to have a natural frequency as low as possible, with negligible damping. Such a spring is often referred to as a soft spring, and its natural frequency should be only a fraction of the forcing or disturbing frequency, i.e. the value for the ratio (ω/p) should be as high as possible, a value in the region of 4 or more being suggested. When running a machine from rest up to its operating speed, it should be accelerated sufficiently to prevent the build-up of amplitude at the resonant condition. If this cannot be achieved, the amplitude of vibration at resonance must be kept to a safe value by the provision of sufficient damping in the form of buffers or other means.

Vibration isolators usually take the form of springs, or mountings made from rubber or cork, and many different types are commercially available. In the first instance, however, the amount of disturbing force which may reasonably be transmitted to the foundation (or alternatively, the problem might be to isolate a delicate piece of equipment or machinery from the motion occurring in the surrounding structure, as in a moving vehicle) must be assessed, and frequently this will involve some experimentation. Having established a value for T_F (either by estimation or experimentation, or possibly by previous experience), and knowing the forcing frequency, the natural frequency for the system can be determined using equation 2.17. Associated with this natural frequency will be a certain amount of deflection, and this may determine the type of vibration isolator to be used, e.g. a large deflection can often only be conveniently obtained by the use of metallic springs. A reduction in the amount of deflection required can be obtained at the expense of an increase in the transmissibility; if the disturbing forces are small, then a higher transmissibility will probably be permissible.

The use of vibration isolation will be demonstrated elsewhere in the series, but for a more thorough treatment of the subject, the reader is referred to the work of Crede[5, 6]. Crede[6] presents a survey of commercially available isolators in the United States. The designer who is called upon to consider vibration isolation should consider using commercially available means, and a useful guide to manufacturers' in the United Kingdom can be found in The Engineer Buyers Guide[7]. The various manufacturers will be only too pleased to supply literature and technical information upon request.

2.6. SYSTEMS WITH MULTIPLE DEGREE OF FREEDOM

The ideal mathematical model illustrated in Fig. 2.1 for a system having a single degree of freedom, is relatively easy to analyse, and for this reason there is great incentive for the designer to use this model whenever possible. Nevertheless, there are many instances where this approach represents an oversimplification of reality, and additional degrees of freedom must be introduced. For example, the vibration isolation of a rigid body supported on springs has, in general, six degrees of freedom.

For systems having multiple degree of freedom, the mathematical analysis becomes somewhat complex, but with the advent of high speed digital computers as a readily available facility[8], such problems are not now so prohibitive. Several methods for obtaining the differential equations of motion from a mathematical model are available, but the system must first of all be reduced to an appropriate mathematical model without significant loss of accuracy in the result. The most common method of obtaining the equations of motion is by the direct application of Newton's second law, as already demonstrated for the single degree of freedom system (Fig. 2.1), but other methods are also available.

2.7. THE LAGRANGE EQUATION

An alternative, and frequently more convenient form, is to use the Lagrange equation. In its fundamental form this may be written as

$$\frac{\mathrm{d}}{\mathrm{d}t}\frac{\partial(KE)}{\partial \dot{q}_i} - \frac{\partial(KE)}{\partial q_i} + \frac{\partial(PE)}{\partial q_i} + \frac{\partial(DE)}{\partial \dot{q}_i} = Q_i \qquad (2.18)$$

where q_i represents the generalised co-ordinate, equal in number to the degrees of freedom of the system, and

KE = total kinetic energy of the system
PE = total potential energy of the system
DE = total dissipative energy of the system
Q_i = generalised external force acting on the system.

In many practical instances the dissipation of energy due to damping is small compared with the other energies, and may consequently be neglected. For a more complete and detailed treatment of Lagrange's equation the reader is referred to Haberman[9] and Thomson[10]. In a conservative system, i.e. a system in which there is no loss of energy, Lagrange's equation may be written in the form[9]

$$\frac{\mathrm{d}}{\mathrm{d}t}\frac{\partial(KE)}{\partial \dot{q}_i} - \frac{\partial(KE)}{\partial q_i} + \frac{\partial(PE)}{\partial q_i} = 0 \qquad (2.19)$$

By way of illustration, consider the three degree of freedom system shown in Fig. 2.4.

Let the displacements of the masses m_1, m_2 and m_3 be x_1, x_2 and x_3

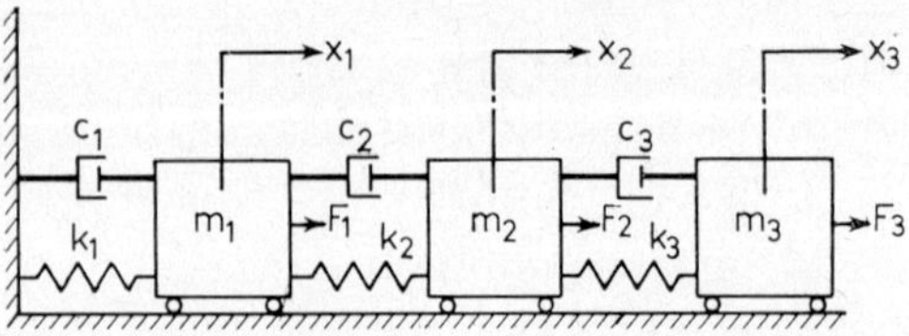

Fig. 2.4. Three degree of freedom system

respectively, as indicated. The equation for kinetic energy, potential energy and dissipative energy are as follows:

$$KE = \tfrac{1}{2}\, m_1\dot{x}_1^2 + \tfrac{1}{2}\, m_2\dot{x}_2^2 + \tfrac{1}{2}\, m_3\dot{x}_3^2$$
$$PE = \tfrac{1}{2}\, k_1 x_1^2 + \tfrac{1}{2}\, k_2(x_2 - x_1)^2 + \tfrac{1}{2}\, k_3(x_3 - x_2)^2$$
$$DE = \tfrac{1}{2}\, c_1\dot{x}_1^2 + \tfrac{1}{2}\, c_2(\dot{x}_2 - \dot{x}_1)^2 + \tfrac{1}{2} c_3(\dot{x}_3 - \dot{x}_2)^2$$

The three simultaneous equations necessary for a solution are obtained by applying equation 2.18 with respect to $q_i = x_1$, $q_i = x_2$, and $q_i = x_3$.
Thus,

$$\frac{\partial(KE)}{\partial\dot{x}_1} = m_1\dot{x}_1; \quad \frac{\partial(KE)}{\partial\dot{x}_2} = m_2\dot{x}_2; \quad \frac{\partial(KE)}{\partial\dot{x}_3} = m_3\dot{x}_3;$$

$$\frac{\partial(KE)}{\partial x_1} = 0; \quad \frac{\partial(KE)}{\partial x_2} = 0; \quad \frac{\partial(KE)}{\partial x_3} = 0;$$

$$\frac{\partial(PE)}{\partial x_1} = k_1x_1 + k_2(x_1 - x_2); \quad \frac{\partial(PE)}{\partial x_2} = k_2(x_2 - x_1) + k_3(x_2 - x_3);$$

$$\frac{\partial(PE)}{\partial x_3} = k_3(x_3 - x_2);$$

$$\frac{\partial(DE)}{\partial\dot{x}_1} = c_1\dot{x}_1 + c_2(\dot{x}_1 - \dot{x}_2); \quad \frac{\partial(DE)}{\partial\dot{x}_2} = c_2(\dot{x}_2 - \dot{x}_1) + c_3(\dot{x}_2 - \dot{x}_3);$$

$$\frac{\partial(DE)}{\partial\dot{x}_3} = c_3(\dot{x}_3 - \dot{x}_2);$$

Substituting into equation 2.18 the following equations of motion are obtained:

$$m_1\ddot{x}_1 + (c_1 + c_2)\dot{x}_1 \quad - c_2\dot{x}_2 \quad + (k_1 + k_2)x_1 \quad - k_2x_2 = F_1$$

$$m_2\ddot{x}_2 \quad - c_2\dot{x}_1 + (c_2 + c_3)\dot{x}_2 - c_3\dot{x}_3 \quad - k_2x_1 + (k_2 + k_3)x_2 - k_3x_3 = F_2$$

$$m_3\ddot{x}_3 \quad - c_3\dot{x}_2 + c_3\dot{x}_3 \quad - k_3x_2 + k_3x_3 = F_3$$

The above equations could, of course, have been obtained very easily by the direct application of Newton's second law. The Lagrange method of solution, however, is frequently a more convenient and powerful method of solution than that offered by the application of Newton's second law, but it has no real advantage for the solution of elementary problems.

The above equations of motion can be more concisely expressed by writing them in matrix form, as follows:

$$\begin{bmatrix} m_1 & 0 & 0 \\ 0 & m_2 & 0 \\ 0 & 0 & m_3 \end{bmatrix} \begin{Bmatrix} \ddot{x}_1 \\ \ddot{x}_2 \\ \ddot{x}_3 \end{Bmatrix} + \begin{bmatrix} (c_1+c_2) & -c_2 & 0 \\ -c_2 & (c_2+c_3) & -c_3 \\ 0 & -c_3 & c_3 \end{bmatrix} \begin{Bmatrix} \dot{x}_1 \\ \dot{x}_2 \\ \dot{x}_3 \end{Bmatrix}$$

$$+ \begin{bmatrix} (k_1+k_2) & -k_2 & 0 \\ -k_2 & (k_2+k_3) & -k_3 \\ 0 & -k_3 & k_3 \end{bmatrix} \begin{Bmatrix} x_1 \\ x_2 \\ x_3 \end{Bmatrix} = \begin{Bmatrix} F_1 \\ F_2 \\ F_3 \end{Bmatrix} \qquad (2.20)$$

Using a shorthand notation:

$$[M]\{\ddot{x}\}+[C]\{\dot{x}\}+[K]\{x\} = \{F\} \qquad (2.20a)$$

where $[M]$, $[C]$ and $[K]$ are referred to as the inertia coefficient, damping coefficient and stiffness coefficient respectively, and are obvious from equation 2.20. For some purposes it is convenient to express equation 2.20a in the alternate form indicated below:

$$\{\ddot{x}\}+[M]^{-1}[C]\{\dot{x}\}+[M]^{-1}[K]\{x\} = [M]^{-1}\{F\} \qquad (2.20b)$$

where $[M]^{-1}$ is the inverse or reciprocal of $[M]$, and for the case under consideration

$$[M]^{-1} = \begin{bmatrix} 1/m_1 & 0 & 0 \\ 0 & 1/m_2 & 0 \\ 0 & 0 & 1/m_3 \end{bmatrix}$$

If the motion is free undamped, then $[C]$ and $\{F\}$ in equation 2.20b are each zero, and we have

$$\{\ddot{x}\}+[M]^{-1}[K]\{x\} = 0 \qquad (2.20c)$$

It is important to mention that, unlike determinants, matrices do not have specific values, but are simply a method for grouping numbers into rows and columns to form arrays. They are, however, extremely useful as a means facilitating the handling of simultaneous equations, and in particular, matrix form is very convenient in connection with programming for a digital computer.

2.8. INFLUENCE COEFFICIENTS

A further alternative approach, which is particularly useful for solving complex multiple degree of freedom systems involving structures, is to use *influence coefficients.* A flexibility influence coefficient is simply defined as the static deflection of a system at some position (say i), due to the application of a unit force being applied, in general, at another point (say j), the unit force being assumed the only force acting. Influence coefficients may also be used for rotational systems, where in this case the influence coefficient is defined as the angular displacement at a point i due to the application of a unit twisting moment or torque applied at another point j.

Considering the three degree of freedom illustrated in Fig. 2.4, if the elastic limit is not exceeded, the displacements will be linear functions of the forces acting, and we may write the following flexibility equations:

$$\begin{aligned} x_1 &= \alpha_{11}F_1+\alpha_{12}F_2+\alpha_{13}F_3 \\ x_2 &= \alpha_{21}F_1+\alpha_{22}F_2+\alpha_{23}F_3 \\ x_3 &= \alpha_{31}F_1+\alpha_{32}F_2+\alpha_{33}F_3 \end{aligned} \qquad (2.21)$$

where the coefficients α_{ij} are the flexibility influence coefficients. The first part of the subscript gives the location of the displacement and the second part refers to that force contributing to the displacement. For a system having n degrees of freedom, we may write the flexibility equation in the general form:

$$x_i = \sum_{i=1}^{n} \alpha_{ij}F_i \qquad (2.22)$$

The reader should note that, by Clerk Maxwell's reciprocal theorem[2],

$$\alpha_{ij} = \alpha_{ji}$$

If the motion is free, undamped and takes place with harmonic motion, then the only forces acting will be the inertia forces, and we have:

$$F_1 = -m_1\ddot{x}_1 = m_1p^2x_1$$
$$F_2 = -m_2\ddot{x}_2 = m_2p^2x_2$$
$$F_3 = -m_3\ddot{x}_3 = m_3p^2x_3$$

Substituting for F_1, F_2 and F_3 from the above equations into equation (2.21)

$$\begin{aligned} x_1 &= \alpha_{11}(m_1p^2x_1)+\alpha_{12}(m_2p^2x_2)+\alpha_{13}(m_3p^2x_3) \\ x_2 &= \alpha_{21}(m_1p^2x_1)+\alpha_{22}(m_2p^2x_2)+\alpha_{23}(m_3p^2x_3) \\ x_3 &= \alpha_{31}(m_1p^2x_1)+\alpha_{32}(m_2p^2x_2)+\alpha_{33}(m_3p^2x_3) \end{aligned} \tag{2.23}$$

Or, writing in matrix form:

$$\begin{Bmatrix} x_1 \\ x_2 \\ x_3 \end{Bmatrix} = p^2 \begin{bmatrix} \alpha_{11}m_1 & \alpha_{12}m_2 & \alpha_{13}m_3 \\ \alpha_{21}m_1 & \alpha_{22}m_2 & \alpha_{23}m_3 \\ \alpha_{31}m_1 & \alpha_{32}m_2 & \alpha_{33}m_3 \end{bmatrix} \begin{Bmatrix} x_1 \\ x_2 \\ x_3 \end{Bmatrix} \tag{2.24}$$

Re-arranging equations (2.23) and writing in the alternative form:

$$\begin{bmatrix} (\alpha_{11}m_1-1/p^2) & \alpha_{12}m_2 & \alpha_{13}m_3 \\ \alpha_{21}m_1 & (\alpha_{22}m_2-1/p^2) & \alpha_{23}m_3 \\ \alpha_{31}m_1 & \alpha_{32}m_2 & (\alpha_{33}m_3-1/p^2) \end{bmatrix} \begin{Bmatrix} x_1 \\ x_2 \\ x_3 \end{Bmatrix} = 0 \tag{2.25}$$

This equation is satisfied if the determinant is zero, i.e. if

$$\begin{bmatrix} (\alpha_{11}m_1-1/p^2) & \alpha_{12}m_2 & \alpha_{13}m_3 \\ \alpha_{21}m_1 & (\alpha_{22}m_2-1/p^2) & \alpha_{23}m_3 \\ \alpha_{31}m_1 & \alpha_{32}m_2 & (\alpha_{33}m_3-1/p^2) \end{bmatrix} = 0 \tag{2.26}$$

The resulting equation is termed the *frequency equation*, and in the more general form this may be written

$$\begin{bmatrix} (\alpha_{11}m_1 - 1/p^2) & \alpha_{12}m_2 & \alpha_{13}m_3 & \ldots \alpha_{1n}m_n \\ \alpha_{21}m_1 & (\alpha_{22}m_2 - 1/p^2) & \alpha_{23}m_3 & \ldots \alpha_{2n}m_n \\ \alpha_{31}m_1 & \alpha_{32}m_2 & (\alpha_{33}m_3 - 1/p^2) & \ldots \alpha_{3n}m_n \\ \alpha_{n1}m_1 & \alpha_{n2}m_2 & \alpha_{n3}m_3 & \ldots (\alpha_{nn}m_n - 1/p^2) \end{bmatrix} = 0 \qquad (2.27)$$

Equation 2.27 may be solved using Cramer's rule[11], but this is a rather tedious and inefficient method for a system having many degrees of freedom. It is suggested that Cramer's rule is suitable for up to three, or perhaps four, simultaneous equations, but it is not very convenient for more than this. Even when the frequency equation has been obtained, by expanding equation 2.27, it is then necessary to solve for the roots of the polynomial equation obtained. This may be performed in a number of different ways, but the reader will probably remember the Newton-Raphson iteration method from his mathematics, which may be written as

$$x_{n+1} = x_n - \frac{f(x_n)}{f'(x_n)} \qquad (2.28)$$

where x_n = assumed value for the unknown quantity x,

$f(x_n)$ = equation involving x with assumed value for x substituted and the result calculated,

$f'(x_n)$ = differential of equation involving x, with assumed value for x substituted and the result calculated.

When the result is obtained that x_{n+1} is equal to x_n (the assumed value for x), this indicates a root of the equation. This, and other methods, are dealt with in books on mathematics[11] and will not be further discussed at the present time.

The reader will find a very good discussion relating to methods of approximation, and also the use of matrices and the solution of linear algebraic equations, in the texts by Pipes[11] and Haberman[9].

2.9. THE HOLZER METHOD

An extremely useful technique, which may be used for calculating the natural frequencies of multi-degree of freedom systems, is a trial and error method due to Holzer[12], and which bears his name. The method may be applied to many lumped mass systems, and although of a trial and error nature, it nevertheless permits the calculation of all the natural frequencies of a system. The method

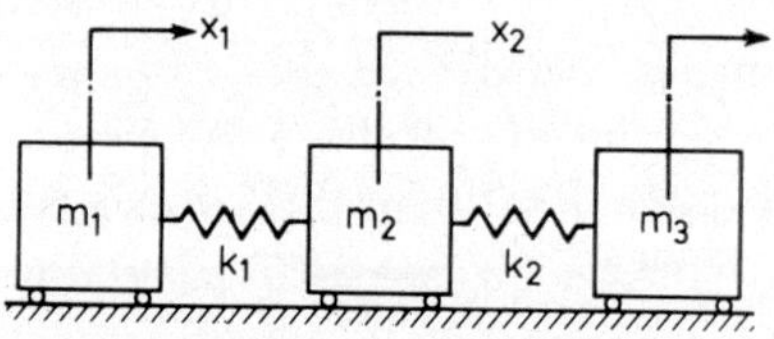

Fig. 2.5. Three degree of freedom system

is particularly well suited for calculating the natural frequencies of torsional vibrations for multi-rotor systems, and this will be discussed presently.

Consider the three degree of freedom system shown in Fig. 2.5. Let the masses m_1, m_2 and m_3 be displaced from their static equilibrium positions by the application of external forces, such that the displacements are x_1, x_2 and x_3 respectively. On the removal of the external forces, the system will try to restore itself to equilibrium and, in the absence of any other forces acting, free vibrations will ensue. Applying Newton's second law, the equations of motion are obtained, thus:

$$\begin{aligned} m_1\ddot{x}_1+k_1(x_1-x_2) &= 0 \\ m_2\ddot{x}_2+k_1(x_2-x_1)+k_2(x_2-x_3) &= 0 \\ m_3\ddot{x}_3+k_2(x_3-x_2) &= 0 \end{aligned} \tag{2.29}$$

Adding these equations of motion results in

$$m_1\ddot{x}_1+m_2\ddot{x}_2+m_3\ddot{x}_3 = 0 \tag{2.30}$$

indicating that the sum of the inertia forces are equal to zero.

This is the basis of the Holzer method. Assuming the motion is harmonic, then $m\ddot{x} = -mp^2x$, and re-writing equation 2.30 in more general terms we have

$$p^2 \sum_{i=1}^{n} m_i x_i = 0 \tag{2.31}$$

where p represents a natural frequency corresponding to a principal mode of vibration. If equation 2.31 is satisfied for any assumed value of p, then the value so assumed is a natural frequency. The method may be most conveniently applied using a tabular solution. It will be apparent that any such solution requires values for the displacements x_1, x_2, x_3, $\ldots x_n$ to be evaluated. From the first of equation 2.29 substituting for $m_1\ddot{x}_1 = -m_1p^2x_1$, we obtain the result

$$x_2 = x_1 - \frac{m_1p^2}{k_1}x_1$$

Similarly,

$$x_3 = x_2 - \frac{p^2}{k_2}(m_1x_1 + m_2x_2)$$

For the general case having n degrees of freedom

$$x_i = x_{i-1} - \frac{p^2}{k_{i-1}} \sum_{i=1}^{i-1} m_i x_i \tag{2.32}$$

The procedure may be summarised as follows. Assume a value for p, and let x_1 be equal to unity. The corresponding values for x_2, x_3, $\ldots$ x_n may then be calculated by the above method, and the value for $p^2 \sum m_i x_i$ can be obtained. If the value is zero, then the assumed value for p is correct and gives one value for the natural frequency. If the value is not zero, an incorrect estimate for p has been made and the calculation must be repeated with a new value for p. A methodical procedure is desirable, and if a range of values for p are selected, these may be plotted against the corresponting values determined for $p^2 \sum m_i x_i$. Where such a curve cuts the axis for $p^2 \sum m_i x_i$ equal to zero, the value for p corresponding to a natural frequency is obtained.

For simplicity, the procedure outlined above has been applied to an unsupported spring mass system, but the method may easily be extended to systems with fixed supports. In this case, however, the displacement at a fixed support must be zero. Consequently, a correct value for p has been assumed only when the displacement at a rigid support is equal to zero.

For the case of *one end fixed and one end free*, the equation to be used for calculating displacements is equation 2.32, and the condition which must be satisfied is

$$x_{n+1} = x_n - \frac{p^2}{k_n} \sum_{i=1}^{n} m_i x_i = 0$$

where the calculation is commenced from the free end, so that the location n_{n+1} corresponds to the fixed support.

For the case of *both ends fixed*, the equation to be used for calculating the displacements is

$$x_i = x_{i-1} + \frac{1}{k_i}\left[k_1 x_1 - p^2 \sum_{i=1}^{i-1} m_i x_i\right] \qquad \ldots(2.33)$$

and this should yield the value zero at the support for a correct value of p, the location at the support being x_{n+1}.

2.10. TORSIONAL VIBRATIONS

Probably one of the most important types of vibration encountered in practice is that of torsional vibrations. Such vibrations may occur in reciprocating engines, rotating electrical machines and many other types of machinery and equipment involving shafting or components subjected to torsion. A reasonably thorough discussion relating to the torsional vibration of shaft systems has already been presented by the author[2], and the present discussion will be applied to the general case of a multi-rotor system.

Suppose that it is required to determine the natural frequencies for the multi-rotor system shown in Fig. 2.6a. The first thing which must be done is to convert the system to a mathematical model, i.e. the real system must be replaced by a lumped parameter approximation, and this is indicated in Fig. 2.6b. In the mathemtical model, the mass of the shaft is replaced by the small rotors at a, *b*, *c*, etc., as shown, such that each of the small rotors has a moment of inertia equivalent to the shaft segment which it represents, and

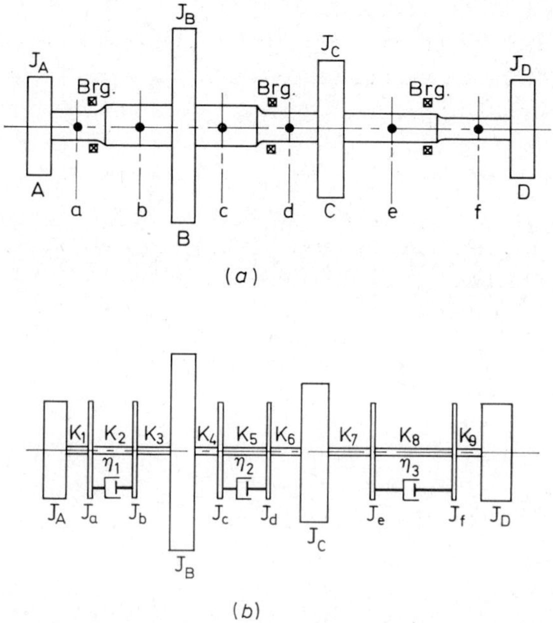

Fig. 2.6. Conversion of physical system into mathematical model (a) Multi-rotor torsional system (b) Equivalent mathematical model

is positioned at the centre of mass of the appropriate section. It must also be mentioned that where a stepped shaft connects two rotors in the original system, Fig. 2.6a, this stepped portion must be replaced by a torsionally equivalent shaft of uniform diameter. It will be assumed at this stage that such a replacement has been carried out before setting up (or as part of setting up) the mathematical model illustrated in Fig. 2.6b; the method for so doing will be presently indicated. For the time being, then, the torsional stiffness between each rotor in the mathematical model is represented by K_1, K_2, K_3, etc.

The effect of the bearings is shown in the mathematical model by the torsional dampers. In a practical system, the damping is usually very small and may quite reasonably be neglected. Note also that if the main rotors are very massive compared with the shaft, the mass effect of the shaft may be small enough to neglect in the analysis. On the other hand, if the shaft mass effect is very significant, then it might be necessary to divide it into a larger number of lumped moments of inertia than that indicated in Fig. 2.6.

The differential equations which govern rotational or torsional motion are similar to those which govern rectilinear motion. Consequently, the mathematical model shown in Fig. 2.6b could be replaced by an equivalent spring-mass system. The analogy between these two types of motion is well known, and for completeness, this is indicated in Table 2.1 below.

Table 2.1

COMPARISON BETWEEN RECTILINEAR AND TORSIONAL SYSTEMS

Physical Term	*Rectilinear*	*Torsional*
Time	t	t
Displacement	x	θ
Velocity	$\dot{x}$	$\dot{\theta}$
Acceleration	$\ddot{x}$	$\ddot{\theta}$
Force of Torque	$F = m\ddot{x}$	$M_t = J\ddot{\theta}$
Mass or Moment of Inertia	m	J
Spring Constant	$k = F/\delta$	$K = G\theta/L$
Damping Coefficient	c	η
Damping Factor	ζ	ζ
Kinetic Energy	$\frac{1}{2}m\dot{x}^2$	$\frac{1}{2}J\dot{\theta}^2$
Potential Energy	$\frac{1}{2}kx^2$	$\frac{1}{2}K\theta^2$

Obviously, then, any of the methods which have previously been discussed in connection with spring-mass systems can be used to solve problems associated with torsional vibrations. As far as design methods are concerned, probably the most convenient of the methods presented, for torsional vibration analysis of multi-rotor systems, is the Holzer method. By analogy, using Table 2.1, we can write down the equivalent torsional equations for the system under consideration. Thus, equation 2.31 becomes (for the case of no torsional restraint at the ends):

$$p^2 \sum_{i=1}^{n} J_i\theta_i = 0 \tag{2.31a}$$

and the angular displacement, by analogy with equation 2.32, is

$$\theta_i = \theta_{i-1} - \frac{p^2}{K_{i-1}} \sum_{i=1}^{i-1} J_i\theta_i \tag{2.32a}$$

The analysis for a shaft with one or both ends torsionally rigid can easily be performed by analogy with the equations previously presented for a rectilinear system.

It has already been mentioned that where a stepped or tapered shaft connects two rotors (as is usual in a practical case), this stepped or tapered section must be replaced by a torsionally equivalent shaft of constant stiffness. It can easily be shown[2] that a stepped shaft, such as that indicated in Fig. 2.7a, may be replaced by an equivalent uniform torsional flexure shaft, such that the length of the equivalent shaft, having diameter d_1, is given by the equation

$$L = l_1 + l_2\left(\frac{G_1J_1}{G_2J_2}\right) + l_3\left(\frac{G_1J_1}{G_3J_3}\right) \tag{2.34}$$

where in this equation J_1, J_2 and J_3 refer to the second moment of areas, and not the mass moment of inertias, of the respective sections. For the usual case of a shaft made from the same material throughout its length, equation 2.34 reduces to

$$L = l_1 + l_2\left(\frac{d_1}{d_2}\right)^4 + l_3\left(\frac{d_1}{d_3}\right)^4 \tag{2.35}$$

Similarly, for a tapered section shaft (as indicated in Fig. 2.7b) it can be shown[2] that the length of the torsionally equivalent shaft, having a uniform diameter d, is given by

$$L = \frac{dl(D^3 - d^3)}{3D^3(D-d)} \tag{2.36}$$

For a shaft consisting of both tapered and stepped sections, first replace the tapered sections by sections of uniform diameter, using equation 2.36, and then replace the stepped shaft system by a torsionally equivalent uniform diameter shaft, having a length calculated from equation 2.34 or 2.35. These latter mentioned

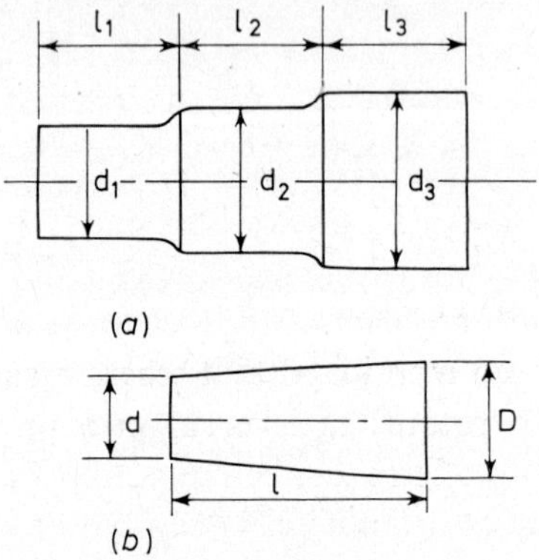

Fig. 2.7. Stepped and tapered shafts (a) Stepped shaft in torsion (b) Tapered shaft in torsion

equations may, of course, be extended to any number of stepped sections.

An excellent and authoritative treatise on the subject of torsional vibrations, which has become accepted as a classic on the subject, is that presented by Dr. W. K. Wilson[13], and the reader who is particularly interested in the subject would do well to study this reference.

Actual design problems involving torsional vibrations are frequently encountered, and the solution of such problems will be considered later in this series.

2.11. TRANSVERSE VIBRATIONS OF BEAMS

Transverse vibrations are those in which each particle vibrates along a straight path perpendicular to the longitudinal axis, i.e. the member is alternately bent and unbent. As a simple example, consider the cantilever shown in Fig. 2.8, firmly fixed at one end, and carrying a heavy mass m at the free end, such that the mass of the cantilever itself may be considered negligible in comparison. If the mass m is displaced by an amount y from the position of

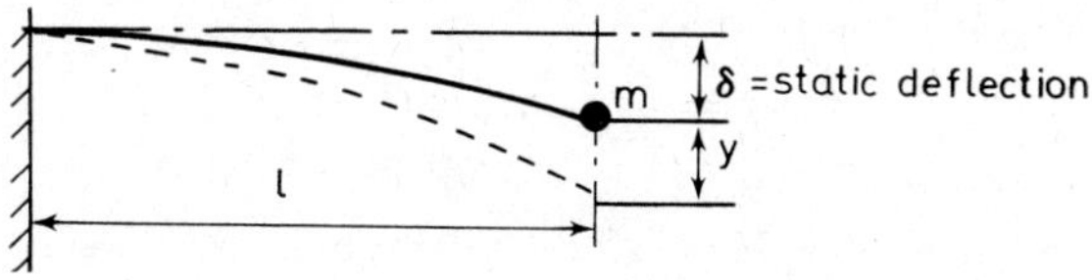

Fig. 2.8. Cantilever beam carrying heavy mass m at the free end

static equilibrium, due to the application of an external force which is subsequently removed, the cantilever will perform transverse vibrations. Applying Newton's second law, the equation of motion is

$$m\ddot{y}+ky = 0$$

i.e.

$$\ddot{y}+p^2y = 0$$

by equation 2.1. Thus,

$$p = \sqrt{\left(\frac{k}{m}\right)}$$

so that

$$f = \frac{p}{2\pi}\sqrt{\left(\frac{k}{m}\right)} \text{ vib/s (Hz)} \qquad (2.37)$$

It is quite usual to express the frequency of transverse vibrations in terms of the static deflection under the mass, and since $k = gm/\delta$, we can write equation 2.37 as

$$f = \frac{1}{2\pi}\sqrt{\left(\frac{g}{\delta}\right)} \text{ vib/s (Hz)} \qquad (2.37a)$$

Using traditional *English units*, $g = 386{\cdot}4$ in/s^2, and measuring δ in inches, gives:

$$f = \frac{3{\cdot}13}{\sqrt{\delta}} \text{ vib/s (Hz)} \qquad (2.38)$$

or

$$N = \frac{187{\cdot}8}{\sqrt{\delta}} \text{ vib/min} \tag{2.39}$$

Using *metric units*, $g = 9{\cdot}81 \times 10^3$ mm/s², and measuring δ in mm, gives:

$$f = \frac{15{\cdot}6}{\delta} \text{ vib/s (Hz)} \tag{2.38a}$$

$$N = \frac{945{\cdot}6}{\sqrt{\delta}} \text{ vib/min} \tag{2.39a}$$

The above equations apply to any one degree lumped mass system, i.e. the beam is considered to be massless but possess elasticity, and any slope of the concentrated mass (as would occur in the case of a large disc) is neglected.

In general, the practical problems with which we shall have to deal will involve beams of non-constant flexure, having continuous

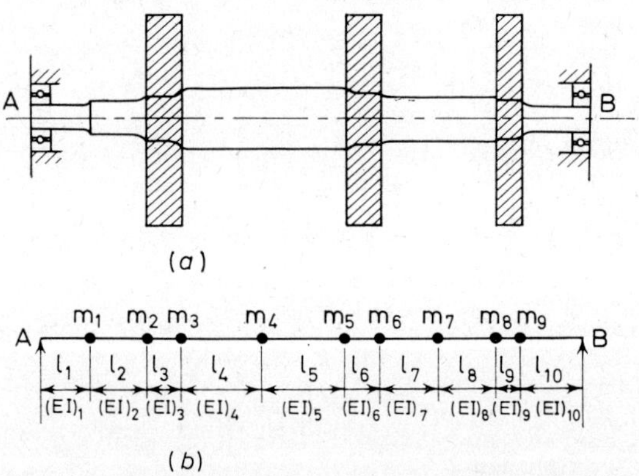

Fig. 2.9. Conversion of physical system into mathematical model (a) Actual multi-rotor shaft system (b) Equivalent mathematical model

mass, and subjected to a number of concentrated masses, somewhat as indicated in Fig. 2.9a. The first thing which must be done is to convert the system to a mathematical model which can be analysed, i.e. the real system must be replaced by a lumped parameter approximation, as suggested by Fig. 2.9b. In the mathematical model, the mass of the beam is replaced by a number of small masses as shown, such that these small masses have a mass equivalent to the beam segment which it represents. For the system shown, the masses have been divided up so that the beam segments connecting them are each of constant flexure. This need not necessarily be the case, and it would be possible to position the mass of each beam segment at its centre of mass. The use of equation 2.27, applied to the mathematical model, will result in the frequency equation being obtained. In this instance, however, the flexibility influence coefficients must be obtained for a beam of non-constant flexure. It will be apparent that such calculations are tedious, and will require the application of a method such as the area-moment method[8] to obtain values.

One of the most commonly used methods for determining the fundamental natural frequency (and to some extent higher frequencies) is that which uses an energy approach and is known as the *Rayleigh or Rayleigh-Ritz method.* In using this approach, it is necessary to assume a shape for the vibrating curve, and Rayleigh has shown[14] that the shape which gives the lowest frequency value is the true one. Nevertheless, any reasonable assumption results in a good approximation to the first or fundamental frequency, and it is commonly assumed in practice that the vibrating curve is similar to the static deflection curve, provided that such an assumption retains the correct mode of vibration. Equating the total maximum strain energy to the total maximum kinetic energy in the system, it can readily be shown[2] that the lowest natural frequency is determined from the following equations: Using traditional *English units:*

$$f = 3{\cdot}13\left(\frac{\sum_{1}^{n} m_i y_i}{\sum_{1}^{n} m_i y_i^2}\right)^{1/2} \quad \text{vib/s (Hz)} \qquad (2.40)$$

$$N = 187{\cdot}8\left(\frac{\sum_{1}^{n} m_i y_i}{\sum_{1}^{n} m_i y_i^2}\right)^{1/2} \quad \text{vib/min} \qquad (2.41)$$

where m_i = mass of appropriate lumped parameter mass, lbf s^2/in (lb),

y_i = deflection corresponding to the i th mass, inches.

Using *metric units:*

$$f = 15{\cdot}76\left(\frac{\sum_1^n m_i y_i}{\sum_1^n m_i y_i^2}\right)^{1/2} \quad \text{vib/s (Hz)} \qquad (2.40a)$$

$$N = 945{\cdot}6\left(\frac{\sum_1^n m_i y_i}{\sum_1^n m_i y_i^2}\right)^{1/2} \quad \text{vib/min} \qquad (2.41a)$$

where m_i = mass of appropriate lumped parameter mass, kg

y_i = deflection corresponding to the i th mass, mm.

As can be seen from the above results, this method necessitates the evaluation of the static deflection curve. Usually, this will involve beams of non-constant flexure, and the methods which have been previously discussed by the author[2, 8] can be used.

The Rayleigh method may also be used to calculate frequencies corresponding to the second and higher modes of vibration, but this necessitates assuming a vibrating curve corresponding to the particular mode of vibration. For example, for a simply supported beam with no overhang, the modes of vibration corresponding to the first (or fundamental), second and third frequencies of transverse vibration will be similar to those indicated in Fig. 2.10. For a complete discussion concerning the application of the Rayleigh method for calculating frequencies above the fundamental, the reader is referred to Duggan[2].

A useful empirical method used to determine the fundamental frequency of transverse vibrations, which should be mentioned, is that known as *Dunkerley's method*, after Professor Stanley Dunkerley[15]. In brief, if a beam of uniform section, or a beam which may be replaced by one of uniform section without serious error (this will be discussed under power transmission shafting), carries masses m_1, m_2, .. m_n, then the fundamental frequency of transverse vibration can be estimated from the equation

$$\frac{1}{f_c^2} = \frac{1}{f_1^2} + \frac{1}{f_2^2} + \ldots + \frac{1}{f_n^2} \tag{2.42}$$

where f_c is the natural frequency for the complete system (Hz), and the frequencies $f_1, f_2, \ldots f_n$ represent the natural frequencies for the system when the masses m_1, m_2, .. m_n act individually. The more

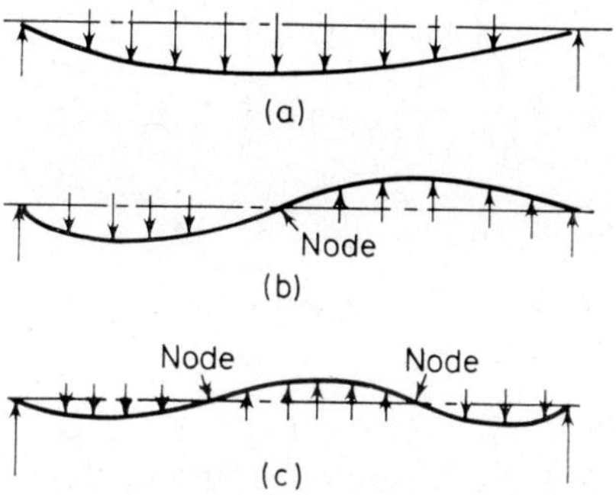

Fig. 2.10. Model of transverse vibration (a) 1st or fundamental mode (b) 2nd mode (c) 3rd mode

usual way of representing this equation is in terms of the frequencies measured in vib/min, in which case

$$\frac{1}{N_c^2} = \frac{1}{N_1^2} + \frac{1}{N_2^2} + \ldots + \frac{1}{N_n^2} \tag{2.42a}$$

Dunkerley's method is widely used and, provided that the beam is of constant flexure or can be replaced by an equivalent beam of constant flexure, and that the loading is not very odd, it may be used to give quite reliable results. It is, however, limited to obtaining an approximation for the first frequency only, and the result expressed by equation 2.42 is based on the assumption that the natural frequencies corresponding to the second and higher modes of vibration are considerably greater than the natural frequency corresponding to the fundamental mode.

A very powerful method for the determination of all the natural frequencies for a beam carrying a number of lumped masses is that known as the *Myklestad*[16] or *Prohl*[17] method. In many ways this method is similar to the Holzer method discussed in Section 2.9, but the computations are considerably more involved, and ideally a digital computer should be used for numerical solutions. The method is one of trial and error and aims at satisfying the boundary conditions. The actual beam system must first be replaced by a

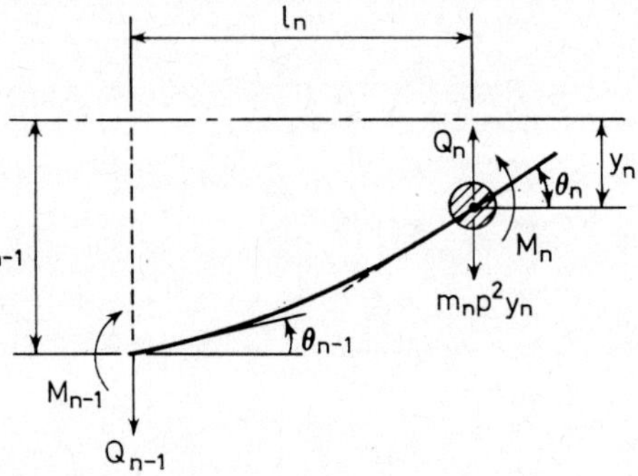

Fig. 2.11. Free body diagram for segment of vibrating beam

mathematical model, consisting of a number of lumped masses connected by massless beam sections (the beam sections possessing the elastic properties of the actual beam), and the method then calculates progressively the shearing force, bending moment, slope and deflection from one station to the next. Consider again the system shown in Fig. 2.9, representing a simple shaft and rotor system, which indicates the mathematical model obtained by distributing the masses in the real system by a number of lumped masses, connected by beam sections of constant flexure. In particular, consider the free body diagram for one of these sections, say from station $(n-1)$ to station n, when the beam is vibrating with harmonic motion at one of its natural frequencies p; this is indicated in Fig. 2.11. The length of the beam section considered is l_n, it has constant flexure $(EI)_n$, and carries a mass m_n at the station n, i.e. the right-hand end. The shear force, bending moment, slope and deflection at station $(n-1)$ are Q_{n-1}, M_{n-1}, θ_{n-1} and y_{n-1} respectively; at the station n (the right-hand end), they are Q_n, M_n, θ_n and y_n respectively, and in addition, there is an inertia force due to the mass m_n (of magnitude $m_n p^2 y_n$). For equilibrium, the shearing force and bending moment relationships are:

$$Q_n = Q_{n-1} + m_n p^2 y_n \tag{2.43}$$
$$M_n = M_{n-1} - Q_{n-1} l_n \tag{2.44}$$

The equations for slope and deflection can be written using influence coefficients (Section 2.8); thus:

$$\theta_n = \theta_{n-1} + (Q_n - m_n p^2 y_n)\gamma_Q + M_n \gamma_m \tag{2.45}$$
$$y_n = y_{n-1} - \theta_{n-1} l_n - (Q_n - m_n p^2 y_n)\alpha_Q - M_n \alpha_m \tag{2.46}$$

where γ_Q and γ_m are the slopes at station n due to the application of unit force and unit moment respectively, and α_Q and α_m are the deflections at station n due to the application of unit force and unit moment respectively, referred to the tangent at the station $(n-1)$. From simple beam theory, treating the segment shown in Fig. 2.11 as a cantilever built-in at station $(n-1)$, the values for the influence coefficients are:

$$\gamma_Q = \alpha_m = \frac{l_n^2}{2(EI)_n}; \quad \gamma_m = \frac{l_n}{(EI)_n}; \quad \alpha_Q = \frac{l_n^3}{3(EI)_n}$$

Substituting these values for the influence coefficients, and for Q_n and M_n in terms of Q_{n-1} and M_{n-1} from equations 2.43 and 2.44 into equations 2.45 and 2.46 we obtain the following equations for slope and deflection:

$$\theta_n = \theta_{n-1} - \frac{Q_{n-1} l_n^2}{2(EI)_n} + \frac{M_{n-1} l_n}{(EI)_n} \tag{2.45a}$$

$$y_n = y_{n-1} - \theta_{n-1} l_n + \frac{Q_{n-1} l_n^3}{6(EI)_n} - \frac{M_{n-1} l_n^3}{2(EI)_n} \tag{2.46a}$$

If a value for p is assumed, then equations 2.43, 2.44, 2.45a and 2.46a enable the shear force, bending moment, slope and deflection to be calculated for successive stations along the beam, the calculations commencing from either a free end or a simply supported end (generally). The method may be further amplified by referring to the mathematical model for the simply supported beam shown in Fig. 2.9. In this instance, if we assume a value for p, and commence our calculations from the left-hand bearing support (assumed simply supported), we know that at this point A, $M_A = 0$ and $y_A = 0$. If then we assume that the shear force Q_A at this point is equal to unity, equations 2.43, 2.44, 2.45a and 2.46a may be used to calculate the values for Q, M, θ and y for all other stations along the beam, in terms of the unknown slope θ_A (say), at the left-hand bearing support. Finally, when we reach the right-hand bearing support, we will have the following four equations

$$Q_B = C_1\theta_A + C_2 \qquad (2.47a)$$

$$M_B = C_3\theta_A + C_4 \qquad (2.47b)$$

$$\theta_B = C_5\theta_A + C_6 \qquad (2.47c)$$

$$y_B = C_7\theta_A + C_8 \qquad (2.47d)$$

where the coefficients $C_1, C_2 \ldots C_8$ have numerical values. Further, we know that at the support B (assuming it to be simply supported), $M_B = 0$ and $y_B = 0$ for the case under consideration. Hence, θ_A can be determined from equation 2.47b above, and if the value so obtained yields the result $y_B = 0$ when substituted into equation 2.47d, then the value for p assumed gives a true value for one of the natural frequencies. A methodical procedure is obviously desirable, and this is established by using a range of values for p, and calculating the corresponding values for Q, M, θ and y for all points along the beam. When the known boundary conditions are satisfied, a true value for one of the natural frequencies is obtained. Because of the somewhat complex nature of the numerical calculations involved, this method ideally requires the use of a digital computer.

In this discussion relating to the transverse vibrations of beams, the mathematical models used have assumed lumped masses concentrated at points. In reality such masses have physical dimensions, and this fact may have an influence on the frequencies of vibration. For example, if the lumped mass carried by the end of the cantilever shown in Fig. 2.8. has been used to replace a heavy disk, then an oversimplification has been made in assuming a single degree of freedom. In actuality, reference to Fig. 2.12 indicates that, due to the disk slope, two parameters are necessary in order to define the system, namely y and θ, indicating that the system has two

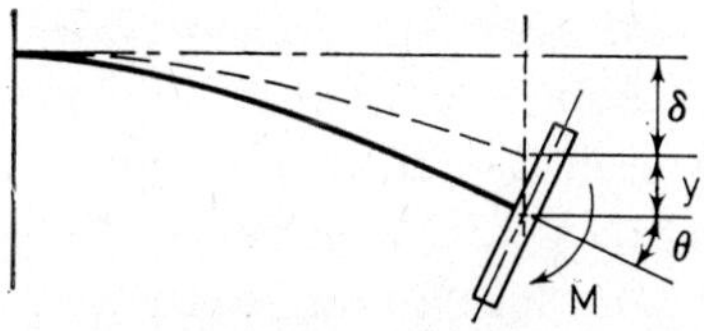

Fig. 2.12. Effect of disc slope

degrees of freedom; the frequency equation can be derived from the flexibility equations.[18] Consequently, the effect of neglecting the disk slope is to reduce the number of degrees of freedom and alter the natural frequencies. Such a simplification might be justified in many instances, but in some the disk slope is important and cannot be neglected without introducing significant errors. In fact, for a beam which is vibrating in a transverse plane at a natural frequency, a rotary inertia effect is involved, which introduces a bending moment at the disk. This moment acts in a direction which increases the deflection at the disk and lowers the values for the natural frequencies. The magnitude of the moment is calculated as follows:

$$M = Ip^2\theta \tag{2.48}$$

where M = moment at disc due to rotary inertia effect,
I = mass moment inertia of disc about its diameter,
θ = slope of disk,
p = estimated value for the natural frequency, rad/s.

The process is an iterative one, requiring p to first be estimated before M can be calculated from equation 2.48, and then included in the beam loading. The procedure is repeated as many times as necessary.

It has also been tacitly assumed in determining the natural frequencies of transverse vibrations of beams that the effect of deflection due to shear is small in comparison with the deflection due to bending. In the majority of instances this is probably quite reasonable, and the shear deflection can be neglected without introducing serious error. There are instances, however, where the shear deflection might be important, such as in very short beams where the length to section depth ratio is comparitively small, and for various hollow sections. In such instances, the deflection due to shear should be added to the deflection due to bending, before computing the values for the natural frequencies. The determination of shear deflection has already been discussed[8].

Another factor which effects the values for natural frequencies is the application of an end load. If the end load is tensile, this has the effect of decreasing the amplitude of vibration (and static deflection) at all points along the beam, and consequently increasing the values for the natural frequencies. If the end load is compressive, the opposite effect occurs, the amplitude of vibration (and static deflection) at all points along the beam are increased, and consequently the values for natural frequencies are decreased. A simple analysis is presented[2] and a discussion is given[19] in the references.

The frequency calculations for a so-called continuous beam, i.e. a beam on many supports, can be performed using the methods which have been discussed[2,3].

2.12. CRITICAL OR WHIRLING SPEED OF SHAFTS

When a shaft rotates at a speed of rotation which coincides with a natural frequency of transverse vibration, the system becomes unstable and violent vibrations may occur. At this condition the shaft whirls about the bearing axis and, if allowed to continue, failure may occur. Such a speed is termed a *critical* or *whirling speed.*

The phenomenon of whirling will be illustrated by considering the case of a massless shaft supported in frictionless bearings (the bearings being assumed inelastic), and carrying a heavy disk of mass M, as illustrated in Fig. 2.13. The deflection under the disk when the shaft rotates at an angular speed of ω rad/s is denoted by y. Any tilting of the disk and consequential dynamic effects are considered negligible at the present time. Due to manufacturing difficulties and the non-homogeneity of the shaft, the centre of

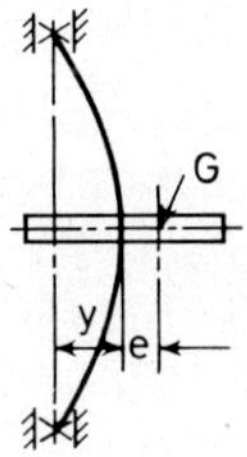

Fig. 2.13. Whirling of single disc system

gravity of the disk will not coincide exactly with the axis of rotation, Consequently, the system will be subjected to centrifugal forces which tend to increase bending of the shaft, and this is opposed by the elastic restoring forces. If the centre of gravity of the disk, denoted by G in Fig. 2.13, lies off the shaft axis by an amount e. termed the eccentricity, the centrifugal force acting along the disk will be given by

$$CF = M\omega^2(y+e)$$

If the shaft has a stiffness k (force per unit deflection), then the restoring force due to the elasticity of the shaft is ky. Applying Newton's second law

$$M\omega^2(y+e) = ky$$

from which

$$y = -\frac{e}{[1-(k/M\omega^2)]} \qquad (2.49)$$

When the term $(k/M\omega^2)$ is equal to unity, the denominator becomes zero, and the deflection tends to infinity. It is when this situation arises that the shaft is said to be running at its critical spped, and to evaluate this speed we must solve the equation

$$\frac{k}{M\omega_c^2} = 1$$

so that

$$\omega_c = \sqrt{\frac{k}{M}} \tag{2.50}$$

where ω_c has been used to represent the angular frequency of rotation corresponding to the critical speed. Equation 2.50 is identical with the result obtained for the natural frequency of transverse vibration for a beam carrying a single mass. Thus, the critical speed is identical with the fundamental frequency of transverse vibrations. This is true for any system of loading and any manner of supports, from which it may be concluded that the theory pertinent to transverse vibrations of beams is also applicable to the whirling of shafts. When the speed of rotation of a shaft coincides with a critical speed, it has been demonstrated that, in the absence of damping, the deflection tends to infinity. This is termed a condition of resonance, and is to be avoided, otherwise the increased bending stresses may cause fatigue failure of the shaft, or the amplitude of vibration may cause repeated impact between rotor and stator. Resonance can occur at any speed of rotation which coincides with a natural frequency of transverse vibration, not only the fundamental, and this fact is of great importance in the design of high speed machines which are to run through the first critical speed into a region of stability. When it is necessary to run through a critical speed, this should be done as quickly as possible to prevent resonance building up to a destructive nature. In any event, the shaft should operate sufficiently remote from a critical speed to prevent the vibrating stresses from causing fatigue failure. Usually, the first critical speed is of major importance, but it is sometimes necessary to calculate higher modes; this is generally limited to the second or third critical speed in practise.

In the above oversimplified analysis, the dynamic effect associated with the disk slope was considered negligible, but under some circumstances this needs to be taken into account. Actually, when a shaft and rotor assembly rotates at speed, any mass on the shaft, such as a disk or wheel, will rotate through a small angle θ, relative to the bearing centre-line, as illustrated in Fig. 2.14. At the cricical speed, the angular velocity of the whirling shaft may be either in the same direction as that of the disk about its axis, or in the opposite direction. Usually, the rotation of the disk and the whirling shaft

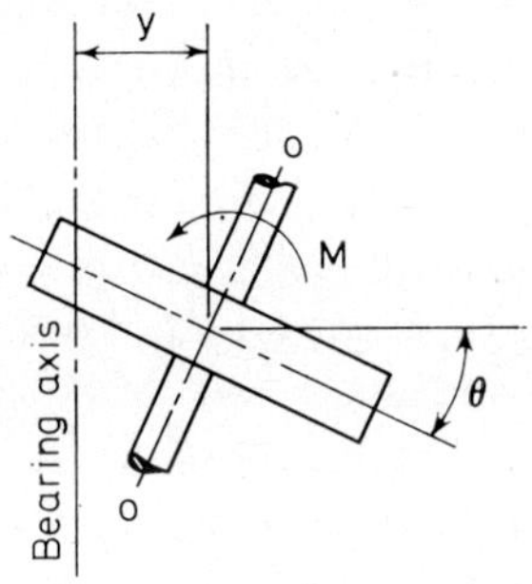

Fig. 2.14. Gyroscopic effect of disc on whirling speed

are in the same direction, giving rise to a gyroscopic couple which acts in a manner which tends to reduce the deflection and increase the critical speed. The magnitude of this gyroscopic couple can be shown[3] to be given by the equation

$$M = (J - I)\omega_c^2\theta \qquad (2.51)$$

where M = gyroscopic couple,
ω_c = radial frequency corresponding to critical speed, rad/s,
J = polar mass moment of inertia of the disc,
I = mass moment of inertia of disk about its diameter.

Equation 2.51 enables an estimate of the gyroscopic couple to be made once the radial frequency of the critical speed has been estimated. In other words, the critical speed of the rotor assembly must first of all be determined neglecting the gyroscopic effects of the disks. This value for ω_c is then used to determine the gyroscopic couple acting on each disk using equation 2.51. The critical speed may now be re-calculated including the gyroscopic couple effects in the shaft loading system. This process may be continued for as many successive approximations as are necessary. Because of the added complexity of including the dynamic effects associated with disk slope, it is not uncommon for it to be neglected, but this can lead to serious errors, particularly if a heavy disk is carried at the end of an overhung shaft or close to the bearing supports, where the disk slope might be very significant. A digital computer programme can be written for the Myklestad-Prohl method of analysis, such that the disk slope is taken into account. One final point which needs to be mentioned again is, that although it is somewhat exceptional, it is possible for the whirling shaft to rotate in a manner opposite to the direction of disk rotation. This condition could occur if disturbing forces, of the same frequency as the critical speed, act to make this possible. Under this condition, the gyroscopic couple would act to increase the deflection and consequently lower the critical speeds. In general, this latter condition does not occur, and the usual effect of the gyroscopic couples is to increase the stiffness of the shaft.

The effect of unbalance and initially bent shafts is discussed by Bishop[20], as is the effect of damping on a rotating shaft[21]. The reader is also referred to the interesting papers by Mitchell[22], Downham[23, 24] and Tuplin[25], and to the book by Dimentberg[26] for further and more complete discussions on the various aspects of shaft whirl.

2.13. CLOSURE

In practice, the engineer may encounter two types of problem. In the first instance, he may be called upon to design a system or machine which uses vibratory motion or forces as a deliberately introduced design feature; alternatively, vibratory motion may be undesirable, and the engineer must then design against vibration. The latter requirement is usually more likely, but in any case the procedures presented in this chapter should prove adequate for dealing with either situation.

Vibration Considerations in Design

When designing against vibration, there are certain fundamental considerations which the engineer should consider. Vibratory forces created by unbalanced rotating or unbalanced reciprocating machines may be objectionable by virtue of the forces transmitted to the foundation or substructure. Alternately, vibratory forces from a structure or machine foundation might be damaging to delicate or precision instruments, and some protection may be necessary. In either instance, the proper design approach is to reduce the excitation of vibration at the source, ideally eliminating it completely. However, for practical reasons concerning cost and manufacturing capabilities, complete elimination is not usually possible, and a compromise is necessary. Where the source of excitation has been reduced to a minimum, but the vibratory forces which remain are still objectionable, then the designer should consider the use of vibration isolators, as discussed in Section 2.5.

Much emphasis has been given in this chapter to the setting up of a mathematical model to replace a physical system, and this is often a most difficult part of design procedure. Such mathematical models must be sufficiently representative of the physical system to ensure that meaningful results can be obtained. From the analysis of mathematical models, the natural frequencies of vibrations can be predicted using the methods discussed in this chapter, and precautions taken to ensure that a condition of resonance is avoided. The designer should also remember that where a machine or structure is to be subjected to forces of a vibratory nature, then fatigue as a design criterion must also be considered[8, 27].

As a final point, it must be mentioned that frequently a machine or structure may be subjected to an external force applied with such rapidity that the consequential displacements and stresses may be significantly larger than the displacements and stresses which would normally be obtained if the same force was applied statically. Such forces are variously referred to as sudden, impact or shock loading. If such a force is repeatedly applied, it is referred to as repeated impact, and the fatigue strength of the part under consideration becomes of vital importance. It should not be thought, however, that any rapid force is an impact force, since this is very closely associated with the lowest natural frequency of the system. Whenever an external force is applied to a structure or element of a machine, no matter how rapidly, there will be some time lag between the application of the force and the response of the system reaching a maximum (i.e. the time involved between the force being applied and the element feeling the full effect); in some instances, this response time may be so small that a square wave impulse can be assumed, but impact loading is not restricted to any particular shape of curve. When the response time is less than half the periodic time associated with the lowest natural frequency of the system or structure, then such loading must certainly be considered as dynamic and, by definition, is termed impact loading. On the other hand, if the response time is greater than three times the periodic time associated with the lowest natural frequency, such loading may be considered statically applied without significant error. There are certain undesirable effects associated with impact loading which require careful consideration. In the first instance, impact loading has the effect of causing greater stresses and displacements than would normally be associated with a static load of the same magnitude, and for structures it is convenient to use so-called stress or deflection *impact factors* to account for these differences. Impact loading will be discussed in more detail in a later volume, and for the present time it is suggested that Juvinall[28], Marin[29] and Goldsmith[30] be consulted for further details.

REFERENCES

1. CREDE, CHARLES E., *Shock and Vibration Concepts in Engineering Design*, Prentice-Hall Inc., Englewood Cliffs, N.J. (1965)
2. DUGGAN, T. V., *Stress Analysis and Vibrations of Elastic Bodies*, Iliffe Books Ltd., London (1964)
3. TIMOSHENKO, S., *Vibration Problems in Engineering* D. Van Nostrand Company, Inc., Princeton, N.J., 3rd edn. (1955)
4. SHIGLEY, J. E., *Dynamic Analysis of Machines*, McGraw-Hill Book Company, Inc., New York (1961)
5. CREDE, C. E., *Vibration and Shock Isolation*, John Wiley and Sons, Inc., New York (1951)
6. CREDE, C. E., *Application and Design of Isolators*, Chapter 32 of *Shock and Vibration Handbook*, Vol. 2, Harris, C. M. and Crede, C. E. (Eds.), McGraw-Hill Book Company, New York (1961)

7. THE ENGINEER BUYERS GUIDE, *The Engineer*, 28 Essex Street, Strand, London, W.C.2 (annually).
8. DUGGAN, T. V., *Applied Engineering Design and Analysis*, Design Technology Series, Iliffe Books Ltd., London (1970)
9. HABERMAN, C. M., *Vibration Analysis*, Charles E. Merrill Publishing Company, Columbus, Ohio (1968)
10. THOMSON, W. T., *Vibration Theory and Applications*, George Allen and Unwin Ltd., London (1965)
11. PIPES, L. A., *Applied Mathematics for Engineers and Physicists*, McGraw-Hill Book Company, New York, 2nd edn. (1958)
12. HOLZER, H., *Die Berechnung der Drehschwingungen*, Springer, Berlin (1922)
13. WILSON, W. K., *Practical Solution of Torsional Vibration Problems*, (3 volumes), Chapman and Hall, London, 3rd edn. (1963)
14. TEMPLE, G. and BICKLEY, W. G., *Rayleigh's Principle*, Dover Publications, Inc., New York (1956)
15. DUNKERLEY, S. 'On the Whirling and Vibrations of Shafts', *Phil. Trans. Soc., Roy.* A., **185** 269–360 (1895)
16. MYKLESTAD, N. O., 'A New Method of Calculating Natural Modes of Uncoupled Bending Vibrations of Airplane Wings and Other Types of Beams, *Jour. Aero. Sci.*, 153–162, (April, 1944) (see also the paper by *R. E. D. Bishop*, 'Myklestad's Method for Non-Uniform Vibrating Beams', *The Engineer*, (14th and 21st Dec 1956))
17. PROHL, M. A., 'A General Method for Calculating Critical Speeds of Flexible Rotors', *Trans. ASME, Jour. of Applied Mechs.*, A-142 (Sept. 1945)
18. HIRSCHHORN, J., *Dynamics of Machinery*, Thomas Nelson and Sons, London (1967)
19. STOKEY, W. F., 'Vibrations of Systems Having Distributed Mass and Elasticity', Chapter 7 of *Shock and Vibration Handbook*, Vol. 1, Harris, C. M. and Crede, C. E. (Eds.), McGraw-Hill Book Company, New York (1961)
20. BISHOP, R. E. D., 'Unbalanced and Initially Bent Shafts', *Engineering*, 735, London, (1960)
21. BISHOP, R. E. D., 'Damping a Rotating Shaft', *Engineering*, 312, London, (1961)
22. MITCHELL, J. M., 'The Critical Speeds of Rigidly Coupled Rotors', *G. E. C. Journal*, **23**, No. 2, April (1956)
23. DOWNHAM, *E.*, 'The Critical Whirling Speeds and Natural Vibrations of Rotating Shaft Rotor Systems', University of London (1956)
24. DOWNHAM, E., 'Theory of Shaft Whirling' (Pts. 1 to 5), *Engineering* 204, 518, London (1957)
25. TUPLIN, W. A., 'Notes on the Whirling of Shafts,' *The Engineer*, 216, No. 5618 (1963)
26. DIMENTBERG, F. M., *Flexural Vibrations of Rotating Shafts*, Butterworths, London (1963)
27. DUGGAN, T. V., 'Fatigue as a Design Criterion', *Engineering Materials and Design*, Part 1, 521–5, April (1967), Part 2, 875–8, June (1967)
28. JUVINALL, R. C., *Engineering Considerations of Stress, Strain, and Strength*, McGraw-Hill Book Company, New York (1967)
29. MARIN, J., *Engineering Materials*, Prentice-Hall, Inc., Englewood Cliffs, N. J. (1952)
30 GOLDSMITH, W. *Impact*, Edward Arnold (Publishers) Ltd., London (1960)

Chapter 3

Design of Shafts and Couplings

The design of machine shafts and shafts for the transmission of power requires the consideration of many factors, culminating in 'the selection of a suitable material and geometry which satisfies the specified and implied functional requirements while remaining within the confines of inherently unavoidable limitations' (see Section 1, Chapter 1 of Duggan[1]). Any shaft design must satisfy certain requirements relating to strength (including fatigue), rigidity, vibration, space limitations, and sometimes factors such as resistance to impact or corrosion are also important. The final design should be such that a satisfactory shaft may be manufactured from a suitable material at the lowest possible cost. Because any real shaft will not generally be of constant flexure, due to variations in diameter, keyways, splines, serrations, etc., the proper design of a shaft can be quite a complex exercise.

Frequently, the shafts of two or more separate shaft systems need to be connected together for one reason or another, and this may be accomplished by the use of flexible or rigid couplings. The type of coupling used, i.e. whether flexible or rigid, will greatly influence the characteristics of the overall connected shaft system, and the decision concerning the type of coupling is dependent on a number of factors.

3.1. METHODICAL PROCEDURE FOR SHAFT DESIGN

As with many other machine or structural elements, the first stage in designing a shaft is to determine the loading, both in magnitude, position and direction, to which the shaft is subjected. This may be due to the transmission of power, forces from belts, chains, gears, etc., or due to the masses associated with heavy components mounted on the shaft. The forces acting may not all be in the same plane, and in general both torsional and bending stresses will be induced, both of which may vary along the length of the shaft system. In some instances, a shaft may also be subjected to forces due to axial loading. The positioning of various elements such as pulleys and gears will frequently be determined from physical factors not directly associated with the shaft system under consideration, and in such instances the bearing supports should be positioned as close to these elements as possible with a view to obtaining the greatest transverse rigidity. In any event, any shaft system whether a machine shaft or a power transmission shaft should be designed to be as compact as possible, whilst still satisfying physical and functional requirements.

Once the positions of the bearings and the various elements mounted on the shaft have been established, and the forces acting have been determined, the resultant bending moment diagram can be obtained from applied mechanics. When the forces acting are not co-planar, it will be necessary to resolve the forces into two mutually perpendicular planes, and hence to determine the bending moment diagrams for these two planes (usually referred to as the horizontal and vertical bending moment diagrams, since these are the planes normally used); the resultant bending moment diagram can then be easily obtained. Just as the bending moment varies along the shaft, so also may the power transmitted vary, depending upon the various positions of power input and power output to and from the shaft system. From a consideration of the variation in resultant bending moment diagram and the variation in the torque transmitted, the critical shaft section is found. If the critical stress at this section is determined, in terms of the unknown shaft diameter, then the selection of a suitable material will enable a numerical value to be calculated for the diameter, based on the known material properties, but certain assumptions will usually be necessary. Alternatively, it may be possible to estimate a suitable shaft diameter based upon physical or other considerations, and the actual numerical value for the critical stress can then be calculated. Using a desired reserve factor (*RF*)[1], an appropriate material may be selected based upon the required material properties. It is important to mention at this stage that fatigue may be the important criterion as far as strength is concerned[1], and this will be further discussed later. Having estimated the shaft diameter at the critical section, the required diameters at the bearings should next be estimated based upon the bearing design or the required bearing life (this will be discussed in detail in a later chapter). The variation in diameter along the shaft length can then be established, taking due regard of the requirements for any of the components which are to fit on the shaft at various points, and the appropriate key sizes, splines, etc,. should be defined using the appropriate national standard. It will usually be necessary to check the stresses in the shaft at various other points before the shaft design is finalised from a strength point of view. At this stage, a tentative shaft design with the material and geometry fully defined has been obtained based on strength requirements.

The next factor which requires consideration is that relating to rigidity. The amount of shaft deflection (either torsional wind-up or transverse deflection) and slope permissible at any point depends upon the type of sytem. Just how much deflection can be allowed in any particular instance is often difficult to establish, but on machine tools, where accuracy is of paramount importance, the permissible deflection will obviously be considerably less than on power transmission shafting involving only flexible elements such as belt and chain drives. The deflection and slope at gears and bearings can have an effect on the correct functioning of these elements, and consequently must be limited to acceptable values.

If the calculated deflections are in excess of permissible values, the tentatively designed shaft must be stiffened, either by increasing the shaft diameters, by decreasing the bearing centres, or possibly both.

If the rigidity requirements of the shaft system are satisfactory, the next consideration is that relating to vibration characteristics, and the natural frequencies of transverse and torsional vibrations may need to be calculated to ensure that they do not coincide with operating speeds. Vibration characteristics will usually be of greatest importance on shafts which are to operate at high speeds of rotation.

Finally, before a detail drawing of the shaft is issued for production, some attention should be given to the possibility of reducing the estimated cost. At the design stage, this is known as value engineering, as distinct from value analysis, which relates to methods of reducing the cost of a component already in production. Any reduction in the cost of production at this stage, without adversely affecting the functional requirements of the shaft, is well worth consideration, and value engineering should always be a part of the design process, not only for shafts, but for any component.

The procedure for the methodical design of a shaft can be briefly summarised by the following steps:

1. Determine the position of the bearings and various elements such as pulleys and gears, and calculate the loading on the shaft.
2. Determine the shaft geometry and select a suitable material based on strength considerations.
3. Check rigidity requirements and modify shaft design as necessary.
4. Check vibration characteristics and modify design as necessary,
5. Carry out value engineering.
6. Issue detail drawing for production of shaft.

3.2. STRENGTH CONSIDERATIONS AND MATERIAL SELECTION

The design of a shaft based on strength considerations establishes a shaft size and material, such that failure by excessive yielding or fatigue within the required working life are avoided, whilst minimising the cost of the shaft and associated components. The use of a material capable of withstanding high working stresses enables the shaft size to be kept small, which in turn enables the dimensions of the shaft bearings and housings, fitted hubs, etc., to be kept to a minimum, thus permitting compact design and generally lower cost. A high strength material will usually cost more than one of lower strength, but this may be more than offset by the considerable reduction in cost possible due to a decrease in the size of the shaft and associated components. Other factors, however, often influence the selection of a material, such as the minimum size of bearings required for a desired life, or the rigidity and vibrational characteristics of the shaft system.

The types of stress which occur in a shaft may be due to (*a*) torsion, (*b*) bending, (*c*) shear force, (*d*) axial loading or (*e*) any combination of the above, and the design must be such that the shaft is capable of resisting the applied stresses.

Any shaft which is required to transmit power will be subjected to simple twisting or torsion, which results in a torsional shear stress being induced in the shaft. At this stage it is worth mentioning that in many shaft systems, different sections of the shaft may be subjected to different amounts of twisting, due to the fact that the total power delivered to the shaft may be taken off at various points and transmitted by means of belt, chain or gear drives. If the power transmitted is established for any section of the shaft, the twisting moment M_t for that section can be calculated from the usual relationships:

$$hp = \frac{2\pi N M_t}{33\,000} \tag{3.1}$$

or

$$W = \frac{2\pi N M_t}{60} \quad \text{(J/s or watts)} \tag{3.2}$$

where hp = power transmitted in horse power (1 hp = 33 000 ft lbf/min)

N = speed of rotation, rev/min

M_t = torque or twisting moment (lbf ft in equation 3.1 and Nm in equation 3.2),

W = power transmitted in watts (J/s).

The above equations may be written in a more convenient form for design purposes, as follows:

$$M_t = \frac{63\,000\,hp}{N} \quad \text{lbf in} \tag{3.1a}$$

or

$$M_t = \frac{9.55W}{N} \quad \text{Nm(J)} \tag{3.2a}$$

Once the twisting moment has been calculated using one of the above relationships, the torsional shear stress can be calculated from the usual equation for a circular shaft[2] i.e.

$$\tau = \frac{M_t}{J} r \tag{3.3}$$

where τ = torsional shear stress,

J = polar second moment of area,

r = radius from axis of bar to fibre considered.

For a solid circular shaft $J = \pi d^4/32$ (d being the shaft diameter) and for a hollow circular shaft having an outer diameter D and an internal diameter d, $J = \pi(D^4 - d^4)/32$. Usually, it is the maximum torsional shear stress at any particular section which is of interest to the designer, and since the torsional shear stress is directly proportional to the radius, the maximum value will occur at the outer fibres, i.e. $r = d/2$ for a solid shaft or $r = D/2$ for a hollow shaft. Thus, equation 3.3 may be written as follows:

$$\tau = \frac{16M_t}{\pi d^3} \quad \text{for a solid shaft} \tag{3.4}$$

or

$$\tau = \frac{16M_t D}{\pi(D^4 - d^4)} \quad \text{for a hollow shaft} \tag{3.5}$$

to give the maximum value (i.e. at the outer fibres) of torsional shear stress at the section considered.

The bending stress at any section of the shaft may be calculated, once the bending moment diagram has been obtained, from the usual bending stress equation[2], i.e.

$$\sigma = \frac{M}{I} y \tag{3.6}$$

where σ = bending stress,
M = bending moment at section under consideration,
y = distance from axis of shaft to fibre considered,
I = second moment of area,
= $\pi d^4/64$ for a solid shaft of diameter d,
= $\pi(D^4 - d^4)/64$ for a hollow shaft of external diameter D and internal diameter d.

Thus, the maximum stress due to bending can be obtained from equation 3.6 in terms of the shaft diameter, i.e.

$$\sigma = \frac{32M}{\pi d^3} \quad \text{for a solid shaft} \tag{3.7}$$

or

$$\sigma = \frac{32MD}{\pi(D^4 - d^4)} \quad \text{for a hollow shaft} \tag{3.8}$$

The importance of direct shear has already been discussed[1], and it can be shown[2] that the direct shear stress across the section of a beam is given by the equation

$$\tau = \frac{Q}{bI} \sum A_1 y \tag{3.9}$$

Since the maximum value of shear stress obtained from equation 3.9 occurs at the neutral axis where both the bending stress and the torsional shear stress are zero, it is not usually necessary to consider the effect of direct shear in shaft design.

Sometimes a shaft may be subjected to an axial force P (say), and if this acts with no eccentricity, the direct stress (either tension or compression, depending upon the direction of P) may be calculated from

$$\sigma = \frac{P}{A} \tag{3.10}$$

$$= \frac{4P}{\pi d^2} \quad \text{(for a solid shaft)} \tag{3.11}$$

$$= \frac{4P}{\pi(D^2 - d^2)} \quad \text{(for a hollow shaft)} \tag{3.12}$$

As already indicated, in general a shaft may be subjected to a combination of the above types of stresses, and it then becomes necessary to apply an appropriate failure theory. A detailed discussion of failure theories will be presented in a future volume in this series, and for the present time the reader is referred to a standard text on strength of materials or to the excellent work by Juvinall[3]. Various investigations regarding the application of a suitable failure theory to power transmission shafting, indicates that the two major theories most applicable are the maximum shear stress theory and the distortion energy theory. However, since the stresses in a shaft system are of a fluctuating nature, fatigue as a design criterion must be considered. A review of failure theories and the application to fatigue has been presented in a paper by the author[4], and in more detail in a series of lectures[5] to be published; these detailed discussions will also be presented in another volume in this series. For the present time the equations derived by the so-called equivalent damage concept[4,5] will be presented without proof for the maximum shear stress theory and the maximum shear strain energy theory (also referred to as the distortion energy theory or the Mises criterion).

Combining the *maximum shear stress theory* of failure with the Goodman relationship yields the result[4,5]

$$\left(\frac{\sigma_{sx} - \sigma_{sy}}{S_u}\right)^2 + 4\left(\frac{\tau_s}{S_u}\right)^2 = 1 \tag{3.13}$$

where σ_{sx} and σ_{sy} may be thought of as equivalent normal static stresses in the x and y directions respectively, and τ_s an equivalent static shear stress, which have the same damaging effect as the defined fluctuating stresses. These stress values may be calculated from the following relationships[4, 5]

$$\sigma_{sx} = \sigma_{mx} + c_1\sigma_{\text{alt }x} \tag{3.13a}$$

$$\sigma_{sy} = \sigma_{my} + c_2\sigma_{\text{alt }y} \tag{3.13b}$$

$$\tau_s = \tau_m + c_3\tau_{\text{alt}} \tag{3.13c}$$

where σ_m and τ_m represent normal and shear mean stresses respectively and σ_{alt} and τ_{alt} represent normal and shear alternating stresses respectively. These quantities are calculated in the way previously discussed[1], i.e. in general terms

$$\sigma_m = \tfrac{1}{2}(\sigma_{\max} + \sigma_{\min}) \tag{3.13d}$$

$$\sigma_{\text{alt}} = \tfrac{1}{2}(\sigma_{\max} - \sigma_{\min}) \tag{3.13e}$$

$$\tau_m = \tfrac{1}{2}(\tau_{\max} + \tau_{\min}) \tag{3.13f}$$

$$\tau_{\text{alt}} = \tfrac{1}{2}(\tau_{\max} - \tau_{\min}) \tag{3.13g}$$

The maximum and minimum stress levels are calculated using the equations previously given in this chapter. The quantities c_1, c_2 and c_3 represent combined material and geometrical constants, and are defined[4, 5] as follows:

$$c_1 = \frac{S_{u1}}{S'_{ax}}; \quad c_2 = \frac{S_{u2}}{S'_{ay}}; \quad c_3 = \frac{S_{su}}{S'_{sa}} \tag{3.13h}$$

where S_{u1} and S_{u2} represent the ultimate tensile strengths of the material in the directions x and y respectively, S_{su} represents the ultimate strength of the material in shear or torsion; S'_{ax} and S'_{ay} represent the corrected fatigue strengths in the x and y directions respectively, and S'_{sa} represents the corrected torsional fatigue strength. For a component which is isotropic, $S_{u1} = S_{u2}$ and $S'_{ax} = S'_{ay}$. In any event, for the usual type of shaft loading the stresses in the y direction are zero, and equation 3.13 reduces to

$$\left(\frac{\sigma_{sx}}{S_u}\right)^2 + 4\left(\frac{\tau_s}{S_u}\right)^2 = 1 \tag{3.13j}$$

Application of the *distortion energy theory* or *Mises criterion*, combined with the Goodman relationship, yields the result[4, 5]

$$\left(\frac{\sigma_{sx}}{S_u}\right)^2 + \left(\frac{\sigma_{sy}}{S_u}\right)^2 - \left(\frac{\sigma_{sx}\sigma_{sy}}{S_u^2}\right) + 3\left(\frac{\tau_s}{S_u}\right)^2 = 1 \qquad (3.14)$$

where all the symbols are as previously defined. In the usual case for a shaft σ_{sy} is zero, and equation 3.14 reduces to

$$\left(\frac{\sigma_{sx}}{S_u}\right)^2 + 3\left(\frac{\tau_s}{S_u}\right)^2 = 1 \qquad (3.14a)$$

A reserve factor (*RF*) may be incorporated in the above equations by replacing S_u by S_u/RF. An alternative to the above is to use the Soderberg relationship[1] instead of the Goodman, in which case S_u in the above equations is replaced by the yield or proof strength S_p. The application of the above equations will be illustrated later in this chapter (Example 3.1).

Of the above two theories of failure, the distortion energy theory generally gives better agreement with test results, but the maximum shear stress is more conservative and for this reason is frequently favoured by many designers.

One cannot determine the geometry of a shaft without considering the material from which the shaft is to be manufactured. The precise properties required of the material will depend upon the particular application, but in general, shafts need to be produced from a tough ductile material, and the greater its mechanical strength, the smaller will be the required shaft size. Since rigidity is also frequently a requirement, a material having a high elastic modulus is also a desirable feature. Consequently, shafts for machines or power transmission will invariably be produced from steel of one kind or another. When considering the design of a shaft on the basis of strength, other factors affecting the material selection must also be taken into account at the same time. For example, resistance to corrosion might be an important requirement, particularly for such applications as marine engineering. When a high strength steel is considered desirable in order to reduce the size of the overall system, in order to satisfy the rigidity or vibrational characteristics it may be necessary to introduce additional bearings. Where this is not possible for one reason or another, and the size of the shaft has to be increased in order to satisfy rigidity requirements, there is no point in using a high strength steel if a cheaper one will satisfy the requirements. In other words, like many other instances in design, the process may be of an iterative nature.

From these few brief comments it may be concluded that the selection of a suitable material is not always a simple matter, since there are a number of factors which need to be considered, not the least of which is the cost of the complete system. As previously mentioned[1], a design decision concerning a particular component does not only affect that component, but may have a significant influence on the associated components in the system; such might be the case concerning the selection of a suitable shaft material. The manner in which a shaft is manufactured will depend largely upon the type of material which is selected and also upon the shaft size. For example, shafting produced from carbon steels may be cold drawn up to diameters of about 100 mm (somewhat less than 4 in), but larger sizes require to be produced from hot-drawn material.

The engineer has at his disposal a wide ranger of suitable shaft materials from which to make his selection, including both carbon and alloy steels. For shafts requiring only low strength, carbon steels may be quite suitable, since they are cheaper than alloy steels and are relatively easy to machine. A typical carbon steel suitable for shaft design is En 8 having a tensile strength in the region of 540 MN/m^2 (78 400 lbf/in^2). For higher strength shafts, it is necessary to consider alloy steels, and a wide range of suitable materials for various applications are available. Typical alloy steels which may be used in the production of shafts include En 14 (B) (SAE 1027, SAE 1330) and En 15 (SAE 1036, SAE 1335) which are moderately priced materials possessing excellent ductility and shock resistance when hardened and tempered to give tensile strengths in the range 620–770 MN/m^2 (89 600–112 000 lbf/in^2); En 16 and En 17 developing tensile strengths up to about 1 000

MN/m^2 (145 600 lbf/in^2); En 19 (SAE 4140) which is a 1 per cent chromium molybdenum steel having useful creep resistance up to about 500 °C, and possessing good shock resistance in the medium tensile range, i.e. up to about 926 MN/m^2 (134 400 lbf/in^2), although tensile strengths in excess of this figure are obtainable if a decrease in shock resistance can be tolerated; En 21 (SAE 2340) is a 3 per cent nickel oil hardening steel used for applications requiring tensile strengths up to 695–850 MN/m^2 (100 800–123 000 lbf/in^2) combined with high shock resistance; En 22 having tensile strengths in the range 770–1544 MN/m^2 (112 000–134 400 lbf/in^2) and 850–1 000 MN/m^2 (123 200–145 600 lbf/in^2) depending upon the limiting ruling section; En 24 (SAE 4340) having good resistance to wear and shock and possessing a wide range of mechanical properties, tensile strengths within the range 700–1 544 MN/m^2 (112 000–224 000 lbf/in^2) being possible, depending upon heat treatment and size; En 25 having tensile strengths in the range 850–1 544 MN/m^2 (123 200–224 000 lbf/in^2) and possessing good shock resistance. For applications requiring resistance to corrosion, stainless steels such as En 57 (SAE 51 431) and En 58, which have tensile strengths in the region of 850 MN/m^2 (123 200 lbf/in^2) and 648 MN/m^2 (94 000 lbf/in^2) respectively, may be suitable. Note that the tensile strengths quoted above are very much dependent upon heat treatment and size. The list of steels used to manufacture shafts for different applications could be considerably extended, but the list given here is sufficient indication of the range available. For more detailed and specific information it is suggested that the reader consult the literature available from the various steel manufacturers. A useful booklet containing information relating to shafts in alloy steels has been produced by the International Nickel Company Ltd.[6] and is available upon request.

3.3. RIGIDITY AND VIBRATIONAL CHARACTERISTICS

Once the geometry and material has been tentatively established based upon strength considerations, as discussed above, the next stage in the design is to consider the rigidity and vibrational requirements. Since the natural frequencies of vibration are directly related to the shaft stiffness (either flexural or torsional depending upon type of vibration) rigidity and vibrational characteristics are closely related, i.e. the stiffer we make the shaft, the higher will be the natural frequencies of vibration and the smaller the deflections.

In determining the geometry and material, i.e. defining the physical system, it is necessary to make *design decisions*. On the other hand, in analysing the physical system from the point of view of rigidity and vibrations, it will be necessary to make some *design assumptions* in order to convert the physical system to a mathematical model. The significance of decisions and assumptions in design have already been discussed[1].

From the point of view of calculating the transverse deflection of shafts having non-constant flexure, the methods previously presented[1] can be used; these include graphical methods of integration and the area moment method. Other methods are, however, available[2], and the method of replacing a stepped shaft by an equivalent shaft of uniform constant flexure is quite a useful technique. The derivation of the method has been discussed elsewhere[2], and will be presented here without proof. Fig. 3.1 should be self explanatory, and indicates how a tabular solution may be applied in order to replace a stepped shaft by an approximately equivalent shaft of uniform constant flexure. An estimate is first made of the anticipated point of maximum deflection, designated 0–0 in Fig. 3.1, and the table is completed using the known shaft geometry. Although the table in Fig. 3.1 uses symbols to indicate the entries, in a real problem these entries will be numerical values. It can be shown[2] that if the estimate of the point of maximum deflection is a reasonable one, the ratio of K_1 to K_2 (defined in the table of Fig 3.1) should lie within the range 0.77–1.3, and the equivalent second moment of area for the uniform flexure shaft is given approximately by the equation

$$I_m = \frac{L^3}{4(K_1+K_2)} \tag{3.15}$$

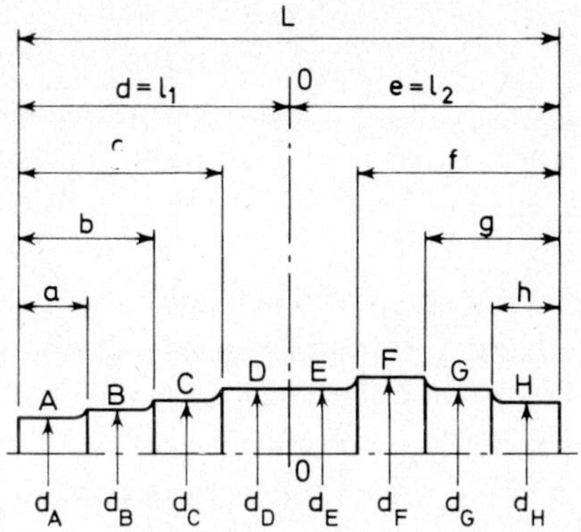

Section	A	B	C	D	E	F	G	H
Dia. 'd'	d_A	d_B	d_C	d_D	d_E	d_F	d_G	d_H
I	I_A	I_B	I_C	I_D	I_E	I_F	I_G	I_H
x	a	b	c	d	e	f	g	h
y	0	a	b	c	f	g	h	0
x^3	a^3	b^3	c^3	d^3	e^3	f^3	g^3	h^3
y^3	0	a^3	b^3	c^3	f^3	g^3	h^3	0
$x^3 - y^3$	a^3	$b^3 - a^3$	$c^3 - b^3$	$d^3 - c^3$	$e^3 - f^3$	$f^3 - g^3$	$g^3 - h^3$	h^3
$\frac{x^3 - y^3}{I}$	$\frac{a^3}{I_A}$	$\frac{b^3 - a^3}{I_B}$	$\frac{c^3 - b^3}{I_C}$	$\frac{d^3 - c^3}{I_D}$	$\frac{e^3 - f^3}{I_E}$	$\frac{f^3 - g^3}{I_F}$	$\frac{g^3 - h^3}{I_G}$	$\frac{h^3}{I_H}$
$\Sigma \frac{x^3 - y^3}{I}$	$K_1 = \Sigma_A^D \frac{x^3 - y^3}{I}$				$K_2 =. \Sigma_E^H \frac{x^3 - y^3}{I}$			

Fig. 3.1. Tabular solution for equivalent uniform shaft

or, more accurately, by the result

$$I_m = \frac{L^3}{(L/l_1)^2K_1+(L/l_2)^2K_2} \tag{3.16}$$

If the first calculation yields a ratio of K_1 to K_2 outside the range indicated, the calculation must be repeated until an acceptable value is obtained; this will involve shifting the position 0–0 in a direction indicated by the first calculation, and modifying the columns headed *D* and *E* in Fig. 3.1. A small point which is worth mentioning is that the modifications are more easily carried out if the entries are initially made in pencil, since they can then be erased, thus eliminating the necessity for completing a new table. With printed sheets and a desk top calculating machine, the calculations required for determining the equivalent uniform flexure shaft to replace the original stepped shaft without significant error for most cases, can be readily performed by unskilled staff. Any of the usual methods of deflection derivation can then be applied to calculate the deflection curve.

Having determined the deflection curve by any suitable method, it is then necessary to make an assessment of the acceptibility or otherwise of the calculated values. In brief, if the calculated values of deflection are such that they indicate malfunctioning of the system may occur, then the shaft must be stiffened by one of the methods previously mentioned. It is difficult to make any generalisations regarding limiting transverse or angular deflections which may be permissible, since each case must be considered on its merits. Faires[11] gives some guidance, and for power transmission shafting quotes that an old rule of thumb is that the deflection should not exceed about 0·8 mm/m (0·01 in/ft) of length between the supports, whilst for machine shafts the permissible deflection is more likely to be in the region of about 0·17 mm/m (0·002 in/ft). Whenever torque is transmitted by a shaft there is bound to be some angular displacement between the ends. In some circumstances, it is desirable to limit the change in angular displacement which accompanies a change in torque, and this may be achieved by ensuring that the shaft has adequate torsional stiffness. The permissible angular twist for the maximum torque condition is dependent upon the particular circumstances, and will be more important for machine tool shafts than for line shafts. For the former, a figure in the region of about 0·26 degree/m (0·08 degree/ft) is a frequently accepted permissible value, whilst for the latter permissible values may be as much as about 3 degrees/m (1 degree/ft).

The above figures are presented only as an indication of some commonly accepted values, and as already mentioned, each case must be considered on its merits. The allowable deflections in any particular system are not magical figures, and need to be considered in relation to other factors. As an extreme example, if the calculated transverse deflection of a shaft is in excess of the clearance between the rotor and stator, then quite clearly the deflection is excessive. In fact, on rotating electrical machines, because of the unbalanced magnetic pull, the deflection must be limited to only a small percentage of the air gap[2], a figure in the region of 20 per cent being commonly used. The slopes at bearings and mating gears must be limited to values which do not cause interference, binding or non-uniform loading. For the successful operation of good quality spur gears, the deflection of the shafts relative to each other (at the section where the gears mesh) should not, in general., exceed about 0·13 mm (about 0·005 in), and the relative slope of one shaft to the other should not exceed about 0·5 mm/m (0·0005 in/in). Naturally, the face width of the gears will also influence the above figures to some extent.

The vibrational characteristics of shaft systems have been quite thoroughly discussed in Chapter 2, and the methods of analysis there presented can be applied to determine the natural frequencies of transverse or torsional vibrations of a rotor system. A condition of resonance should be carefully avoided, by ensuring that the range of operating speeds do not coincide with any of the natural frequencies. In this respect, it is worth repeating that the position of the supporting bearings do not influence the torsional frequencies of vibrations to any extent, whereas the natural frequencies of transverse vibrations are considerably influenced by the number and position of the supporting bearings. If the calculated values of natural frequencies are too close to the operating speeds, then the shaft system will need to be modified by one of the methods previously discussed. In actual fact, of course, the natural frequency must be sufficiently remote from operating speeds to ensure that an undesirable magnification factor does not cause unpleasant vibrations. If a shaft is to operate below the first natural frequency (either transverse or torsional), the first or fundamental frequency should be about 1·3–1·5 times the operating speed.

3.4. KEYS AND SPLINES

When designing a shaft system, it is necessary to consider the various means by which components can be attached to the shaft. The methods which are used vary depending upon the application. For example, a rolling contact bearing is usually fitted so that the inner ring has an interference fit on the shaft. In other instances, it may be desirable to attach a component so that it may slide longitudinally along the shaft, but is fixed rigidly in the torsional direction. Frequently a component must be rigidly attached to a shaft so that it moves neither longitudinally nor torsionally relative to the shaft.

Perhaps the most simple and common method of attaching a disc or hub to a shaft is by means of a key. A key is simply a piece inserted between a shaft and a hub, the major function of which is to prevent relative rotation between the shaft and hub. Keys may be considered as standard components, and their dimensions should conform to the relevant national standard. In the United Kingdom, BS 46: Part 1: 1958[7] has been, and continues to be, the national standard for keys and keyways in inch units. However, with the adoption of metric dimensions, a more recent British Standard Specification has been prepared, namely BS 4235: Part 1: 1967[8], which is the first of a series for metric keys and keyways. The essential data for BS 4235: Part 1: 1967, has been extracted from the ISO Recommendation *Rectangular or Square Parallel*

Keys and their Corresponding Keyways and *Taper Keys and their Corresponding Keyways (with or without gib head)*, and the metric sections stated have been in satisfactory use in most European countries for many years. In the United States, the national standard generally used is ASA Standard B 17·1-1943[9], which is in inch units.

Apart from special applications, the usual types of keys may be divided into the following categories:

1. Parallel keys — either rectangular or square;
2. Taper keys — rectangular or square, plain or with gib-head;
3. Tangential keys;
4. Woodruff keys.

Parallel keys find wide application for conditions where the torque to be transmitted is unidirectional, and the transmission is not likely to be subjected to heavy starting loads. These keys are usually side fitting with top clearance, and normally retained in the shaft rather more securely than in the hub. BS 46: Part 1: 1958, providies for only one class of fit, whereas BS 4235: Part 1: 1967 deals with three classes of fit, namely (a) *free*, where the hub is required to slide over the key when in use, (b) *normal*, where the key is to be inserted in the keyway with minimum fitting, as is required for mass production assembly, and (c) *close*, where an accurate fit of key is required. BS 46: Part 1: 1958 recommends the use of square keys for shaft sizes up to and including 1 in diameter, or for shafts up to 6 in where it is desirable to have a greater key depth than that provided by rectangular keys; usually, however, the general purpose key for shaft sizes in excess of 1 in diameter is the rectangular key. BS 4235: Part 1: 1967 recommends the use of square keys for shaft sizes up to and including 22 mm, and rectangular keys for shaft diameters over 22 mm. The recommendations for the dimensioning of parallel keys and keyways are given in BS 46: Part 1: 1958 (inch dimensions) and BS 4235: Part 1: 1967 (metric dimensions), extracts of which are reproduced in Appendix A by permission of the British Standards Institution.

Taper keys are used in preference to parallel keys for applications requiring the transmission of heavy torques, either unidirectional or reversing, and where periodic withdrawal of the key may be a necessity. They cannot, however, be used in applications requiring a sliding hub member. Where gibheaded keys can be readily accommodated, the provision of such facilitates the easy withdrawal of keys. The recommendations for the dimensioning of parallel keys and keyways are given in the above mentioned British Standard Specifications, extracts of which are reproduced by permission of the British Standards Institution in Appendix A.

Tangential keys are sometimes used for the trasmission of reversing torque under conditions of heavy loading. Details for inch dimensions are to be found in BS 46: Part 1: 1958, or in other relevent national standards.

Woodruff keys are used for the angular location of associated parts on tapered shafts, and for light applications only. Details for inch dimensions can be found in BS 46: Part 1: 1958, or in other relevent national standards.

Keys complying with BS 46 and BS 4235 should be manufactured from steel to En 6A, En 8 or En 9 of BS 970 *Wrought Steels*[10], except for rectangular parallel keys for marine tailshafts, for which BS 46 recommends the use of En 4 or En 5.

Reference to the appropriate standards indicates that the width of key for parallel and taper keys is made approximately equal to about one quarter of the shaft diameter (see Appendix A). However, although the cross-sectional dimensions of keys are specified for particular shaft sizes, the designer must usually determine the length of key. This may be fixed as a matter of convenience in some instances, but in any case, the key should be of sufficient proportions to resist failure under the applied load. Preferred lengths of keys, both plain and gib-headed, are indicated in the appropriate standards.

A typical sunk key used to transmit torque from a shaft to a hub is shown in Fig. 3.2. Let the length of the key be l. Failure of the key may be due to either shearing or crushing. It is usual to assume that the force transmitted by the key is uniformly distri-

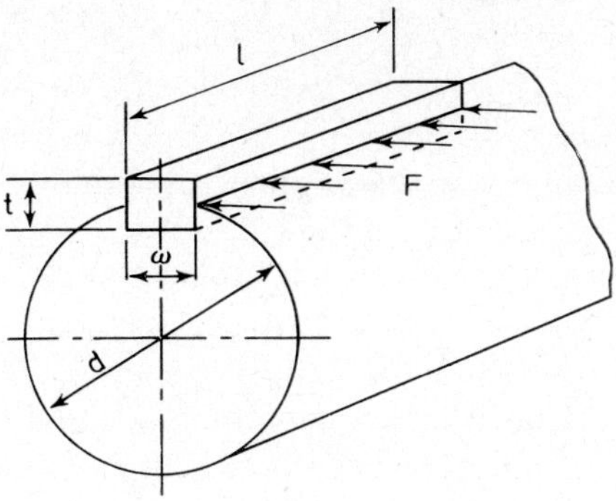

Fig. 3.2. Typical sunk key

buted along the effective length, and acts at the surface of the shaft. Thus, if the torque to be transmitted by the key is M_t, the total force acting on the key is given by

$$F = \frac{2M_t}{d} \tag{3.17}$$

Assuming that the shear force is uniformly distributed across the shear area, the shear stress is given by

$$\tau = \frac{F}{wl} = \frac{2M_t}{dwl} \tag{3.18}$$

The stress due to crushing is similaı ly given by the force divided by the area in crushing, i.e.

$$\sigma_c = \frac{2F}{tl} = \frac{4M_t}{dtl} \tag{3.19}$$

where w = width of key,
t = thickness of key.

Thus, for a given material, if the limiting values for τ and σ_c are known, the required length of key can be determined. On the other hand, if a desirable value for the key length l is known, a material may be selected to give the necessary reserve factor. The limiting value for τ is usually taken as the shearing yield strength divided by a reserve factor having a value within the range

1·5–4·5, depending upon the nature of the load. Faires[11] suggests a value of $RF = 1{\cdot}5$ for smooth loads, $RF = 2\text{–}2{\cdot}25$ for minor shock loading, and up to 4·5 for severe shock loads. In addition to the crushing of the key, there will also be the associated possibility of crushing of the shaft (keyway) or hub (keyway). Consequently, three conditions need to be considered in connection with crushing strength, namely that associated with the key itself, that associated with the shaft and that associated with the hub. The lowest value should be used in determining the required length l from equation 3.19. The limiting value for σ_c is the lowest of the three yield strengths in compression divided by the appropriate reserve factor (as indicated above). Frequently, a hub will be fitted with a press fit, so that some frictional forces act between the surfaces of the shaft and hub. The above simple stress analysis assumes that this is negligible. In some instances, a grub screw may be used to locate a hub axially, in which case care should be taken in the detail drawings to ensure that the grub screw is screwed down on the key rather than on the shaft.

Keys are used to connect hubs and shafts in the manner discussed above by machining keyways in the members to be connected. This not only removes material from the shaft and hub, but also has the adverse effect of introducing stress concentration, thus reducing the fatigue strength. In this connection, fatigue tests carried out by Peterson[12] indicate that the shape of the keyway end can have a considerable influence on the resistance to fatigue from reversed bending stresses, the so-called sled-runner end having considerable advantage over the somewhat more common profiled keyway; these two types of keyway end are illustrated in Fig. 3.3. The actual values of fatigue strength reduction factors are dependent upon a number of other factors, such as material and heat-treatment, but values of $K_f = 1{\cdot}6$ for sled-runner keyways and $K_f = 2$ for profiled keyways have been suggested [13] as reasonable and conservative to be used with nominal bending stresses; these values, of course, apply at the endurance limit, and may be much less than this for shorter fatigue lives. The fatigue strength reduction factor for keyways under torsion depends very much upon the geometry,

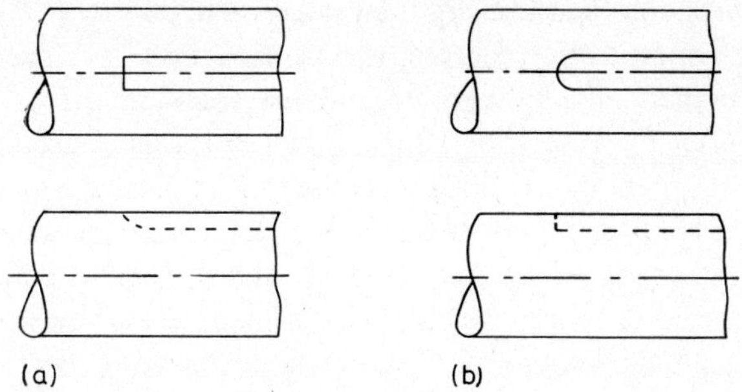

Fig. 3.3. Sled-runner and profiled keyways (a) Sled-runner end (b) profiled end

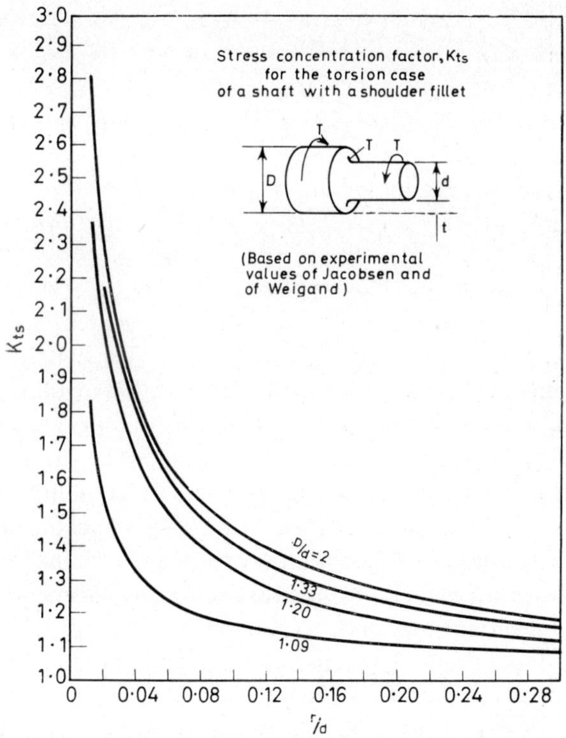

Fig. 3.4. Stress concentration Factor for Keyway under torsion (Reproduced from 'Stress Concentration Design Factors' by kind permission of John Wiley and Sons)

and in particular the radius at the bottom of the keyway. Theoretical stress concentration factors for this condition have been obtained mathematically by Leven[14] and presented in the form of a design curve by Peterson[15]; this design curve is reproduced by Duggan[1] (Appendix C) and is repeated in Fig. 3.4 for convenience.

When a shaft is highly stressed, the stresses in the usual types of sunk keys may be excessively high, and in such circumstances the use of two or more keys may be considered. However, modern

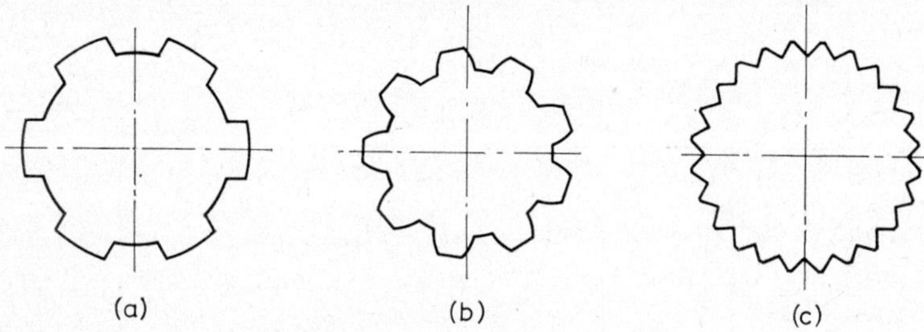

Fig. 3.5. Splined and serrated shafts (a) Straight sides splines (b) Involute splines (c) Straight sided serrations

design favours the use of splines or serrations. Splines may be thought of as multiple keys spaced around the circumference, and machined as an integral part, of the shaft. These may be either straight sided, as indicated in Fig. 3.5a or of involute form as shown in Fig. 3.5b. Splines are used in many applications where the strength of keys are inadequate, particularly in the automobile industry. Straight sided splines are produced by milling away the spaces between the splines on the shaft, and by broaching the mating recesses in the hub. Involute splines are produced by similar methods to those used for machining internal and external gears. The British Standard recommendation for straight sided splines and serrations (in inch dimensions) is given in BS 2059: 1953[16]. Part 1 of this specification includes six straight sided splines only (Fig. 3.5a) and provides for three different types of fit. *Fit 1* represents the closest fit and is used where the minimum amount of backlash is necessary; *Fit 2* is used for conditions where ease of assembly is required; *Fit 3* has larger allowances than Fit 2 and is used when such clearances can be accepted in the assembly. Extracts from Part 1 of BS 2059 are reproduced in Appendix A by permission of the British Standards Institution. In the United States, straight sided splines with 4, 6, 10 and 16 splines have been standardised by the Society of Automotive Engineers[17], and in other countries the appropriate national standards should be consulted. Part 2 of BS 2059: 1953 deals with straight sided serrations (Fig. 3.5c) and again provides for three different types of fit. *Fit 1* which is an interference fit, is intended for permanent or semi-permanent assemblies, and to permit assembly it is necessary to heat the internally serrated member or cool the externally serrated member; *Fit 2* is a transition fit, and is intended for assemblies requiring accurate location of serrated members, but which must be readily capable of being dismantled; *Fit 3* is a clearance or sliding fit, intended for general applications. Serrations should generally be confined to assemblies which require no relative axial movement in operation. Extracts from BS 2059, Part 2 are reproduced in Appendix A.

The British Standard recommendations for involute splines is given in BS 3550: 1963[18], the basic dimensions of which are the same as those given in the American Standard ASA Standard B5·15–1960[19]. For further details, it is suggested that the reader refer to the above standards, or in other countries the appropriate national standards which apply.

3.5. EFFECT OF SHRINK FITS ON FATIGUE STRENGTH

Another important factor concerning the fatigue life of a shaft assembly is that of the fatigue strength reduction factor due to press or shrink fits between the hubs of wheels, gears, collars, etc., and the shaft. It is suggested by Horger[20] that the fatigue strength of a press- or shrink-fitted assembly approaches that of a member machined from one piece of material which simulates the geometry of the shaft and hub assembly, and further that this weakness due to shape may be aggravated by *fretting corrosion*, i.e. the small relative movement which may occur between the mating surfaces of the shaft and hub. A commonly quoted value for the fatigue strength reduction factor for a press fitted assembly under bending is in the region of $K_f = 2$, although widely differing values may be obtained. The effect on fatigue strength of a press- or shrink-fitted assembly is more severe for larger sizes than for smaller sizes, a factor of great practical importance[21]. Where a press-fit is used in conjunction with a keyway, the fatigue strength reduction factor may typically be in the region of $K_f = 3$. The above factors are quoted as being typical, but the many parameters involved mean that factors very different may apply in particular cases. The results obtained from fatigue tests carried out on about 170 solid shaft and interference-fit collar assemblies have been analysed and presented in the form of fatigue curves by the Engineering Sciences Data Unit[22], and further information on methods of improving the fatigue strength of such assemblies is given in 'Engineering Science Data, Item No. 68002[23] .

3.6. SHAFT COUPLINGS

The decision concerning the selection of a suitable shaft coupling involves the designer in the consideration of a number of factors. Apart from the direct use in connecting shafts of separately manufactured units, couplings may be used to provide for misalignment of the shafts or to introduce mechanical flexibility (or stiffness) into the system, thus altering the deflection and vibrational characteristics. In addition, the proper selection of a coupling can reduce the transmission of shock from one shaft to another and provide some protection against overload. There is no shortage of commercially available couplings as standard components and, in general, the selection can usually be made from manufacturers' catalogues which describe and tabulate design characteristics, operating capabilities, etc.

Shaft couplings are usually divided into two categories according to whether they have flexibility or not; those having no flexibility are termed *rigid couplings* and those having some flexibility are termed *flexible couplings.*

Rigid couplings are usually only suitable for connecting accurately aligned shafts which rotate at relatively slow speeds, or very flexible shafts such as may be used in power transmission systems. If rigid couplings are used to connect shafts which have significant misalignment, either lateral or angular (by angular is meant the slope of one shaft relative to the other), this might quite well have the effect of imposing excessively heavy loads on the couplings, bearings and shafts. Further, shafts which are initially in alignment may develop misalignment under service conditions. Consequently, the big disadvantage associated with rigid couplings is that indiscriminate use can lead to forces and resultant stresses of unknown magnitude, leading to fatigue failures in some instances, and great care must be exercised in their use. Nevertheless, it should be mentioned that rigid couplings have been used successfully, and to advantage, in connecting the shafts of high speed rotors[24].

From the designer's point of view, the simplifying assumptions which are so often made when calculating strength, deflection and vibration characteristics for a system incorporating a flexible coupling, cannot be made for the same system in which a rigid coupling is utilised, even neglecting the unknown forces due to misalignment. Whenever a rigid coupling connects two shafts and the total number of supporting bearings is more than two, the system is statically indeterminate.

One method of rigidly connecting shafts is to incorporte flanged ends on the shafts themselves, and machine one face with a spigot and the other with a recess, as indicated in Fig. 3.6. A series of equally spaced bolts positioned on a common pitch diameter are then used to connect the two flanges. Frequently, the bolts are made a good fit in the holes in order that the torque transmitted may be assumed uniformly distributed amongst them. In actuality, if the bolts are properly tightened to produce the required pre-tension[1], under the action of the torque the frictional forces between the faces will be sufficient to transmit the torque, and the holes for the bolts may be made clearance holes; this is of obvious advantage from the point of view of cost and also from the standpoint of interchangeability. The relationship between the torque to be transmitted and the pre-tension in the bolts is given by

$$M_t = f F_i n \frac{D+d}{4} \tag{3.20}$$

where M_t = torque to be transmitted,
f = coefficient of friction between the faces (a conservative value is about 0·1),
F_i = initial pre-tension in each bolt,
n = number of bolts,
D = outer diameter of friction face,
d = inner diameter of friction face.

There are obvious disadvantages associated with the above method of rigidly connecting shafts, although large forged machine shafts sometimes use this method.

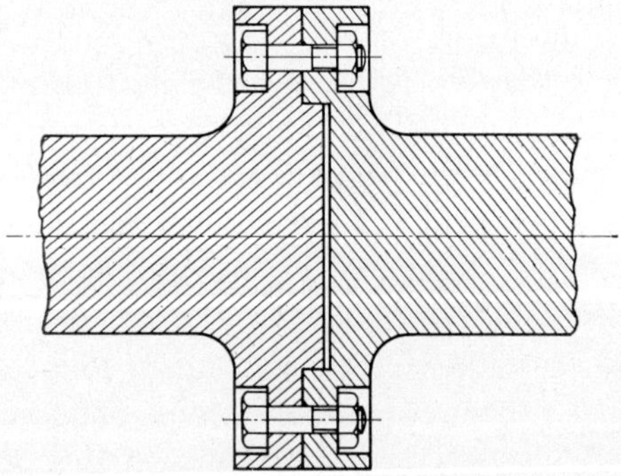

Fig. 3.6. Flanged shaft ends

A rigid coupling which utilises the above principle is a *flange coupling*. This consists essentially of two half couplings, each of which is keyed to the end of the respective shaft, and the two halves are bolted together by a series of equally spaced bolts on a common pitch diameter. Usually, one half coupling is machined with a spigot which locates in a recess machined in the other half, thus providing for alignment of the two halves. Such couplings are readily available as standard components, as illustrated in Fig. 3.7.

Fig. 3.7. Rigid pulley coupling (Manufactured by Crofts (Engineers) Ltd., Bradford)

Flexible couplings are preferred to rigid couplings, because they are capable of operating satisfactorily under conditions of misalignment, or are capable of reducing the transmission of shock from one shaft to another. A wide variety of flexible couplings are available to the designer as standard components, and selection will be made on the specific requirements, including load conditions, speed requirements, relative motion (i.e. transverse or angular misalignment of the shafts), equipment protection, environment, installation and maintenance, and cost; these factors are discussed in detail by the American Society of Mechanical Engineers[25].

A flexible coupling consists essentially of two half couplings with an intermediate member (or members), which may be either flexible or floating, providing the flexibility.

The simplest type of flexible coupling is the *flexible pin coupling*, consisting of two half couplings provided with flanges, and connected together by means of flexible pins, as illustrated in Fig. 3.8. This type of coupling permits relative axial movement but has small torsional flexibility.

Fig. 3.8. Crown pin flexible coupling (Manufactured by Crofts (Engineers) Ltd., Bradford)

Fig. 3.9. Multiflex flexible spring coupling (Manufactured by Crofts (Engineers) Ltd., Bradford)

The *flexible spring coupling* shown in Fig. 3.9, consists of two slotted members connected by a number of steel springs, the springs providing the flexibility. This type of coupling, which is capable of accommodating loss of alignment and damping shock and vibration, is particularly suitable for high power transmission. A desirable feature is that as the torque increases, the coupling becomes stiffer. The springs are packed with grease and the assembly is enclosed in grease-tight covers. In some designs, a continuous spring is used instead of the multi-springs shown here, a design which is frequently termed a Bibby coupling.

Gear couplings transmit torque through the teeth or splines of the member attached to the driving shaft and having external teeth, to the

(a)

(b)

Fig. 3.10. Internal gear couplings (a) NTS *type (b)* MB *type (Manufactured by Crofts (Engineers) Ltd., Bradford)*

mating member with internal teeth; this outer member, which may be in one piece or separable and connected by bolts, then transmits the torque through the other pair of mating teeth to the driven shaft. Such couplings, Fig. 3.10, are used primarily to accommodate axial misalignment, and have practically no torsional flexibility, other than that which may be provided by clearance. Gear couplings are oil lubricated, the oil being retained within the coupling by means of oil-tight covers which are an integral part of the design, and the action of centrifugal force supplies oil to the gear teeth.

Cushion type flexible couplings, such as those illustrated in Fig. 3.11, are relatively inexpensive and are used extensively for light

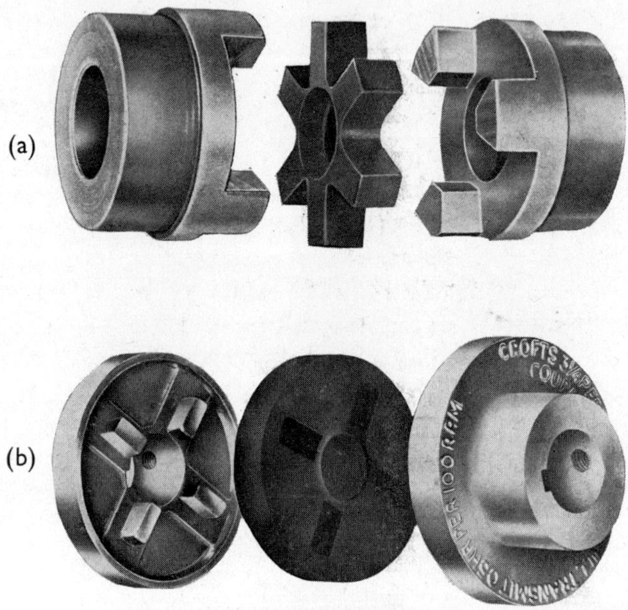

Fig. 3.11. Cushion type flexible couplings (a) SP type (b) Perfect coupling (Manufactured by Crofts (Engineers) Ltd., Bradford)

power transmission requirements. The flexibility is obtained from the intermediate cushion, which is usually made from oil resistant synthetic rubber.

This brief description of couplings by no means exhausts the variety of available types, but simply serves to indicate some of the more common commercial units. For more specific information the reader should consult the manufacturers' catalogues, such as Crofts (Engineers) Ltd.[26]

3.7. CONCLUDING REMARKS

In this chapter the factors involved in the design of shaft systems have been considered, and particular attention has been given to the selection of a suitable material and the determination of the shaft geometry, based on considerations of strength, rigidity and vibration characteristics.

It has been shown that in order to arrive at a physical system it is necessary to make some design decisions; the physical system, i.e. the geometry and material, being usually established based on strength considerations.

Having determined the geometry and selected a suitable material, it is then necessary to convert the physical system into a mathematical model; this mathematical model may then be analysed to yield the rigidity and vibrational characteristics. The methods relating to vibrational analysis have been thoroughly discussed in Chapter 2, and the reader should have no difficulty in applying these methods to shaft systems.

Finally, some discussion relating to shaft couplings has been undertaken, with a brief discussion of a variety of different types.

3.8. EXAMPLE

A schematic layout of a system for a forage harvester is indicated in Fig. 3.12. This consists of two shafts, each supported in its own bearings, connected by means of a flexible coupling. The speed of operation is 450 rev/min.

Power input to the system is 25 kW (33·5 hp) by means of a double ply flat belt through a 750 mm (somewhat less than 30 in) diameter pulley positioned at *A*. A double chain sprocket having a pitch diameter of 250 mm (somewhat less than 10 in) is positioned at *B*, and provides 20 kW (approximately 27 hp) to drive a cutter. At *C*, a *V*-belt pulley with a pitch diameter of 250 mm (somewhat less than 10 in) provides 5 kW (approximately 6·5 hp) to a blower.

It is necessary to carry out a design study which enables the selection of a suitable material and the determination of shaft diameters at the critical sections.

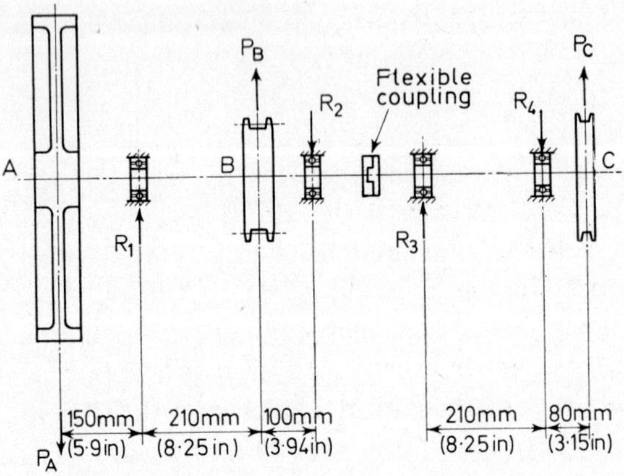

Fig. 3.12. Schematic layout for power transmission shaft

Solution

In determining the total pull at the pulleys and sprockets, the semi-empirical equation presented in Section 5.5 will be used, namely:

$$P = \frac{19\,100\,kW\,F_x}{DN} \quad \text{Newtons (N)} \qquad (3.21)$$

where P = total pull, N,

kW = power transmitted, kW or kJ/s,
F_x = belt or chain factor (see Section 5.5),
D = effective diameter of pulley or sprocket, m,
N = speed of rotation, rev/min.

Reference to Chapter 5 indicates the following belt and chain factors:

For double ply flat belt, $F_x = 2{\cdot}5$

For double chain, $F_x = 1{\cdot}25$

For *V*-belts, $F_x = 1{\cdot}5$

Thus, from equation 3.21 the effective pulls at A, B and C are found.

$$\downarrow P_A = \frac{19\,100 \times 25 \times 2{\cdot}5}{0{\cdot}750 \times 450} = 3\,540 \text{ N (796 lbf)}$$

$$\uparrow P_B = \frac{19\,100 \times 20 \times 1{\cdot}25}{0{\cdot}250 \times 450} = 4\,240 \text{ N (955 lbf)}$$

$$\uparrow P_C = \frac{19\,100 \times 5 \times 1{\cdot}5}{0{\cdot}250 \times 450} = 1\,275 \text{ N (287 lbf)}$$

For the purpose of calculating the bearing reactions, it will be assumed that the flexible coupling offers negligible transverse stiffness, so that each shaft may be considered as a separate system. If this assumption is not considered reasonable, or if a rigid coupling connected the two shafts, the problem becomes somewhat more involved and the analysis entails that of a multi-supported beam[2].

By the usual method of statics,

$$\uparrow R_1 = 3\,885 \text{ N (868 lbf)}$$
$$\downarrow R_2 = 4\,585 \text{ N (1 025 lbf)}$$
$$\uparrow R_3 = 485 \text{ N (108 lbf)}$$
$$\downarrow R_4 = 1\,760 \text{ N (393 lbf)}$$

The bending moment diagrams are obtained in the usual way, and twisting moment diagrams are obtained by noting the power which is transmitted along the various sections of the shafts. Thus, the loading, bending moment and twisting moment diagrams are as indicated in Fig. 3.13. In general, the power transmitted is given by

$$W = \frac{2\pi N M_t}{60} \tag{3.22}$$

From A to B $$M_t = \frac{60 \times 25\,000}{2 \times \pi \times 450} = 530 \text{ Nm (4 680 lbf in)}$$

From B to C $$M_t = \frac{60 \times 5\,000}{2 \times \pi \times 450} = 106 \text{ Nm (940 lbf in)}$$

For the shaft carrying pulleys A and B, the critical section clearly occurs at the bearing reaction R_1, where the bending moment and twisting moment respectively have the values

$$M = 530 \text{ Nm } (4\,680 \text{ lbf in})$$
$$M_t = 530 \text{ Nm } (4\,680 \text{ lbf in})$$

The critical section on the shaft carrying pulley C occurs at the bearing reaction R_4. Attention will be confined to the critical section R_1.

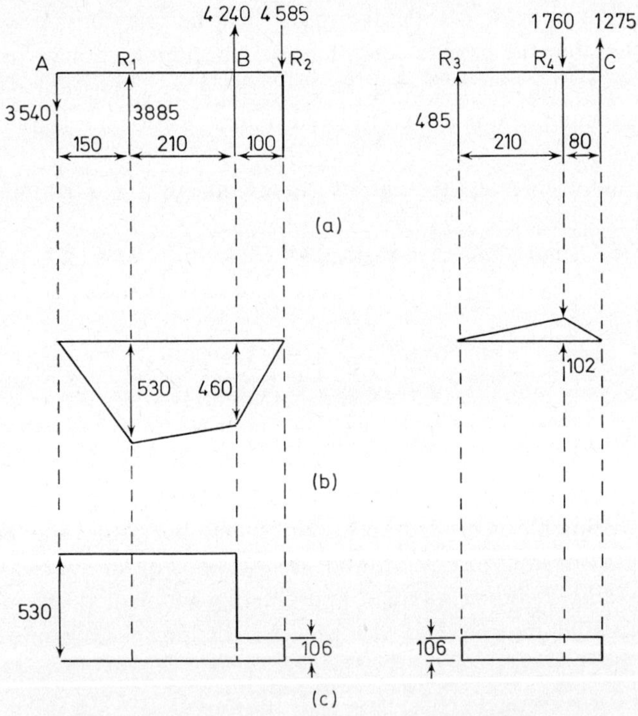

Fig. 3.13. Loading, bending moment and twisting moment diagrams (a) Loading diagram (b) Bending moment diagram (c) Twisting moment diagram

Since for each revolution of the shaft the outer fibres are subjected alternately to tension and compression, the criterion of failure will be fatigue. Where the bearing fits there will be a stress concentration due to the fillet radius in the corner, and the interference fit between the bearing inner ring and the shaft. Typically, at this section the shaft would have a geometry similar to that indicated in Fig. 3.14.

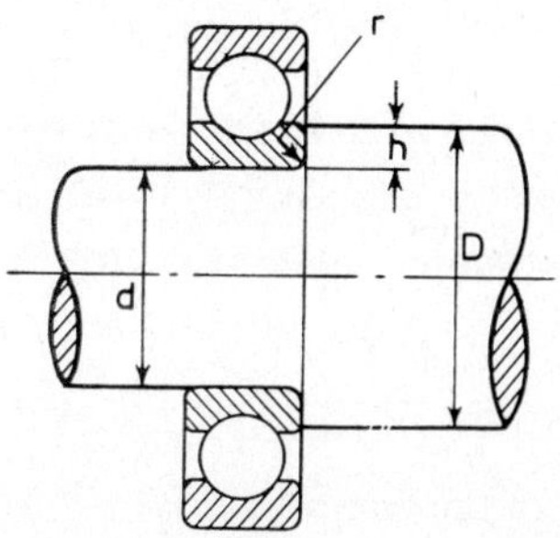

Fig. 3.14. Typical geometry at critical section

At this stage, the size of bearing required for the desired life could be considered, and the geometry fixed accordingly. A fatigue analysis would then be required to establish the integrity of the design based on the selection of a suitable shaft material. The selection of rolling contact bearings is discussed in Chapter 4, and it is shown (see Section 4.19) that a suitable bearing for this application is that of a single row deep groove ball bearing. Further, corresponding to a life of 5 000 h at a speed of 450 rev/min, suitable bearings may be selected from manufacturers' catalogues. Thus, for this application, suitable bearing sizes would be

BRE 055 (55 mm bore)
BRL 035 (35 mm bore)
BRM 030 (30 mm bore)
BRH 020 (20 mm bore)

depending upon compatibility with the required shaft size based on strength requirements. It will now be necessary to make a tentative decision regarding a suitable shaft material, and estimate the component fatigue strength using some assumptions. From the previous discussion relating to shaft materials, a suitable choice of a moderately high strength steel might be En 24 (V) which, by reference to Appendix B of Duggan[1], has the following mechanical properties:

$$S_u = 1\,000 \text{ MN/m}^2 \ (146\,000 \text{ lbf/in}^2)$$

$$S_p = 800 \text{ MN/m}^2 \ (116\,000 \text{ lbf/in}^2)$$

Designing the shaft for a life in excess of 10^6 cycles (i.e. assumed infinite life), and assuming the approximate relationship between endurance limit in rotating bending S_e and the tensile strength S_u (see Duggan[1]),

$$S_e = 0.5S_u = 500 \text{ MN/m}^2 \ (73\,000 \text{ lbf/in}^2)$$

The fatigue strength of the component at 10^6 cycles may be approximated[1] by the equation

$$S'_e = \frac{S_e}{K_f} C_s C_L K_s \tag{3.23}$$

where S'_e = component fatigue strength at 10^6 cycles,
S_e = intrinsic fatigue strength at 10^6 cycles,
K_f = fatigue strength reduction factor,
C_s = size factor,
C_L = loading factor,
K_s = surface finish factor.

These factors, and a procedure for determining the appropriate numerical values, have been thoroughly discussed in the first book in this series[1], and have the following values for this situation:

$$C_s = 0{\cdot}86$$

$$C_L = 1{\cdot}00$$

$$K_s = 0{\cdot}75$$

$$K_f = 1 + q(K_t - 1)K_s$$

The value for K_f can only be obtained when the geometry has been defined, and at the present time it is necessary to make a conservative estimate of its probable value. A reasonable assumption would be to let $K_f = 2$.

Substituting numerical data into equation 3.23,

$$S'_e = \frac{500}{2} \times 0{\cdot}85 \times 1{\cdot}0 \times 0{\cdot}75$$

$$= 160 \text{ MN/m}^2 \text{ (13 300 lbf/in}^2\text{)}$$

The required shaft diameter can now be determined using the Goodman relationship[1] in conjunction with an appropriate theory of failure. Using the Mises criterion, this gives the result previously presented by equation 3.14a, i.e.

$$\left(\frac{\sigma_{sx}}{S_u}\right)^2 + 3\left(\frac{\tau_s}{S_u}\right)^2 = 1$$

where the symbols are as defined.

Assuming that the torque remains sensibly constant (neglecting starts and stops), the alternating torsional shear stress component will be zero, and

$$\tau_s = \tau_m = \frac{M_t}{Z_t} = \frac{M_t}{2Z}$$

where Z_t = section modulus in torsion = $2Z$
Z = section modulus in bending.

$$\therefore \quad \tau_s = \frac{265}{Z} \text{ N/m}^2 \left(\frac{2\,340}{Z} \text{ lbf/in}^2\right)$$

Since the outer fibres of the shaft are subjected alternately to equal numerical values in tension and compression, the mean bending stress component is zero, and

$$\sigma_s = c_1 \sigma_{\text{alt}}$$

where

$$c_1 = \frac{S_u}{S'_e} = \frac{1\,000}{160} = 6{\cdot}25$$

$$\sigma_{\text{alt}} = \frac{M}{Z} = \frac{530}{Z} \text{ N/m}^2 \left(\frac{4\,680}{Z} \text{ lbf/in}^2\right)$$

$$\therefore \quad \sigma_s = \frac{6{\cdot}25 \times 530}{Z} = \frac{3\,312}{Z} \text{ N/m}^2 \left(\frac{29\,300}{Z} \text{ lbf/in}^2\right)$$

Substituting for τ_s and σ_s into equation 3.14a, and introducing a reserve factor *RF* of 1·5,

$$\left(\frac{\sigma_s RF}{S_u}\right)^2 + 3\left(\frac{\tau_s RF}{S_u}\right)^2 = 1$$

$$\left(\frac{3\,312\times 1{\cdot}5}{Z\times 10^6\times 1\,000}\right)^2 + 3\left(\frac{265\times 1{\cdot}5}{Z\times 10^6\times 1\,000}\right)^2 = 1$$

From which

$$Z = 5\times 10^{-6}\ \text{m}^3 = 5\times 10^3\ \text{mm}^3$$

$$= \frac{\pi d^3}{32}$$

$$d^3 = \frac{32\times 5\times 10^3}{\pi} = 50{\cdot}93\times 10^3\ \text{mm}^3$$

$$d = 37\ \text{mm (1{\cdot}456 in)}$$

This suggests that if a slightly smaller value than 1·5 can be accepted for the RF, a suitable choice of bearing would be BRL 035. This requires a certain minimum shoulder height *h*, and a shaft corner radius *r*, both of which can be obtained from the manufacturers' catalogues (see Chapter 4). Corresponding to BRL 035, and referring to Fig. 3.14, the following geometry is established.

$$d = 35\ \text{mm}$$
$$h = 3{\cdot}5\ \text{mm}$$
$$D = d + 2h = 42\ \text{mm}$$
$$r = 1\ \text{mm}$$

A more reliable estimate of K_f may now be obtained from the equation

$$K_f = 1\times q(K_t - 1)\,K_s$$

where K_t = theoretical stress concentration factor (see Duggan[1]) = 2·3

K_s = surface finish factor = 0·75

q = notch sensitivity index (see Duggan[1]) = $1/(1+a/r)$

a = material constant = 0·098 mm

r = notch radius = 1 mm.

Substituting this data into the equation for K_f gives a value of 1·89, which is somewhat less than the value originally assumed.

REFERENCES

1. DUGGAN, T. V., *Applied Engineering Design and Analysis*, Design Technology Series, Iliffe Books Ltd., London (1970)
2. DUGGAN, T. V., *Stress Analysis and Vibrations of Elastic Bodies*, Iliffe Books Ltd., London (1964)
3. JUVINALL, R. C., *Engineering Considerations of Stress, Strain and Strength*, McGraw-Hill Book Company, New York (1967)
4. DUGGAN, T. V. 'Fatigue as a Design Criterion', *Engineering Materials and Design*, Part 2, 875–877, (June, 1967) **10,** No.6
5. DUGGAN, T. V., *Fatigue as a Design Criterion* (A Post-Graduate Design Course, 4th–15th November, 1968), Department of Mechanical Engineering and Naval Architecture, Portsmouth Polytechnic (to be published).
6. WEBB, C. R., *Shafts in Alloy Steels*, The International Nickel Company (Mond) Ltd., Thames House, London, 2nd edn. (1959)
7. B. S. 46: PART 1: 1958, *Keys and Keyways*, British Standards Institution, 2, Park Street, London, W.1.
8. B. S. 4235: PART 1: 1967, *Metric Keys and Keyways*, British Standards Institution, 2 Park Street, London, W.1.
9. A. S. A. STANDARD B 1.71, *Keys and Keyseats*, American Socitey of Mechanical Engineers, New York, (1967)
10. B. S. 970: 1955, *Wrought Steels in the Form of Bars, Billets and Forgings, for Automobile and General Engineering Purposes, En Series*, British Standards Institution, 2 Park Street, London, W.1.
11. FAIRES, V. M., *Design of Machine Elements*, The Macmillan Company, New York, 4th edn. (1965)
12. PETERSON, R. E. 'Fatigue of Shafts having Keyways', *Proc. Amer. Soc. Test Mat.* 32, **11** 413, (1932)
13. LIPSON, C., NOLL, G. C. and CLOCK, L. S., *Stress and Strength of Manufactured Parts*, McGraw-Hill Book Company, New York (1950)
14. LEVEN, M. M., 'Stresses in Keyways by Photoelastic Methods and Comparison with Numerical Solutions', *Proc. of Soc. for Expt. Stress Analysis*, **7,** No.2, 141 (1950)
15. PETERSON, R. E., *Stress Concentration Design Factors*, J. Wiley and Sons, Inc., New York (1953)
16. B.S. 2059: 1953, *Straight Sided Splines and Serrations*, British Standards Institution, 2, Park Street, London, W.1.
17. S.A.E. HANDBOOK, 'Parallel Side Splines for Soft Broached Holes in Fittings', SAE J 499, *SAE Handbook*, 603–604, New York (1968)
18. B.S. 3550: 1963, *Involute Splines*, British Standards Institution, 2, Park Street, London, W.1.
19. A.S.A. B. 5–15 –1960, *Involute Splines, Serrations and Inspection*, American Society of Mechanical Engineers. New York (1960)
20. HORGER, O. J., 'Press- and Shrink-Fitted Assembly', *A. S. M. E. Handbook, Metals Engineering Design*, Chapter 8.13, 340, McGraw-Hill Book Company, 2nd edn. (1965)
21. COYLE, M. B. and WATSON, S. J., 'Fatigue Strength of Turbine Shafts with Shrunk-on Discs, *Proc. Inst. Mech. Engrs.*, **178**, 147–183, London (1964)
22. ENGINEERING SCIENCE DATA, ITEM NO. 68005, *Shafts with Interference-Fit Collars, Part IV:* Fatigue Strength of Plain Shafts, January (1968) Engineering Sciences Data Unit, 251–259 Regent Street, London, W.1.

23. ENGINEERING SCIENCE DATA ITEM NO. 68002, *Shafts with Interference-Fit Collars, Part 1:* Some Factors Affecting the Design of an Assembly, Engineering Sciences Data Unit, 251–269 Regent Street, London, W.1.
24. MITCHELL, J. M., 'The Critical Speeds of Rigidly Coupled Rotors', *G.E.C. Journal*, **23**, No. 2 (April 1956)
25. A. S. M. E., *New Design Standards for Flexible Couplings*, American Society of Mechanical Engineers, New York, (1966)
26. 'CROFT FLEXIBLE COUPLINGS', Crofts (Engineers) Ltd., Bradford, England.

Chapter 4

Bearings, Wear and Lubrication Considerations in Design

When two surfaces in contact move relative to each other under the action of an applied force, there will be some resistance to motion due to friction; this frictional resistance generates heat and causes wear of the surfaces. When the surfaces are free from contaminating films, the frictional resistance is said to be due to dry or Coulomb friction; even if the contacting surfaces are produced to a high degree of accuracy and are apparently quite smooth to the eye and touch, if examined under a microscope, such surfaces will be seen to consist of peaks and troughs, as illustrated in Fig. 4.1. Relative

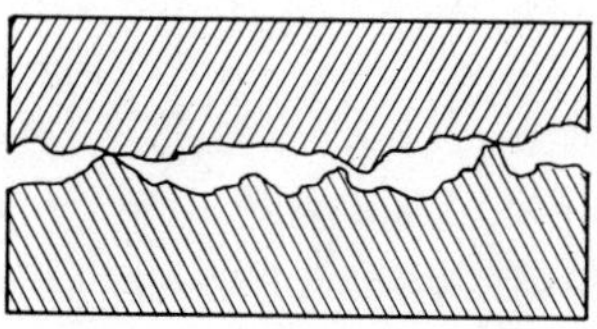

Fig. 4.1. Appearance of contacting surfaces

motion of such surfaces in contact may have the effect of causing the peaks to be welded together, and the consequential force required to shear these welds apart may be considerable. In any event, the frictional forces under conditions of dry or Coulomb friction are most undesirable. If this condition is allowed to continue, the surfaces will very quickly wear or break down, and if the temperature increase due to the heat generated is sufficiently high, complete failure may occur due to seizure of the parts. Thus, whenever two surfaces in contact move relative to each other, as for example the teeth of mating gears or the profile of a cam and its contacting follower, it is usual to reduce the frictional forces by the use of a lubricant. For the present purposes any substance which, when inserted between the moving surfaces, has the effect of reducing friction and wear, will be considered a lubricant. A lubricant does not eliminate friction entirely, but has the effect of greatly reducing it.

In this chapter, some elements of wear and lubrication will be discussed, and the results applied to the design of journal bearings. The various types of rolling contact bearings will also be considered, and attention will be given to the proper selection of such bearings.

Finally, in this brief introduction, it should be mentioned that the modern name for the subjects of wear and lubrication is *tribology*, and this term has now been widely accepted.

4.1. LUBRICATION AND WEAR

Perhaps one of the most important and costly hazards with which the engineer has to contend is that of wear. Lipson[1] defines wear as the deterioration of surfaces due to use, and suggests that it is often the major factor limiting the life and performance of machine components. Since wear occurs whenever force and motion are involved, its importance as a design factor is evident. The phenomenon of wear is, however, quite complex, since it may occur due to a variety of different causes and is dependent upon many factors, such as the nature of the load; relative velocity of the surfaces; nature of the surfaces themselves; type of lubricant; temperature of operation, and so on. Lipson[1] has classified wear processes into the following categories:

(a) galling, scuffing, scoring and seizing;

(b) abrasion;
(c) pitting;
(d) frettage;
(e) cavitation erosion and
(f) galvanic corrosion.

These wear processes will not be discussed at the present time, and the reader is recommended to consult the above mentioned reference for further details.

Even when a lubricant is present, although it is desirable to maintain a film thickness sufficient to ensure that actual contact of the moving surfaces does not occur, solid contact is sometimes inevitable. Under these conditions, *boundary lubrication* is said to occur. Boundary lubrication may be considered to be a condition somewhere between that of *thick film* or *perfect* lubrication, and that of dry or Coulomb friction. With thick film lubrication, if it can be maintained at all times, wear is hardly a problem at all, because the frictional forces are quite low. Since in many instances, however, dry or boundary lubrication may occur at least part of the time, it is necessary to consider means of combating wear. This is largely a matter of the correct selection of materials and surfaces. By far the most important factor for reducing wear is high hardness, but good surface finish of the parts and the selection of materials which have low solid-solubility also help.

Wear and lubrication has become more important as a design criterion in modern times, due to the development of high speed machinery, and it is true to say that the successful operation of gas turbines, high speed aircraft and the like, is largely dependent upon the application of lubrication and bearing technology. Since wear, as it is discussed here, is largely connected with the mechanism of friction, one obvious way of combating it is to reduce the frictional forces as much as possible. This means that dry or Coulomb friction must be prevented as far as possible, by the use of a lubricant which separates the surfaces and replaces the dry or Coulomb friction by the molecular friction of the lubricant.

One of the most important characteristics of a lubricant is its *viscosity*, which may be defined as that property which determines the amount of resistance which a fluid has to a shearing force. This may be illustrated by referring to Fig. 4.2, which indicates two parallel surfaces separated by a lubricant having a film thickness h; one of the surfaces is considered stationary whilst the other moves with a velocity U. If it is assumed that the fluid adheres to each surface so that it has zero velocity at the surface of the stationary

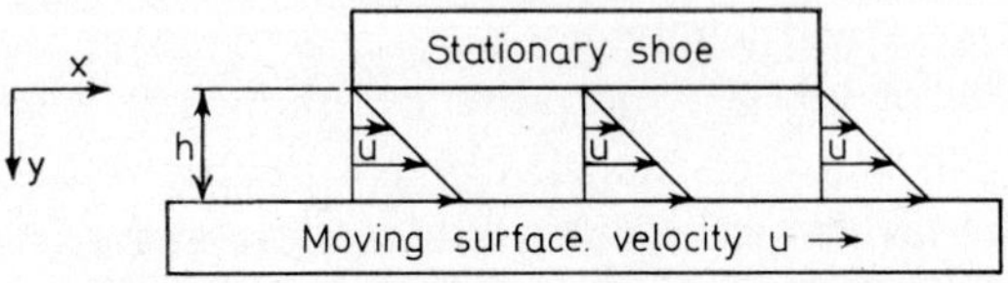

Fig. 4.2. Velocity gradient across parallel film

shoe and velocity U at the surface of the moving plate, the velocity at some intermediate point between the two surfaces will depend upon the distance from the stationary shoe. From Newton's law of viscous flow, the shear stress is proportional to the velocity gradient, and stated mathematically:

$$\tau = \mu \frac{\partial u}{\partial y} = \mu \frac{U}{h} \tag{4.1}$$

where $\partial u/\partial y$ is the velocity gradient, and μ is the absolute viscosity of the fluid.

4.2. MECHANISM OF FLUID LUBRICATION AND REYNOLDS EQUATION

It will be of interest, before considering the design of journal bearings, to consider the mechanism of fluid lubrication in a little more detail. By fluid lubrication is meant that condition where the surfaces are separated by a film of sufficient thickness to prevent contact of the parts (obviously the rougher the surface finish of the parts the greater will be the film thickness necessary to cause complete separation), the only friction then acting being that which occurs within the fluid itself. Fluid lubrication is also called thick film, perfect or hydrodynamic lubrication. Following the celebrated experiments of Beauchamp Tower[2, 3] in the 1880's, Osborne Reynolds deduced that fluid lubrication was the result of hydrodynamic action[4], and depended upon the viscosity of the fluid.

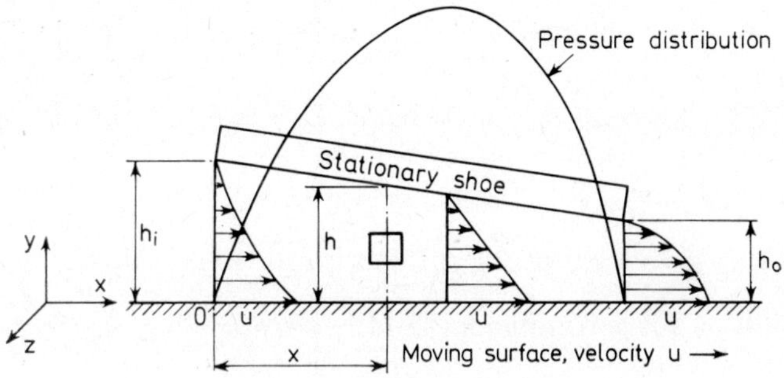

Fig. 4.3. Velocity gradient across converging film

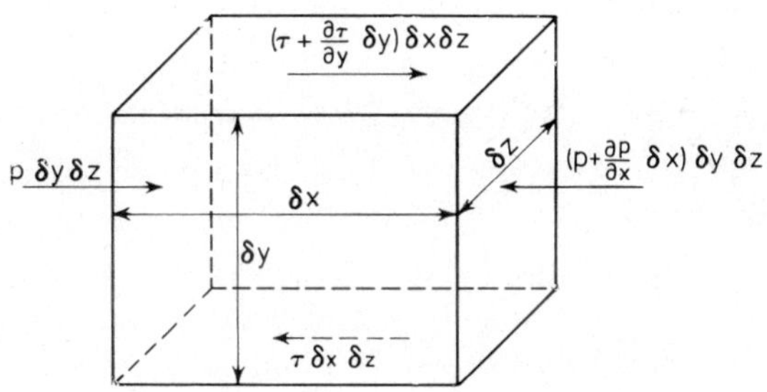

Fig. 4.4. Forces acting on element of fluid

Consider the case where the film separating the surfaces is not of uniform thickness, but rather converges as shown in Fig. 4.3. For continuity of flow, it will be evident that the velocity clearance diagrams cannot always be triangular as was assumed in the case of parallel surfaces, but will have different gradients at different sections, as indicated in Fig. 4.3. In particular, consider an element at some section in the film where the film thickness is h, distance x from the origin, having the elemental lengths δx, δy and δz in the directions Ox, Oy and Oz respectively. Such an element is shown in Fig. 4.4, which indicates the forces acting for equilibrium.

It is assumed that the fluid under consideration completely fills the space and follows Newton's law of viscous flow; that the pressure and viscosity may be assumed constant throughout the film thickness; that the fluid is incompressible and any inertia effects of the forces in the oil film are negligible; that motion of the fluid is laminar; that flow in the direction Oz is negligible.

Writing the equation for static equilibrium:

$$p\,\delta y\,\delta z - \left(p + \frac{\partial p}{\partial x}\,\delta x\right)\delta y\,\delta z + \left(\tau + \frac{\partial \tau}{\partial y}\,\delta y\right)\delta x\,\delta z - \tau\,\delta x\,\delta z = 0$$

which reduces to

$$\frac{\partial p}{\partial x} = \frac{\partial \tau}{\partial y}$$

Substituting for τ from equation 4.1, i.e. assuming Newton's law of viscous flow:

$$\frac{\partial p}{\partial x} = \mu\,\frac{\partial^2 u}{\partial y^2} \tag{4.2}$$

From this equation it may be concluded that if the velocity gradient $(\partial u/\partial y)$ is constant along the length, since $(\partial^2 u/\partial y^2)$ would then be zero, the change in pressure would also be zero, indicating that no pressure could be built up in the film, and consequently no external load could be supported. Thus, a necessary condition for hydrodynamic lubrication is that of a converging film.

The equation for velocity may be obtained by integrating equation 4.2 twice with respect to y, thus:

$$\frac{\partial u}{\partial y} = \frac{1}{\mu}\,\frac{\partial p}{\partial x}\,y + C_1$$

$$u = \frac{1}{2\mu}\,\frac{\partial p}{\partial x}\,y^2 + C_1 y + C_2$$

The constants of integration may be obtained from the boundary conditions, i.e. $u = U$ at $y = O$, and $u = O$ at $y = h$. Sustituting these conditions,

$$C_1 = -\frac{1}{2\mu}\,\frac{\partial p}{\partial x}\,h - \frac{U}{h}$$

$$C_2 = U$$

Thus,

$$u = \frac{1}{2\mu}\,\frac{\partial p}{\partial x}\,(y^2 - hy) + U\left(\frac{h-y}{h}\right) \tag{4.3}$$

The volume of fluid flowing in the x direction, per unit time, through a width of unity in the Oz direction, is given by

$$Q = \int_0^h u\,\mathrm{d}y$$

$$= \int_0^h \left[\frac{1}{2\mu}\,\frac{\partial p}{\partial x}\,(y^2 - hy) + U\left(\frac{h-y}{h}\right)\right]\mathrm{d}y$$

$$= \frac{Uh}{2} - \frac{h^3}{12\mu}\,\frac{\partial p}{\partial x} \tag{4.4}$$

Assuming that the fluid is incompressible, for continuity of flow

$$\frac{\partial Q}{\partial x} = 0$$

$$\therefore \frac{\partial Q}{\partial x} = \frac{U}{2}\frac{\partial h}{\partial x} - \frac{\partial}{\partial x}\left(\frac{h^3}{12\mu}\,\frac{\partial p}{\partial x}\right) = 0$$

$$\frac{\partial}{\partial x}\left(\frac{h^3}{12\mu}\,\frac{\partial p}{\partial x}\right) = \frac{U}{2}\,\frac{\partial h}{\partial x} \tag{4.5}$$

This equation is commonly referred to as Reynolds equation for two dimensional flow. The pressure varies along the length, i.e. in the direction of x, and has its maximum value when $\partial p/\partial x$ is zero. Let the film thickness corresponding to the maximum value of p be $\bar{h}$, so that at this section, from equation 4.4, substituting for $\partial p/\partial x = 0$, we have

$$Q = \frac{U\bar{h}}{2} \tag{4.6}$$

But for continuity of flow, we may equate equations 4.4 and 4.6, giving

$$\frac{U\bar{h}}{2} = \frac{Uh}{2} - \frac{h^3}{12\mu}\,\frac{\partial p}{\partial x}$$

from which

$$\frac{\partial p}{\partial x} = 6\mu U\left[\frac{h-\bar{h}}{h^3}\right] \tag{4.7}$$

Equation 4.7 may be integrated to obtain the pressure distribution, but this requires the evaluation of the film thickness h as a function of x; this relationship depends upon the shape of the film. The converging film shown in Fig. 4.3 is the classical shape of the tilting pad bearing which was developed independently by Michell[5] of Australia and Kingsbury[6] of the United States, in order to provide the converging or wedge shaped film in a thrust bearing. The application of the hydrodynamic theory of lubrication to a tapered wedge is discussed by Fuller[7], to which the reader is referred for further details. The pressure distribution for a converging film is indicated in Fig. 4.3.

From the above discussion it is evident that hydrodynamic lubrication requires motion of the surface relative to the lubricant, and a sufficient supply of lubricant to completely fill the space; further, a converging film is necessary in order to build up sufficient pressure to support the external load.

4.3. BEARING CLASSIFICATION

In Chapter 3, the design of shafts for machines and power transmission systems was studied from the standpoint of strength, rigidity and vibrational characteristics, and throughout the discussions it was assumed that such shafts were supported by bearings. At that time, however, shaft design was considered somewhat in isolation of the supporting bearings, but in a real design the bearings need to be considered an integral part of the system.

Bearings may be defined as machine elements which give the desired restraint to the applied external load, ideally with the minimum amount of friction. They may be divided into two major groups, namely

(a) Sliding bearings, and
(b) Rolling contact bearings.

Phelan[8] uses the term sliding bearings to describe all bearings that do not involve the use of balls or rollers as load supporting elements, and rolling contact bearings as those which depend upon contact between rolling elements to give the desired restraint. This classification is convenient and will be used here.

4.4. SLIDING BEARINGS

Sliding bearings usually operate with thick film lubrication, in which case the surfaces are completely separated by the lubricant, and the performance of the bearing is then characterised mainly by the physical properties of the film. This is a desirable condition of operation and every effort should be made to achieve this condition. Sometimes, however, all the conditions for hydrodynamic or thick film lubrication may not be completely met, and in such cases sliding bearings are said to operate under conditions of *thin-film* or *boundary lubrication*. The exact nature of boundary lubrication and friction is not fully understood, but it seems that in this region viscosity is unimportant, and performance is more dependent upon the chemical properties of the lubricant and the contacting surfaces, and the same factors which affect dry friction also affect boundary friction[7]. In some special cases, sliding bearings may also be required to operate under dry conditions, and the problem becomes one of selecting correct materials to satisfy the product of pressure times velocity and other significant factors. Because sliding bearings are not usually capable of withstanding simultaneously both radial and axial loads, it is desirable to divide them into two groups, namely radial or journal bearings and thrust bearings.

4.5. JOURNAL BEARINGS

Probably the most common application of the hydrodynamic theory of lubrication is encountered in so-called sleeve or journal bearings, used extensively to provide lateral support for rotating shafts. Essentially, journal bearings consist of two elements, namely a bearing bush or surrounding shell, known briefly as the *bearing*, and that part of the shaft which rotates within and is supported by the bearing, known as the *journal*. When the bearing contact angle is 360 degrees, i.e. the bearing completely surrounds the journal, the bearing is termed a full journal type; if the bearing contact angle is 180 degrees or less, it is known as a partial bearing type.

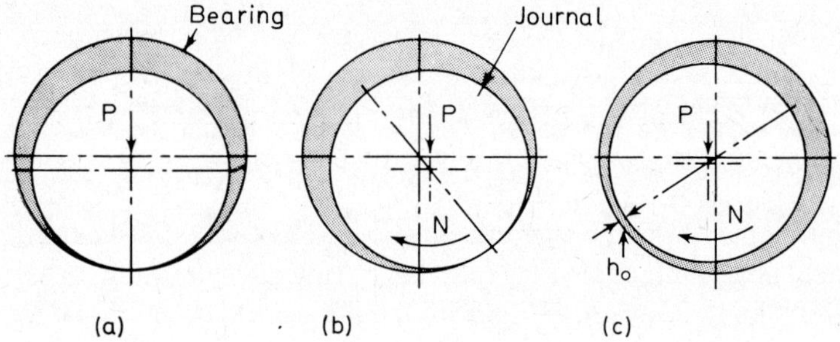

Fig. 4.5. Mechanism of lubrication in a journal bearing (a) Stationary (b) Running-up (c) At Speed

The action of hydrodynamic lubrication in a journal bearing may be illustrated by referring to Fig. 4.5, which indicates three conditions of a journal in its bearing. In Fig. 4.5a, the shaft or journal is at rest and supports an external force *P;* for this condition, solid contact occurs between the journal and its bearing. As the shaft starts to rotate, it will climb up the bearing wall as indicated in Fig. 4.5b, and if a lubricant is present, the conditions of boundary lubrication may apply (in the absence of a lubricant the condition of dry friction will prevail). As the shaft speed is increased, if sufficient lubricant is available, the surface of the rotating shaft will tend to pull lubricant into the wedge shaped clearance area between the shaft and bearing. Thus, the shaft is forced away from the bearing wall due to the oil pressure on the entering side being greater than that on the exit side. Eventually, if all the conditions for hydrodynamic lubrication are met, an equilibrium condition will be reached, as indicated in Fig. 4.5c. At this condition, the surfaces will be separated by a film of oil having a minimum thickness h_0; the actual value of h_0 and the position of the line of centres will depend upon a number of factors, such as the magnitude of the external force, the surface velocity of the rotating shaft, the clearance between the shaft and bearing, the length of the bearing, the properties of the lubricant and temperature of operation, and so on. Fig. 4.6 illustrates the variation of radial and axial pressure in a journal bearing.

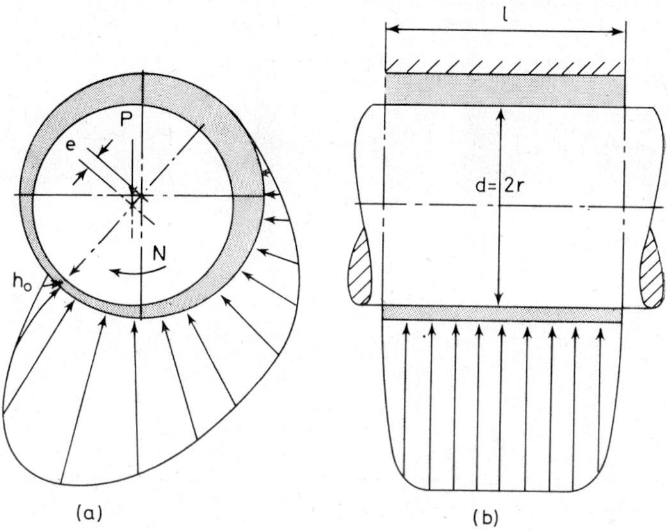

Fig. 4.6. Radial and axial pressure distribution in a journal bearing (a) Radial pressure (b) Axial pressure

The amount of friction generated in a journal bearing can be estimated for lightly loaded bearings if certain assumptions are made. A commonly used method is that proposed by Petroff in 1883[9]. Consider the vertical shaft and bearing indicated in Fig. 4.7, where the shaft rotates at N rev/min. It is assumed that the shaft and bearing are concentric, and that the radial clearance c is completely filled with lubricant; it is further assumed that side leakage is negligible. The shaft diameter is d and radius r, and the bearing length is l. Assuming Newton's law of viscous flow, equation 4.1 we have

$$\tau = \mu \frac{U}{h}$$

$$= \mu \frac{2\pi r N}{60c}$$

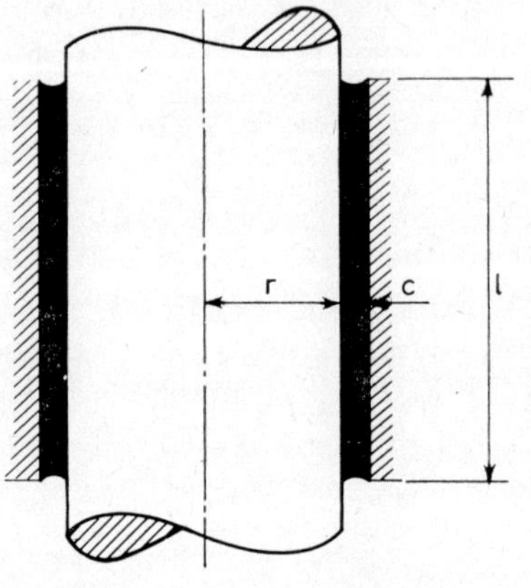

Fig. 4.7. Petroff bearing

The frictional force is simply the shear stress multiplied by the shear area, and the frictional torque, say M_t, is thus given by

$$M_t = \mu \frac{\pi r N}{30c}(2\pi r l)(r)$$

$$= \frac{\pi^2 r^3 l}{c} \frac{\mu N}{15}$$

If the force acting on the bearing is light, having magnitude P, then the frictional torque is equal to

$$M_t = fPr$$

where f is the journal coefficient of friction. Thus, we have that

$$f = \left(\frac{\pi^2}{30}\right)\left(\frac{r}{c}\right)\left(\frac{\mu N}{p}\right) \tag{4.8}$$

where p = pressure intensity of loading = $P/2rl$. Equation 4.8 is known as Petroff's law, and the parameters (r/c) and $(\mu N/p)$ can be seen to be non-dimensional parameters. The assumptions made in the derivation of the Petroff equation make it unsuitable for use with all except very lightly loaded bearings.

Reynolds classical equation for two-dimensional flow, represented by equation 4.5, may be applied to a journal bearing if it is assumed that curvature is negligible. This is tantamount to considering the film to be unwrapped so that the surfaces may be considered flat. The solution to Reynolds equation is complex, and in any case requires certain assumptions to be made.

Perhaps the best known solution which has been widely applied to journal bearings is that due to Sommerfeld[10], who performed the integration using the assumption of an infinitely long bearing. The usual way of expressing Sommerfeld's solution is in the form

$$\frac{r}{c}f = \phi\left[\left(\frac{r}{c}\right)^2\left(\frac{\mu n}{p}\right)\right] \tag{4.9}$$

where r = journal radius,
c = radial clearance,
f = coefficient of friction,
μ = absolute viscosity of the lubricant,
n = speed of rotation, rev/s = $N/60$
p = pressure intensity = $P/2rl$,
l = bearing length,

The symbol ϕ in equation 4.9 simply represents some function. The quantity in parenthesis is termed the *Sommerfeld number* or *bearing characteristic number*, designated S, and is a very important parameter in journal bearing design. Thus,

$$S = \left(\frac{r}{c}\right)^2\left(\frac{\mu n}{p}\right) \tag{4.10}$$

It is convenient to define a new quantity, known as the eccentricity ratio ε, as

$$\varepsilon = \frac{e}{c}$$

and from the geometry of the bearing (Fig. 4.6), it may be seen that the eccentricity ratio is related to the minimum oil film thickness h_0 in the following manner:

$$h_0 = c - e$$
$$= c(1-\varepsilon)$$

so that

$$\varepsilon = 1-(h_0/c) \tag{4.11}$$

In terms of this eccentricity ratio, Sommerfeld's equation for the coefficient of friction for a complete journal bearing is given by[11]

$$f = \left(\frac{c}{r}\right)\left(\frac{1+2\varepsilon^2}{3\varepsilon}\right) \tag{4.12}$$

Another solution to Reynolds equation has been obtained by Ocvirk[12], who made the opposite assumption to Sommerfeld, i.e. that the bearing may be considered infinitely short, and this solution is referred to as the short bearing approximation. In terms of the Sommerfeld number and the eccentricity ratio, Ocvirk's so-called load equation may be written[11] as

$$S\left(\frac{l}{2r}\right)^2 = \frac{(1-\varepsilon^2)^2}{\pi^2\varepsilon(1+0{\cdot}62\varepsilon^2)^{1/2}} \tag{4.13}$$

Substituting for S from equation 4.10 and transposing, we obtain

$$\left(\frac{p}{\mu n}\right)\left(\frac{c}{r}\right)^2\left(\frac{2r}{l}\right)^2 = \frac{\pi^2\varepsilon(1+0{\cdot}62\varepsilon^2)^{1/2}}{(1-\varepsilon^2)^2} \tag{4.14}$$

and expressing the speed of rotation in the more usual engineering unit of rev/min, designated N,

$$\left(\frac{60p}{\mu N}\right)\left(\frac{c}{r}\right)^2\left(\frac{2r}{l}\right)^2 = \frac{\pi^2\varepsilon(1+0{\cdot}62\varepsilon^2)^{1/2}}{(1-\varepsilon^2)^2} \tag{4.15}$$

The left hand side of equation 4.15 is defined as the *load number*, designated N_L, so that,

$$N_L = \left(\frac{60p}{\mu N}\right)\left(\frac{c}{r}\right)^2\left(\frac{2r}{l}\right)^2 \tag{4.16}$$

A consistent set of units must be used with equation 4.16 remembering that the speed of rotation in rev/min has been divided by 60, so that the unit of time used by other quantities should be in seconds. Experimentally determined values for N_L as a function of ε, have been obtained by DuBois, Ocvirk and Wehe[13], and a useful design curve representing this relationship is reproduced in Fig. 4.8, by kind permission of *N.A.S.A.*

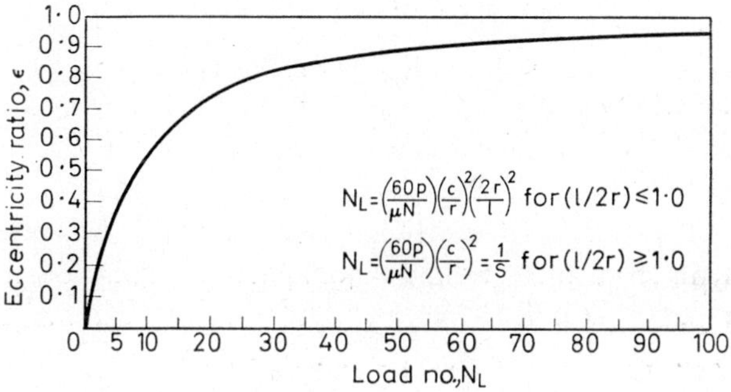

Fig. 4.8. Eccentricity ratio ε against load number, N_L (Reproduced by kind permission of the National Aeronautics and Space Administraion[13])

From equation 4.11, the eccentricity ratio may be obtained if values for h_0 and c are known. In fact, as already mentioned, the value for the minimum oil film thickness h_0, required to ensure complete hydrodynamic lubrication, is a function of the surface finish of the shaft and bearing. Phelan[8] suggests that the minimum oil film thickness in practical design should be at least 0·000 15 mm per mm (0·000 15 in per in) of shaft diameter, and a more complete discussion concerning the minimum allowable film thickness is given by Fuller[7]. Further, the radial clearance will usually be within the range 0·000 5–0.001 mm per mm (0·000 5–0·001 in per in) of shaft diameter. Consequently, the value for ε can be calculated from equation 4.11 by assuming reasonable values for h_0 and c, and then using Fig. 4.8 the corresponding maximum permissible value for N_L to satisfy the minimum oil film thickness requirement can be obtained. A methodical design procedure is desirable, and other factors need to be considered; this will presently be discussed. The application of the Ocvirk method is well illustrated by Phelan[8].

Although equation 4.16 is strictly speaking only applicable to short bearings, i.e. where the ratio of $(l/2r)$ is equal to or less than unity, there is evidence to suggest that within the range 1 to 2, only minor error is incurred by assuming that $(2r/l)^2$ is equal to unity when calculating N_L from equation 4.16. Ocvirk's friction equation is given by[11]

$$f = 2\pi^2 \left(\frac{c}{r}\right) \frac{S}{(1-\varepsilon^2)^{1/2}} \tag{4.17}$$

which reduces to Petroff's equation by substituting for S from equation 4.10 and putting $\varepsilon = 0$.

The coefficient of friction in a journal bearing may be represented generally in terms of non-dimensional parameters, i.e.

$$f = \phi\left[\frac{\mu N}{p}\ \frac{d}{C}\ \frac{l}{d}\right] \tag{4.18}$$

where μ = absolute (or dynamic) viscosity (units to be presently discussed),

d = diameter of journal bearing,

C = diametral clearance = $2c$

and other symbols are as previously defined. Experimental investigations by the McKee brothers[14] on automotive type bearings having various (l/d) ratios and a variety of clearances, indicated that the curve for f as a function of $(\mu N/p)$ has the form indicated in Fig. 4.9. It can be seen that the curve may be divided into three regions, indicating a thin-film or boundary region, a transition region and a thick-film region. For hydrodynamic lubrication, the bearing must operate in the thick film region, which means that the parameter $(\mu N/p)$ must have a value sufficiently high to prevent operation in the thin-film or transition region. Consequently, it has been common practice to design bearings based upon certain minimum values for $(\mu N/p)$, preferably based upon known satisfactory conditions previously obtained in service. However, this has probably led to somewhat over-conservative design in many instances, and note that as the value for $(\mu N/p)$ is increased, the coefficient of friction also increases. The importance attached to $\mu N/p$ cannot be over-emphasised, however, because it does represent very conveniently the condition of the film in a non-dimensional form (Provided that consistent units are used).

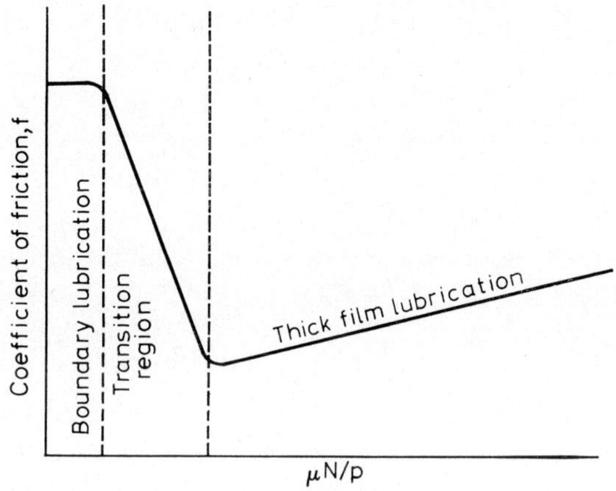

Fig. 4.9. Coefficient of friction against ($\mu N/p$)

One widely used equation for the coefficient of friction in a journal bearing, for the thick film region, is that suggested by McKee, and this has the form

$$f = \left(\frac{\pi^2}{30}\right)\left(\frac{\mu N}{p}\right)\left(\frac{r}{c}\right)+k \qquad (4.19)$$

The first term is the coefficient of friction as calculated from Petroff's equation, and experimental data obtained by McKee and McKee indicated that an average value for k is about 0·002 for (l/d) ratios between 0·75 and 2·8. Thus, equation 4.19 may be written as

$$= \left(\frac{\pi^2}{30}\right)\left(\frac{\mu N}{p}\right)\left(\frac{r}{c}\right)+0{\cdot}002 \qquad (4.20)$$

4.6. THERMAL EQUILIBRIUM OF JOURNAL BEARINGS

The frictional forces in a journal bearing are responsible for generating heat, and consequently an increase in the operating temperature. If this operating temperature is excessive, the oil film may break down, which in turn may lead to complete failure of the bearing. On the other hand, since the viscosity of a fluid decreases with increase in temperature (this will be discussed later), the frictional forces will be smaller for higher operating temperatures (assuming that thick-film lubrication is maintained), and the heat generated will therefore be less.

The heat generated by a self-contained journal bearing will be dissipated through the bearing wall by conduction, and from the bearing outer surface to the atmosphere by convection, until an equilibrium temperature of operation is reached. If the loads and speeds are such that the equilibrium temperature of the oil film is in excess of about 80–95 °C (180–200° F), then some type of external cooling is necessary, either by cooling the bearing directly, e.g. water cooling, or by supplying oil under pressure to the bearing in sufficient quantity to take away the heat, the oil then being returned to a sump for cooling

Heat generated in a journal bearing is determined from the frictional power loss, which in simple terms is the product of frictional resistance multiplied by the peripheral or sliding velocity. Thus, if the heat generated in unit time is designated H_1, then

$$H_1 = fPU \quad \text{J/s or W} \tag{4.21}$$

where P = radial force on bearing, (N),
U = peripheral velocity, (m/s)
$= \pi dn$
d = diameter of bearing or shaft, (m),
n = speed of rotation, rev/s.

If English units are used, since 778 ft lbf of energy are equal to

1 British Thermal Unit, we may write equation 4.21 in the form

$$H_1 = \frac{fPdN}{2\,973} \quad \text{Btu/min} \tag{4.22}$$

where P is measured in lbf; d is measured in inches; and N is in rev/min.

The majority of bearings are usually designed so that the heat generated by friction can be dissipated by natural cooling to the surrounding air without excessive temperature increase. Since heat flow calculations in general cannot usually be performed with great reliability, determining the heat dissipated by a journal bearing is not always a simple matter. The problem may be further complicated by the fact that most of the heat generated occurs at the point of minimum oil film thickness.

Fuller[7] states that the amount of heat carried away by conduction through the shaft or along the pedestal is small, and a good approximation for the heat dissipated is obtained by assuming that the heat generated by friction in a self contained bearing is dissipated entirely from the bearing surface to the surrounding air. For a more comprehensive discussion of this problem, Fuller[7] should be consulted.

Using Newton's law of cooling, the heat dissipated from a journal bearing in which sufficient oil may not be supplied to carry away the generated heat, may be estimated from the equation

$$H_2 = KA\,\Delta\theta \tag{4.23}$$

where H_2 = heat dissipated through the bearing,
K = coefficient of heat transfer,
= heat lost per unit area of heated surface per unit time per degree temperature rise,
A = area of surface which is losing heat,
$\Delta\theta$ = difference between surface temperature of bearing and the ambient temperature.

In equation 4.23, the value for the heat transfer coefficient K is difficult to determine, since this depends upon so many factors, such as the material, finish and colour of the bearing housing, the velocity of air being circulated around the housing, the temperature difference between the housing and the surroundings, etc. Investigations by Karelitz[15] on pedestal bearings 4 in diameter by 8 in long (approximately 100 mm diameter×200 mm long) indicated that the value for K varied from 2.1 Btu/ft^2 h deg. F (11·9 W/m^2 deg. C) in still air to 5·9 Btu/ft^2 h deg. F (33.5 W/m^2 deg. C) in air with a velocity of 500 ft/min (2·54 m/s). An average value for K suggested for general design purposes[16] is in the region of about 2·7 Btu/ft^2 h F deg. (15·3 W/m^2 deg. C). In equation 4.23

$$\Delta\theta = \theta_b - \theta_a \tag{4.24}$$

where θ_b = surface temperature of bearing housing,
θ_a = ambient temperature.

Further, it may be assumed [15] that the surface temperature of the bearing is mid-way between the average operating temperature of the oil and the ambient temperature, i.e.

$$\theta_b = \frac{\theta_0 + \theta_a}{2} \tag{4.25}$$

where θ_0 = operating temperature of the oil.
Since the operating temperature of the oil is initially unknown, and both the heat generated and the heat dissipated depend upon this value, a systemmatic procedure is necessary in estimating the bearing performance.

An alternative and frequently used method for calculating the heat dissipated by a self-contained journal bearing is that due to Lasche[17], and which relates the heat dissipated to the projected area of the bearing. Lasche's equation may be written as

$$H_2 = Kld(\theta_b - \theta_a) \tag{4.26}$$

In this equation, K is the coefficient of heat transfer including the effects of both convection and radiation, and since this is a function of temperature, it is convenient to use a design chart giving values for $K(\theta_b - \theta_a)$ as a function of $(\theta_b - \theta_a)$. Such a curve is shown in Fig. 4.10, which uses SI metric units and English units.

Once the equilibrium temperature has been reached, the heat dissipated from the surface of the bearing will be equal to the heat generated by friction in the oil film. Thus, by assuming a feasible

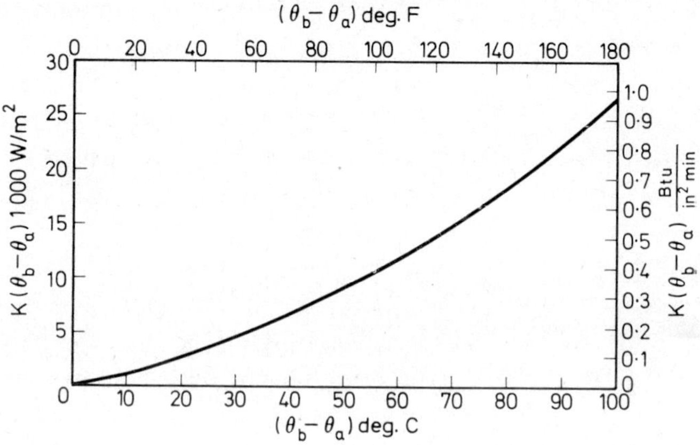

Fig. 4.10. $K(\theta_b - \theta_a)$ *against* $(\theta_b - \theta_a)$ *(Based on data by Lasche[17])*

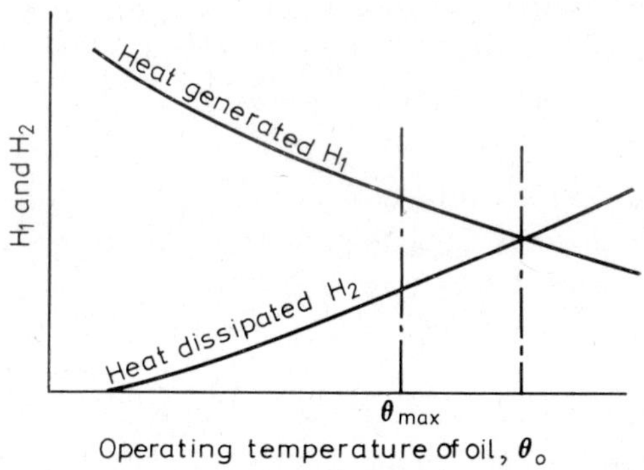

Fig. 4.11. Heat generated and heat dissipated against oil temperature

range of probable operating temperatures for the oil, the heat generated and the heat dissipated can be calculated, and the results plotted graphically, as illustrated in Fig. 4.11. The intersection of these two curves occurs at the temperature corresponding to thermal equilibrium. If this temperature is below the maximum permissible operating temperature of the oil, the bearing as designed will be satisfactory, and it will be unnecessary to consider other methods of cooling the bearing, such as oil circulating systems or water cooling. Since a self-contained bearing is most desirable, it is worth the designer making every effort to achieve this condition.

As previously stated, if the thermal equilibrium operating temperature of the oil is in excess of about 80–95 °C (180–200 °F), in order to avoid deterioration of the oil or weakening of the bearing material, it will be necessary to reduce the temperature by some means of artificial cooling. If the maximum permissible operating temperature of the oil is known (say θ_{max}, as indicated in Fig. 4.11), the difference between H_1 and H_2 at this temperature is the amount of heat required to be dissipated by artificial cooling. One method of getting rid of this heat is to supply sufficient oil under pressure to ensure that the heat is carried away, the oil being circulated to an oil sump where it may be cooled before recirculating to the bearing again. The quantity of oil required is dependent upon the amount of heat which needs to be dissipated, and the temperature rise of the oil as it passes through the bearing. If the heat dissipated in this way is designated H_3, then in equation form

$$H_3 = QC_p(\theta_2 - \theta_1) \tag{4.27}$$

where Q = quantity (mass) of oil flowing in unit time,
C_p = thermal capacity per unit mass (discussed below),
θ_1 = temperature of oil entering the bearing,
θ_2 = temperature of oil leaving the bearing.

It is important when using equations 4.23, 4.26 and 4.27 to ensure that a consistent set of units is used. If the SI metric system is used, the energy or heat dissipated should be expressed in J/s, consistent with the heat generated expressed by equation 4.21. If British units are used, the energy or heat dissipated will generally be expressed in Btu/min, consistent with the heat generated expressed by equation 4.22.

In equation 4.27 C_p is the thermal capacity per unit mass or specific heat capacity of the oil, defined as the amount of heat required to raise unit mass of the substance through one degree of temperature. Using the SI metric system, the units of C_p will be J/kg deg. C, an average value for petroleum oils being in the region of about 1 900 J/kg deg. C. In the British system, C_p will have the units Btu/lb deg. F, the average value for petroleum oils being in the region of about 0·45 Btu/lb deg F. The value for specific heat capacity varies slightly with temperature, and more exact values can be obtained by referring to the properties of specific lubricants. However, the values quoted here are usually sufficiently accurate for design purposes.

4.7. PRACTICAL CONSIDERATIONS OF VISCOSITY

At this stage it is desirable to discuss the units of absolute or dynamic viscosity which should be used with the equations previously presented in this chapter. The absolute or dynamic viscosity in British units is the reyn (after Reynolds), and has the units lbf s/in^2. In practice the reyn is seldom used in lubrication work, since the oil industry has, in the past, used the centimetre-gram-second (c.g.s.) system, in which case absolute or dynamic viscosity has the units dyne s/cm^2, termed the poise (P) (after Poiseuille). The poise is a rather large unit to work with, and consequently the centipoise (cP) has been widely used to represent dynamic viscoity. In the SI metric system, the recommended units for absolute or dynamic viscosity are N s/m^2 which is called a poiseuille (Pl). Since

$$1\text{P} = 0{\cdot}1\ \text{Ns/m}^2 \quad \text{or} \quad 0{\cdot}1\ \text{Pl}$$

and

$$1\ \text{cP} = 1\ \text{Ns/m}^2 \times 10^{-3}$$

the continued use of the centipoise as the unit of absolute viscosity would seem to be perfectly acceptable. The symbol μ has been used throughout this chapter so far, but the reader will find the symbol Z (or z) extensively used in the literature to represent dynamic viscosity when measured in cP.

The so-called McKee equation, i.e. equation 4.20, may be written, using SI metric units, as

$$f = \frac{329}{10^6}\left(\frac{ZN}{p}\right)\left(\frac{d}{C}\right) + 0{\cdot}002 \qquad (4.28)$$

where Z = dynamic viscosity, cP (i.e. $\text{Pl} \times 10^{-3}$),
N = speed of rotation of shaft, rev/min,
p = pressure intensity based on projected area, N/m²,
= P/ld, P = radial bearing load, N,
l = length of bearing, m,
d = shaft diameter, m,
C = diametral clearance, m.

Similarly, in terms of British units, equation 4.20 may be written as

$$f = \frac{473}{10^{10}}\left(\frac{ZN}{p}\right)\left(\frac{d}{C}\right) + 0{\cdot}002 \qquad (4.29)$$

where Z = dynamic viscosity, cP,
N = speed of rotation of shaft, rev/min,
p = pressure intensity based on projected area, lbf/in²,
= P/ld, P = radial bearing lod, lbf,
l = length of bearing, in,
d = shaft diameter, in,
C = diametral clearance, in.

Thus, equations 4.28 and 4.29 provide an approximate method for calculating the coefficient of friction in a journal bearing.

The load number N_L, represented by equation 4.16, may be expressed in the following more convenient forms.
Using the SI metric system,

$$N_L = 60 \times 10^3 \left(\frac{p}{\mu N}\right)\left(\frac{C}{d}\right)^2\left(\frac{d}{l}\right)^2 \qquad (4.30)$$

Using English units,

$$N_L = 414 \times 10^6 \left(\frac{p}{\mu N}\right)\left(\frac{C}{d}\right)^2\left(\frac{d}{l}\right)^2 \qquad (4.31)$$

The units of μ in both the above equations is cP, and other units are consistent with the system used, as indicated immediately beneath equations 4.28 and 4.29 for SI metric and English units respectively.

The viscosity of a lubricant is determined experimentally, and for this purpose various methods are available. The most common method in the United Kingdom is that which uses the Redwood viscometer; in the United States the most common method is the Saybolt Universal viscometer; and on the Continent that known as Engler. Each of these methods measures the time taken for a certain volume of oil to flow through a standard capillary tube at a desired temperature, and the relationship between the recorded time and the *kinematic viscosity* (designated ν), may be represented by the following empirical equation[18]

$$\nu = At - \frac{B}{t} \tag{4.32}$$

where t is the time recorded in seconds; A and B are empirical constants. The values for A and B are indicated in Table 4.1.

Table 4.1. EMPIRICAL CONSTANTS

	Redwood	*Saybolt Universal*	*Engler*
A	0·26	0·22	0·147
B	171	180	374

The dynamic or absolute viscosity is related to the kinematic viscosity by the mass density γ, such that

$$\mu = \nu\gamma = \left(At - \frac{B}{t}\right)\gamma \tag{4.33}$$

where the constants A and B are obtained from Table 4.1 corresponding to the method of measurement used.

Kinematic viscosity has the units of in²/s in the English system, and m^2/s in the SI metric system. In practice, the usual unit used for kinematic viscosity is the stoke, St (after Stokes), but for convenience, the centistoke is preferred. Since

$$1 \ \text{St} = 10^{-4} \ \text{m}^2/\text{s}$$

and $$1 \ \text{cSt} = 10^{-6} \ \text{m}^2/\text{s}$$

this should cause no inconvenience. This means that the units for mass density are $(\text{kg/m}^3) \times 10^3$, which correspond to the units used in the centimetre, gram, second (c.g.s.) system of g/cm^3. Since in the c.g.s. system the mass density corresponds very closely numerically to the specific gravity, this is quite convenient. However, to be

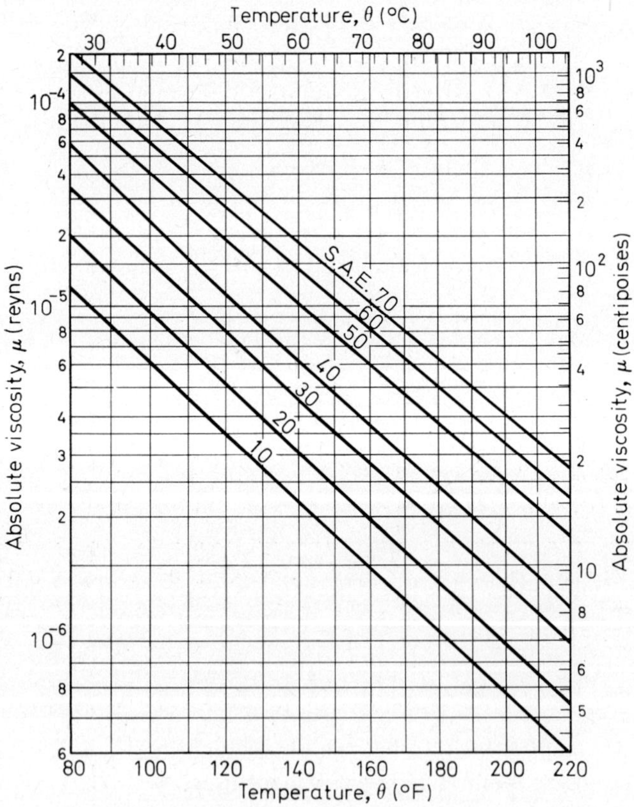

Fig. 4.12. Viscosity-temperature chart for typical SAE numbered oils (Reproduced by kind permission of Raimondi and Boyd[19])

meaningful, the specific gravity needs to be stated corresponding to the temperature under consideration, i.e. at the operating temperature of the oil. This is an appropriate time to mention that viscosity is a function of temperature (and pressure), and in all the previous discussions it has been tacitly assumed that the value for μ (or Z) corresponds to the operating temperature of the oil. In actuality, the operating temperature of a journal bearing is not known initially, and a methodical procedure of design and analysis is necessary. Fig. 4.12 shows a viscosity temperature chart for typical SAE numbered oils, and is reproduced by permission of Raimondi and Boyd[19].

Although the mass density per unit volume γ is dependent upon the operating temperature of the oil, this variation is quite small, and the error introduced by assuming an average value will generally be insignificant. Values sufficiently accurate for most design purposes will be in the region of 0·85–0·90 g/cm^3, or in English units about 0·03–0·033 lb/in^3.

4.8. THE RAIMONDI-BOYD METHOD OF JOURNAL BEARING DESIGN AND ANALYSIS

A somewhat sophisticated method for the analysis and design of journal bearings has been proposed by Raimondi and Boyd of the Westinghouse Electric Corporation Research Laboratories[19]. The method is based on the numerical solution of the fundamental equations, using digital computer techniques; the results so obtained have been presented in the form of charts and tables which permit the determination of friction, film thickness, flow, temperature rise, etc,. and both full and partical bearings have been considered, with corrections for end leakage included. The method proposed greatly simplifies the procedure for assessing the performance of journal bearings, and curves for the 120 degree partial bearing and 360 degree full bearing are reproduced in Fig. 4.14–4.33, by kind permission of A. A. Raimondi and J Boyd, and the Westinghouse Electric Corporation. Some of the symbols used here are not identical with those used in the original papers of Raimondi and Boyd[19], but are consistent with the symbols used throughout this chapter. Fig. 4.13 indicates the bearing nomenclature used.

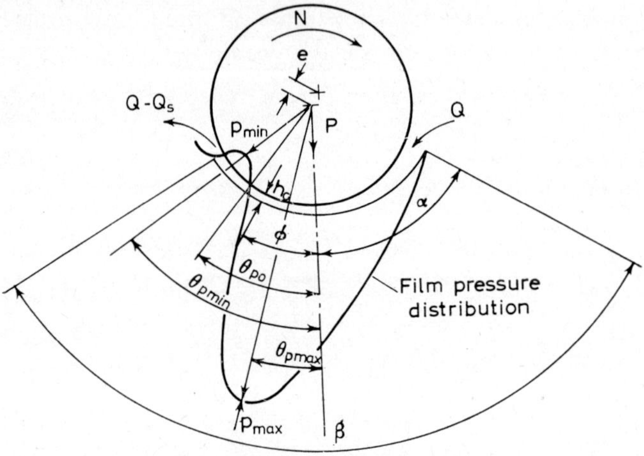

Fig. 4.13. Bearing nomenclature

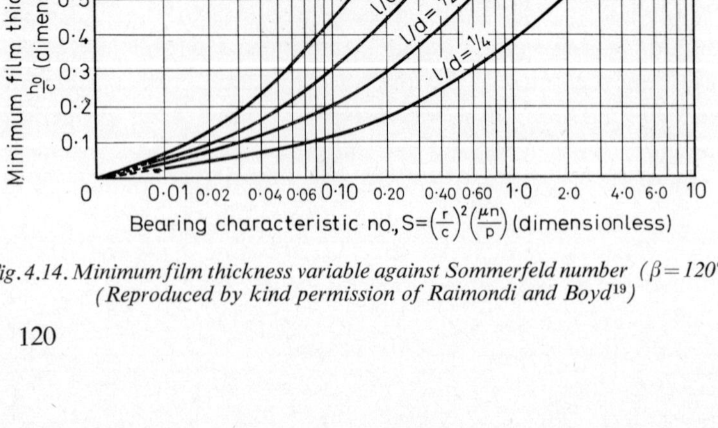

Fig. 4.14. Minimum film thickness variable against Sommerfeld number ($\beta = 120°$) (Reproduced by kind permission of Raimondi and Boyd[19])

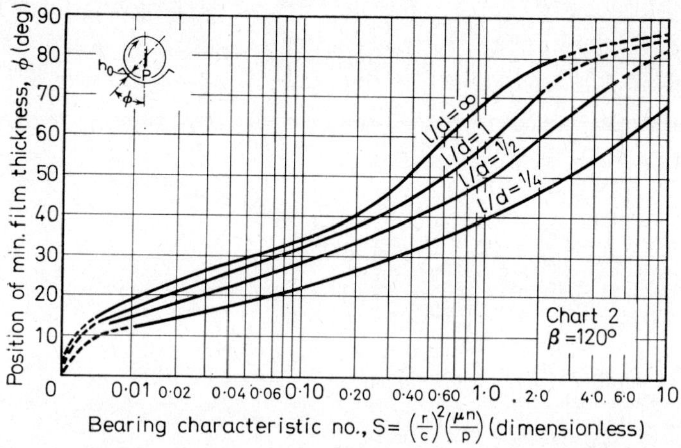

Fig. 4.15. Position of minimum film thickness agains Sommerfeld number ($\beta = 120°$) (Reproduced by kind permission of Raimondi and Boyd[19])

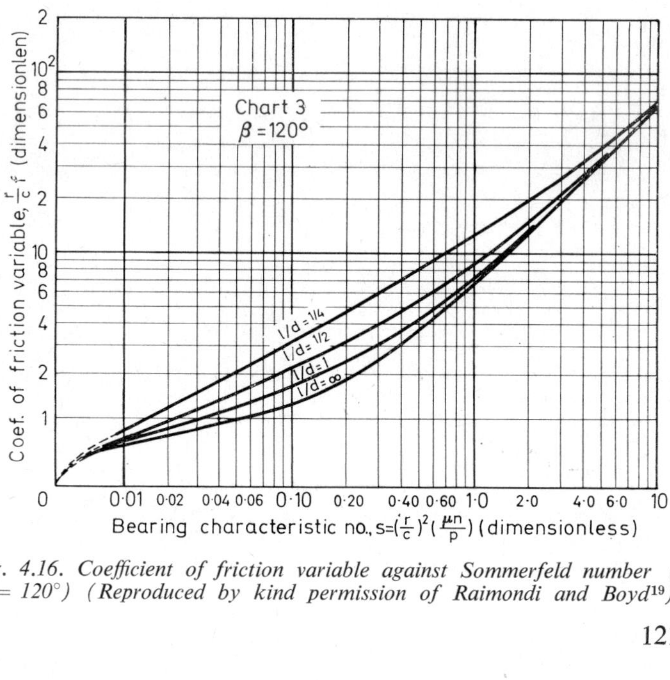

Fig. 4.16. Coefficient of friction variable against Sommerfeld number ($\beta = 120°$) (Reproduced by kind permission of Raimondi and Boyd[19])

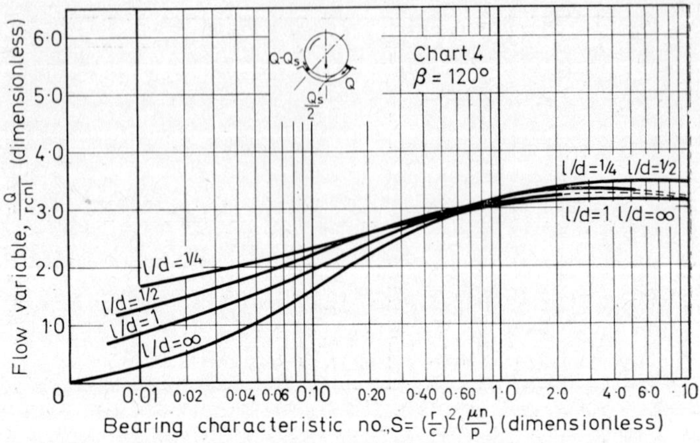

Fig. 4.17. Flow variable against Sommerfeld number (β = 120°) (Reproduced by kind permission of Raimondi and Boyd[19])

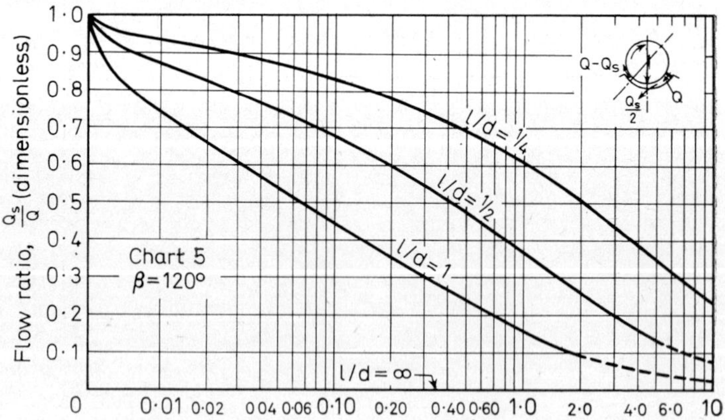

Fig. 4.18. Flow ratio against Sommerfeld number (β = 120°) (Reproduced by kind permission of Raimondi and Boyd[19])

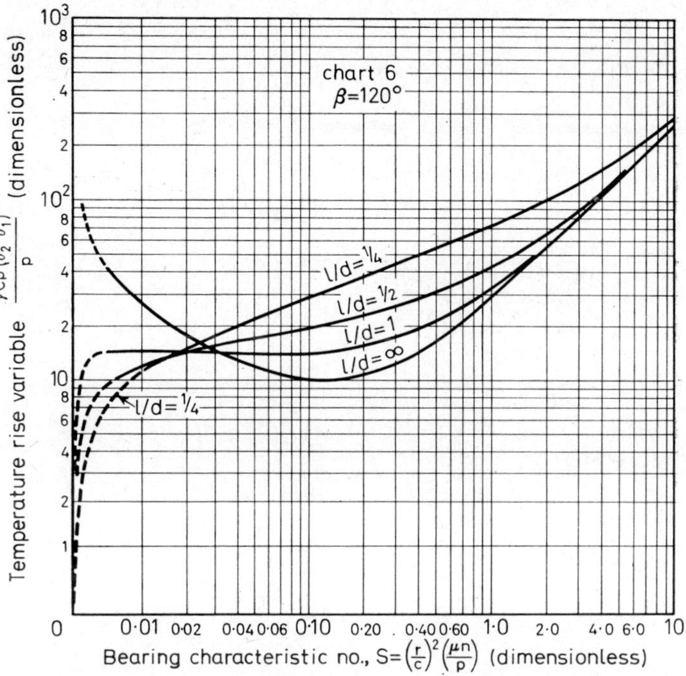

Fig. 4.19. Temperature rise variable against Sommerfeld number ($\beta = 120°$)
(Reproduced by kind permission of Raimondi and Boyd[19])

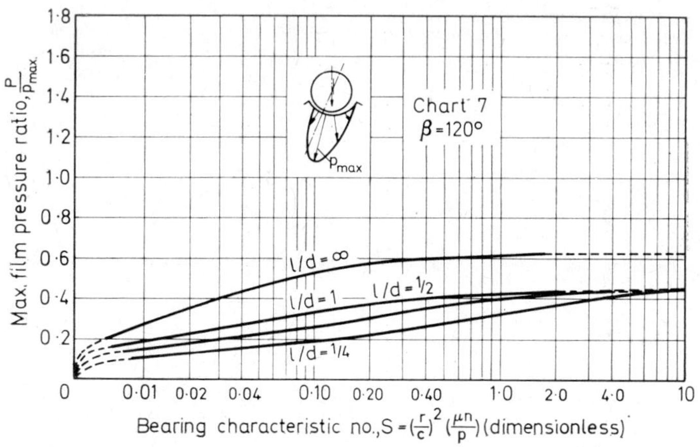

Fig. 4.20. Maximum film pressure ratio against Sommerfeld number ($\beta = 120°$)
(Reproduced by kind permission of Raimondi and Boyd[19])

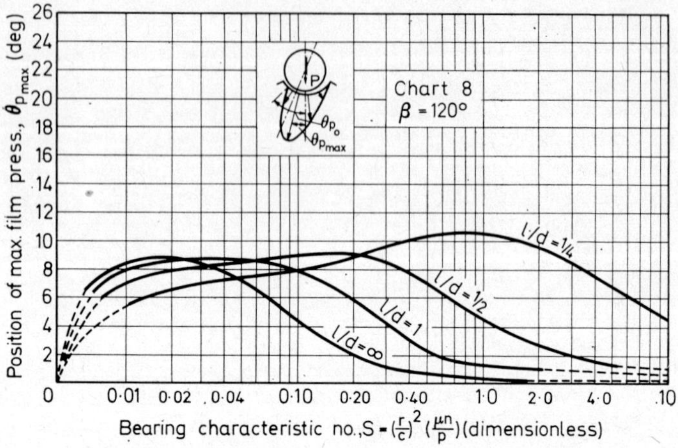

Fig. 4.21. Position of maximum film Pressure against Sommerfeld number ($\beta = 120°$) (Reproduced by kind permission of Raimondi and Boyd[19])

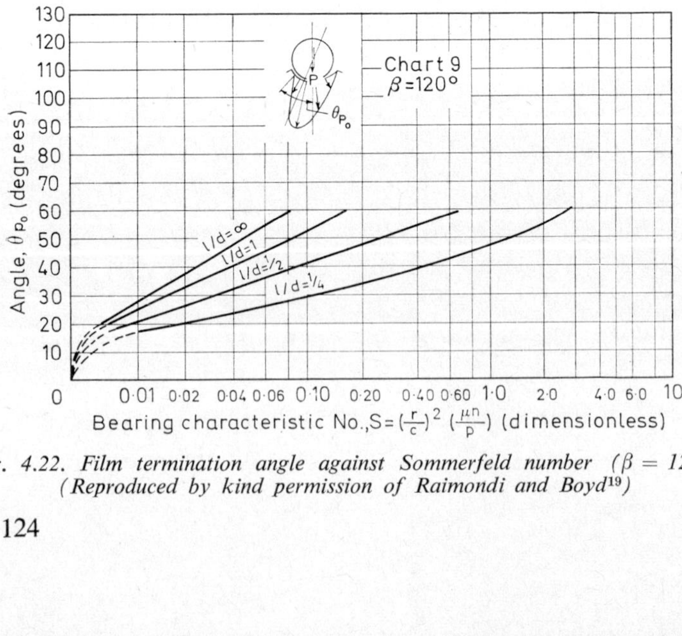

Fig. 4.22. Film termination angle against Sommerfeld number ($\beta = 120°$) (Reproduced by kind permission of Raimondi and Boyd[19])

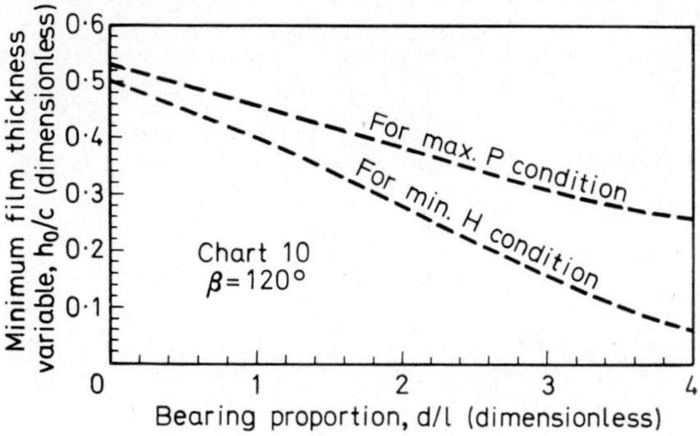

Fig. 4.23. Maximum film thickness variable corresponding to maximum load or Minimum power loss against Sommerfeld number ($\beta = 120°$) (Reproduced by kind permission of Raimondi and Boyd[19])

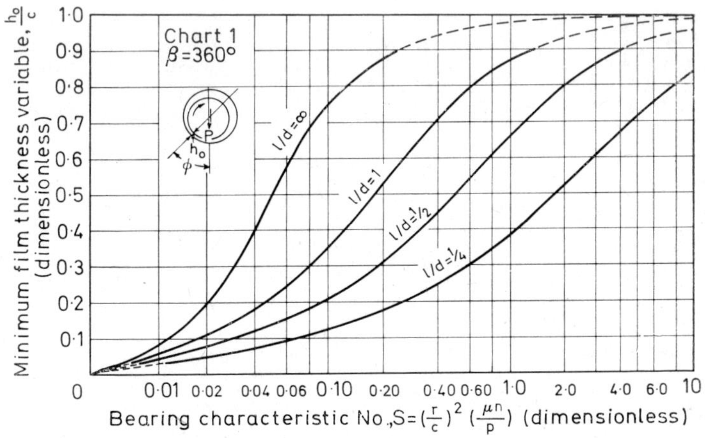

Fig. 4.24. Minimum film thickness variable against Sommerfeld number ($\beta = 360°$) (Reproduced by kind permission of Raimondi and Boyd[19])

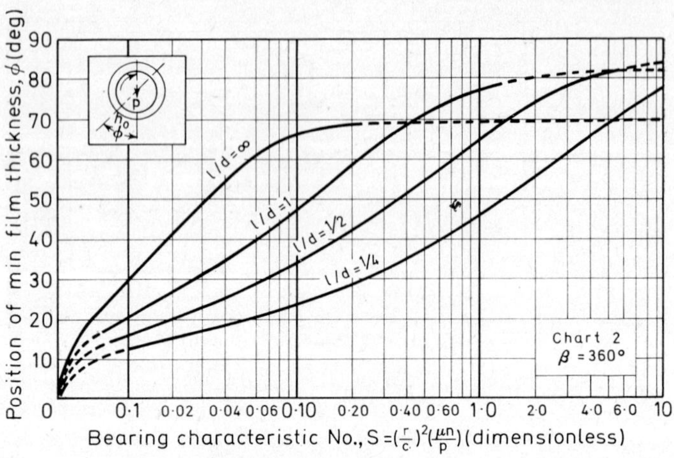

Fig. 4.25. Position of minimum film thickness variable against Sommerfeld Number ($\beta = 360°$) (Reproduced by kind permission of Raimondi and Boyd[19])

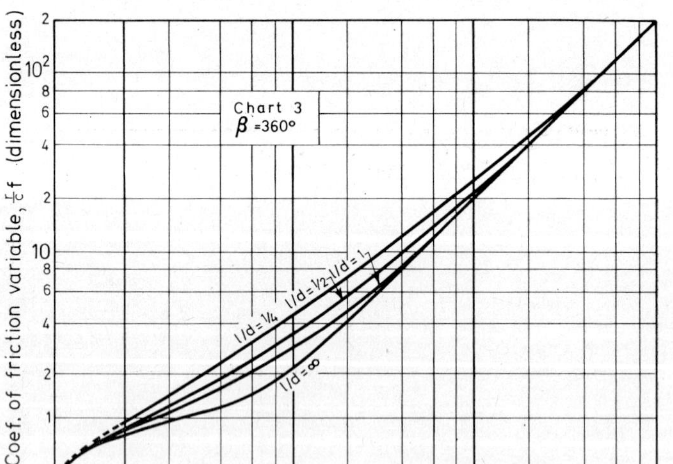

Fig. 4.26. Coefficient of friction variable against Sommerfeld number ($\beta = 360°$) (Reproduced by kind permission of Raimondi and Boyd[19])

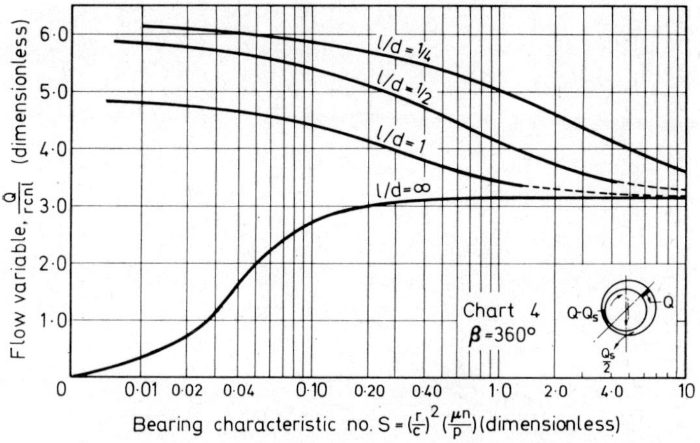

Fig. 4.27. Flow variable against Sommerfeld number ($\beta = 360°$) (Reproduced by kind permission of Raimondi and Boyd[19])

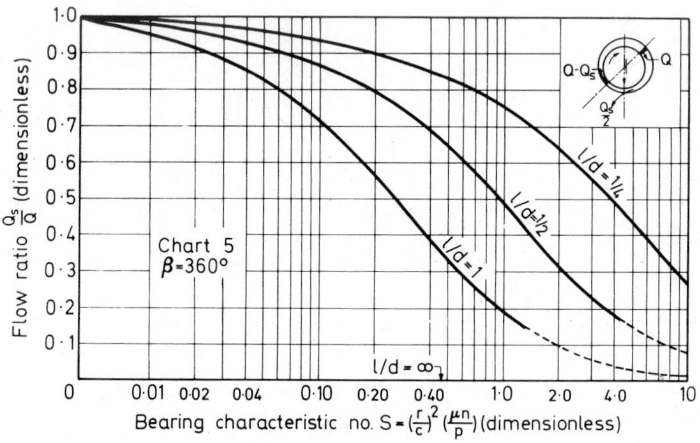

Fig. 4.28. Flow ratio against Sommerfeld number ($\beta = 360°$) (Reproduced by kind permission of Raimondi and Boyd[19])

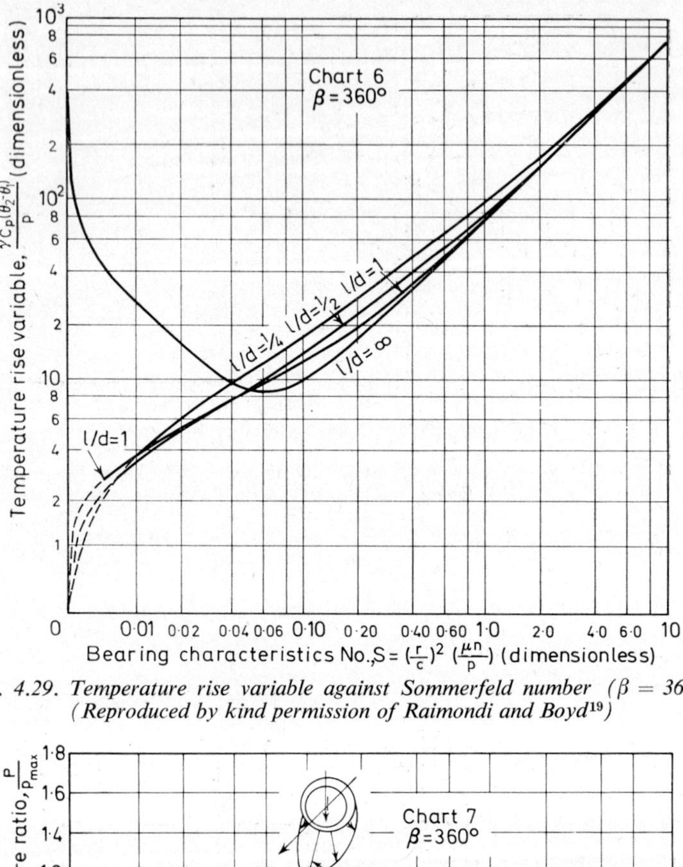

Fig. 4.29. Temperature rise variable against Sommerfeld number ($\beta = 360°$)
(Reproduced by kind permission of Raimondi and Boyd[19])

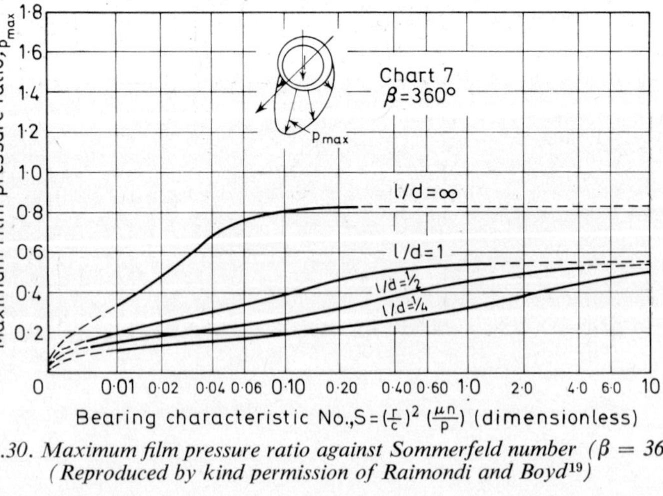

Fig. 4.30. Maximum film pressure ratio against Sommerfeld number ($\beta = 360°$)
(Reproduced by kind permission of Raimondi and Boyd[19])

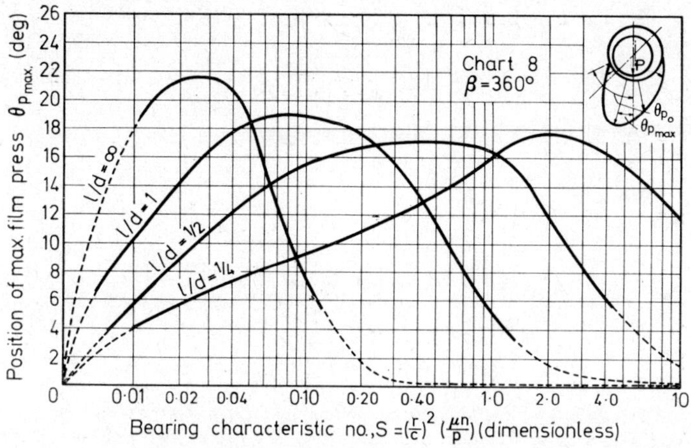

Fig. 4.31. Position of maximum film pressure against Sommerfeld number ($\beta = 360°$) (Reproduced by kind permission of Raimondi and Boyd[19])

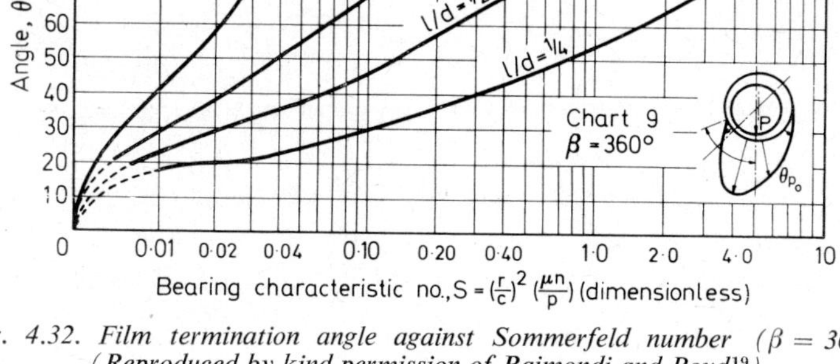

Fig. 4.32. Film termination angle against Sommerfeld number ($\beta = 360°$) (Reproduced by kind permission of Raimondi and Boyd[19])

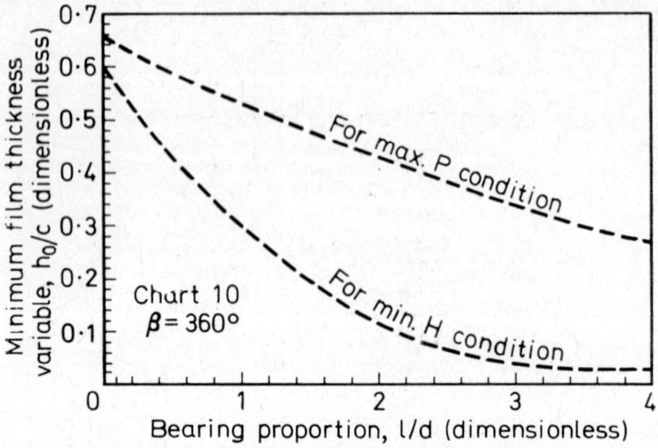

Fig. 4.33. Minimum film thickness variable corresponding to maximum load or minimum power loss against bearing proportion ($\beta = 360°$) (Reproduced by kind permission of Raimondi and Boyd[19])

For convenience in making the calculations, the results of the computer solutions are plotted showing the various performance variables as dimensionless groups as functions of the Sommerfeld number. This has a number of advantages, amongst which is the fact that the curves may be used with any system of units, provided that consistency is maintained. Raimondi and Boyd have produced many more design charts than those presented here, and it is suggested that the reader refer to the excellent work mentioned[19] for a more comprehensive discussion.

4.9. METHODICAL PROCEDURE FOR THE DESIGN OF JOURNAL BEARINGS

The many factors involved in the design of journal bearings makes a methodical procedure most desirable. In the first place, the bearing must be designed and the correct lubricant selected to ensure that the minimum film thickness is sufficient to avoid metal to metal contact or boundary lubrication. Secondly, the amount of friction generated in the bearing must be assessed in order to ensure that the heat generated is not in excess of that which can be dissipated without causing the oil to operate at a too high temperature. Further, the quantity of oil available or supplied must be sufficient to ensure adequate lubrication at all times.

The application of a methodical procedure can best be demonstrated by way of examples, and for this purpose two bearing designs will be considered. In the first case, the Ocvirk short bearing approximation will be used, together with the McKee equation for the determination of the journal coefficient of friction, and the Lasche curve for heat dissipated considerations. In the second example, the design curves of Raimondi and Boyd will be used.

4.10. EXAMPLE 1

A journal bearing for a steam turbine is required to support a radial force of 2 660 N (600 lbf). The shaft diameter at the bearing based on other requirements, is 60 mm (2·36 in), and the estimated bearing length is 90 mm (3·54 in); the speed of rotation is 1 800 rev/min. Make the necessary decisions and calculations required to assess the bearing performance if it is to operate in an ambient temperature of 15 °C (approximately 60 °F), with a maximum operating temperature of not more than 80 °C (176 °F).

Solution

Assuming an average value for the radial clearance of 0·000 75 mm per mm of shaft diameter,

$$c = 0{\cdot}000\,75 \times 60 = 0{\cdot}045 \text{ mm } (0{\cdot}001\,77 \text{ in})$$

The minimum oil film thickness h_0 required to maintain hydrodynamic lubrication is a function of the surface finish, and assuming the recommended minimum value of 0·000 15 mm per mm of shaft diameter,

$$h_0 = 0{\cdot}000\,15 \times 60 = 0{\cdot}009 \text{ mm} = (0{\cdot}000\,35 \text{ in})$$

The eccentricity ratio is obtained from equation 4.11, i.e.

$$\varepsilon = 1-\frac{h_0}{c}$$

$$= 1-\frac{0{\cdot}009}{0{\cdot}045} = 0{\cdot}8$$

Using Fig. 4.8, corresponding to $\varepsilon = 0{\cdot}8$, the load number N_L is found to be 28. This represents the maximum permissible value to give the oil film thickness of 0·009 mm (0·000 35 in). But from equation 4.16, assuming that $(2r/l)$ may be taken equal to unity (refer also to Fig. 4·8) since $(l/2r) > 1.0$,

$$N_L = \left(\frac{60p}{\mu N}\right)\left(\frac{c}{r}\right)^2$$

Substituting known numerical data,

$$N_L = \left(\frac{60\times 2\,660\times 10^6\times 10^3}{60\times 90\times 1\,800\times \mu}\right)\left(\frac{0{\cdot}045}{30}\right)^2$$

$$= \frac{35{\cdot}5}{\mu}$$

Note that the dynamic viscosity μ is in cP, and that this has been divided by 10^3 in order to maintain a consistent set of units. Since the load per projected area is of importance, it will be convenient to calculate this value separately, i.e.

$$p = \frac{2\,660}{60\times 90} = 0{\cdot}491 \text{ MN/m}^2 \text{ (71·2 lbf/in}^2\text{)}$$

which is well within the accepted range for this application (for steam turbines a range of about 690–2 000 kN/m² i.e. about 100 –300 lbf/in²).

The coefficient of friction may be obtained using the McKee equation, i.e. equation 4.28 using SI metric units or equation 4.29 using English units. Thus, by equation 4.28

$$f = \frac{329}{10^6}\left(\frac{\mu N}{p}\right)\left(\frac{d}{C}\right)+0.002$$

Substituting known numerical data

$$f = \frac{329 \times 1\,800 \times 60 \times \mu}{10^6 \times 0{\cdot}491 \times 10^6 \times 0{\cdot}09} + 0{\cdot}002$$

$$= 0{\cdot}803 \times 10^{-3}\mu + 0{\cdot}002$$

The heat generated in friction is calculated from equation 4·21, i.e.

$$\begin{aligned} H_1 &= fPU \\ &= 2\,660 \times \pi \times 0{\cdot}060 \times 30f \\ &= 15{\cdot}05 \times 10^3 f \text{ W (J/s)} \\ &= 856\, f \text{ Btu/min (by equation 4·22)} \end{aligned}$$

The heat dissipated in a self contained journal bearing may be estimated by the application of the Lasche equation, i.e. equation 4.26, namely

$$H_2 = Kld(\theta_b - \theta_a)$$

where $K(\theta_b - \theta_a)$ can be obtained from Fig. 4.10, and the bearing temperature is obtained from equation 4.25.

It is convenient at this stage to tentatively select an oil, and carry out a methodical procedure of solution. This can be most conveniently performed by assuming a range of operating temperatures for the oil, and tabulating the results correponding to these temperatures. The dynamic or absolute viscosity can be obtained from Fig. 4.12.

SAE 10 oil will be initially assumed. The calculations are recorded in Table 4.2.

A plot of H_1 and H_2 against operating temperature of the oil is shown in Fig. 4.34, from which it is clearly seen that for the range of temperatures considered, an equilibrium temperature is not reached. Since the maximum permissible operating temperature of the oil is to be limited to 80 °C (176 °F), it is not possible for the bearing to be self contained, and some means of external cooling will be required. The amount of external cooling required must be sufficient to carry away the difference between the heat generated

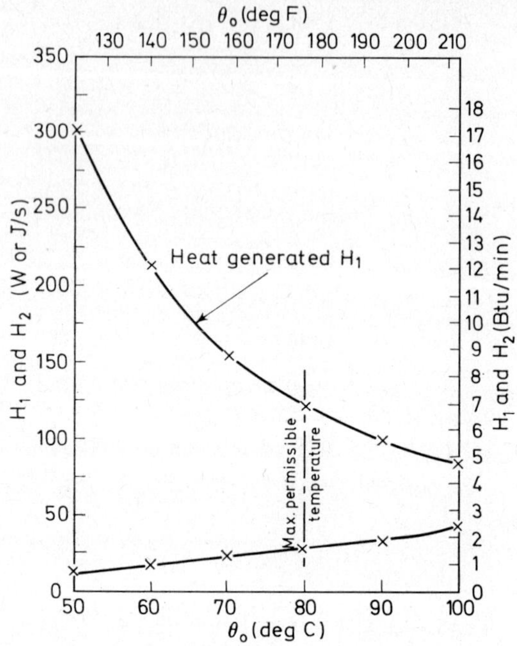

Fig. 4.34. H_1 and H_2 against θ_0 for bearing example 1

and the heat dissipated at 80 °C (176 °F). This is

$$H_1 - H_2 = 122 - 28{\cdot}1 = 93{\cdot}9 \text{ W (5.32 Btu/min)}$$

Note that the tentative selection of SAE 10 oil has now been established as a reasonable choice, since the load number N_L is adequately satisfied, and the choice of a thicker oil would have the effect of increasing friction, which is undesirable.

The oil film thickness will, in fact, be greatly in excess of the minimum permissible value of 0·009 mm (0·00035 in) originally assumed, and the actual thickness may be estimated by working backwards, i.e. determine ε from Fig. 4.8, corresponding to $N_L = 4{\cdot}68$ (from Table 4.2 for an operating temperature of 80 °C), and using equation 4.11 to calculate h_0. Thus, in this way it is found that

$$h_0 = 0.029\,5 \text{ mm (0.001 6 in)}$$

Table 4.2. CALCULATIONS FOR EXAMPLE 1

θ_0 °C	50	60	70	80	90	100
θ_0 °F	122	140	158	176	194	212
μ (cP)	22.7	15.1	10.3	7.7	5.7	4.46
N_L	1.565	2.35	3.45	4.68	6.24	7.96
f	0.02	0.014	0.010	0.008	0.006 6	0.005 6
H_1 (W)	304	211	154	122	99	84
H_1 (Btu/min)	17.3	12	8.7	6.92	5.62	4.76
$(\theta_b - \theta_a)$ deg C	17.5	22.5	27.5	32.5	37.5	42.5
$(\theta_b - \theta_a)$ deg F	31	40	49	58	67	76
$K(\theta_b - \theta_a)$ (W/m²)	2 200	3 100	4 100	5 200	6 200	7 500
$K(\theta_b - \theta_a)$ (Btu/in² min)	0.081	0.114	0.15	0.191	0.227	0.275
H_2 (W)	11.89	16.75	22.1	28.1	33.4	40.5
H_2 (Btu/min)	0.674	0.954	1.26	1.6	1.9	2.3

Let us now estimate the amount of oil which would be required to ensure that the excess heat is carried away, the oil being circulated to an oil sump where it may be cooled before recirculating to the bearing again. By equation 4.27, the heat dissipated in this way is given by

$$H_3 = QC_p(\theta_2 - \theta_1)$$

Assuming a temperature rise through the bearing of 6 deg C (10·8 deg F), and an average value for C_p of 1 900 J/kg deg C (0·45 Btu/lb deg F), the quantity of oil flow required to carry away the whole of the heat generated by friction is

$$Q = \frac{122}{1{\cdot}900 \times 6}$$

$$= 0{\cdot}010\ 8\text{kg/s}\ (0{\cdot}023\ 9\ \text{lb/s})$$

Assuming an average value for mass density of 875 kg/m³ (0·031 5 lb/in³)

$$Q = \frac{0{\cdot}0108 \times 3\,600}{875}$$

$$= 0{\cdot}044\,5 \text{ m}^3/\text{h } (2\,700 \text{ in}^3/\text{h})$$

$$= 4{\cdot}45 \text{ l/h } (9{\cdot}85 \text{ gal/h})$$

Consequently, if the oil drawn into the leading edge by means of the pumping action of the bearing is in excess of this amount, the bearing will operate satisfactorily, provided that the inlet temperature is maintained at less than 74 °C (165 °F). The amount of lubricant drawn into the leading edge by the pumping action of the journal can be estimated from the Raimondi-Boyd lubricant flow chart, i.e. Fig. 4.27.

4.11. EXAMPLE 2

A full journal bearing for supporting a generator rotor has a diameter of 100 mm (3·937 in) and a length of 100 mm (3·937 in). It has to support a central radial force of 8 000 N (1 760 lbf) when the shaft rotates at 3 000 rev/min. It is anticipated that the inlet temperature of the lubricating oil will be about 60 °C (140 °F). Make the necessary decisions and calculations required to assess the bearing performance based on the optimum condition of minimum power loss. Use the Raimondi-Boyd design charts.

Solution

The information relating to this problem is as follows:

$$\beta = 360^\circ,$$

$$r = 50 \text{ mm } (1{\cdot}968\,5 \text{ in})$$

$$d = 2r = 100 \text{ mm } (3{\cdot}937 \text{ in})$$

$$l = 100 \text{ mm } (3{\cdot}937 \text{ in})$$

$$N = 3\,000 \text{ rev/min}$$

$$P = 8\,000 \text{ N}, \; p = P/ld = 0{\cdot}8 \text{ MN/m}^2 \; (116 \text{ lbf/in}^2)$$

$$\theta_1 = 60 \text{ °C } (140 \text{ °F})$$

From Fig. 4.33 corresponding to $d/l = 1$, the optimum value for h_0/c for minimum power loss is found to be

$$\frac{h_0}{c} = 0{\cdot}3$$

It is convenient to consider a range of values for h_0, say from 0·015 mm (0·000 59 in) upwards. Corresponding to each assumed value for h_0, the value for c may be obtained to satisfy the condition for minimum power loss, i.e. from $h_0/(h_0/c)$. Further, from Fig. 4.24, the Sommerfeld number corresponding to the optimum value for h_0/c is determined, i.e.

$$S = 0{\cdot}077$$

From Fig. 4.29 the temperature rise variable corresponding to S = 0·077 is found to be

$$\frac{\gamma C_p(\theta_2 - \theta_1)}{p} = 11$$

Substituting numerical data and assuming average values for $\gamma = 875$ kg/m³ (0·031 5 lb/in³) and $C_p = 1\,900$ J/kg deg C (0·45 Btu/lb deg F),

$$(\theta_2 - \theta_1) = \frac{11 \times 0{\cdot}8 \times 10^6}{875 \times 1\,900}$$

$$= 5{\cdot}29 \text{ deg C } (9{\cdot}52 \text{ deg F})$$

$$\theta_2 = \theta_1 + 5{\cdot}29$$

$$= 65{\cdot}29\ ^\circ\text{C } (149.5\ ^\circ\text{F})$$

The average operating temperature of the oil is approximated[19] by the equation

$$\theta_0 = \theta_1 + \tfrac{1}{2}(\theta_2 - \theta_1) \qquad (4.34)$$

$$= 62{\cdot}64\ ^\circ\text{C } (144.76\ ^\circ\text{F})$$

The viscosity of the oil may be obtained from the Sommerfeld equation, i.e.

$$S = \left(\frac{r}{c}\right)^2 \left(\frac{\mu n}{p}\right)$$

from which

$$\mu = \left(\frac{Sp}{n}\right)\left(\frac{c}{r}\right)^2$$

Substituting numerical data with μ having the units of cP,

$$\mu = \frac{0{\cdot}077\times0{\cdot}8\times10^6\times10^3}{50}\left(\frac{c}{50}\right)^2$$

$$= 492{\cdot}8\ c^2\ \text{(cP)}$$

Thus, the values for μ corresponding to different values of c can readily be obtained. The calculations are conveniently summarised in Table 4.3 for various assumed values of h_0. Note that the assumed values are in excess of the minimum suggested value of 0·000 15 mm per mm of shaft diameter, which is 0·015 mm (0·000 59 in) in this case.

Table 4.3

h_0 (mm)	0.020	0.030	0.040	0.050	0.060
h_0 (in)	0.000 79	0.001 18	0.001 58	0.001 97	0.002 36
c (mm)	0.066	0.100	0.133	0.166	0.200
c (in)	0.002 6	0.003 94	0.005 25	0.006 54	0.007 9
μ (cP)	2.15	4.93	8.71	13.6	19.7

Reference to Fig. 4.12 indicates that for an operating temperature of 62·6 °C (144·8 °F), SAE 10 oil has a viscosity of 13·4 cP. This corresponds quite closely to the viscosity calculated in Table 4.3 for a minimum oil film thickness of 0·050 mm (0·001 97 in). Consequently, if the only available oils are those indicated in Fig. 4.12, the optimum selection will be SAE 10; the use of thicker oils would permit a larger radial clearance to be used, giving a higher value for the film thickness, but with increased friction.

Having chosen SAE 10 oil, and establishing the operating temperature and the Sommerfeld number (S = 0·077), the other bearing performance charcteristics may now be obtained from the Raimondi-Boyd design charts.

From Fig. 4.25, the position of the minimum film thickness is

$$\phi = 44 \text{ degrees}$$

From Fig. 4.26, the coefficient of friction variable is

$$f\frac{r}{c} = 2{\cdot}4$$

$$f = \frac{2{\cdot}4 \times 0{\cdot}166}{50} = 0{\cdot}007\,98$$

By equation 4.21, the heat generated in friction is

$$\begin{aligned} H_1 &= fPU \\ &= 0{\cdot}007\,98 \times 8\,000 \times \pi \times 0{\cdot}1 \times 50 \\ &= 1\,000 \text{ W} \end{aligned}$$

From Fig. 4.27 the flow variable is

$$\frac{Q}{rcnl} = 4{\cdot}5$$

from which

$$\begin{aligned} Q &= 4{\cdot}5\ rcnl \\ &= 4{\cdot}5 \times 0{\cdot}050 \times 0{\cdot}050 \times 10^{-3} \times 50 \times 0{\cdot}1 \\ &= 0{\cdot}187\,4 \times 10^{-3} \text{ m}^3/\text{s } (11{\cdot}4 \text{ in}^3/\text{s}) \\ &= 0{\cdot}674\,6 \text{ m}^3/\text{h } (41\,040 \text{ in}^3/\text{h}) \\ &= 674{\cdot}6 \text{ l/h } (148 \text{ gal/h}) \end{aligned}$$

This represents the amount of lubricant which is drawn into the leading edge by the pumping action of the journal. If the amount supplied is less than Q, the bearing is said to be starved, and its performance characteristics will be altered. Of the oil flowing into the leading edge, the amount which escapes laterally from both sides of the bearing may be found from the flow ratio. From Fig. 4.28 the flow ratio is

$$\frac{Q_s}{Q} = 0{\cdot}76$$

from which

$$Q_s = 0{\cdot}76 \times 674{\cdot}6$$
$$= 512 \text{ l/h } (112{\cdot}6 \text{ gal/h})$$

From Fig. 4.30 the maximum film pressure ratio is

$$\frac{p}{p_{\max}} = 0{\cdot}36$$

from which

$$p_{\max} = \frac{0{\cdot}8}{0{\cdot}36} = 2{\cdot}23 \text{ MN/m}^2 \text{ (323 lbf/in}^2\text{)}$$

From Fig. 4.31 the position of the maximum film pressure is

$$\theta_{\max} = 19 \text{ degrees}$$

From Fig. 4.32 the film termination angle, i.e. the angle to the position of zero pressure, is found to be

$$\theta_{p0} = 64 \text{ degrees}$$

The bearing performance characteristics have thus been assessed, and the decisions required for minimum power loss have been made.

4.12. MATERIALS FOR PLAIN BEARINGS

As stated previously, the two essential elements in a plain bearing are the bearing or bearing material itself, and the shaft or moving member. The bearing or bearing material is located in a housing or structure, and may or may not be integral with it. Separating these two elements is the lubricant, introduced by some suitable means, e.g. manual oiling, wick-feed or drop-feed, oil rings or chains, splash system, or external pressure feeding, depending upon the design and requirements. If the bearing is intended to operate under conditions of hydrodynamic lubrication, i.e. pressure being built up in the oil film by means of the pumping action created by the velocity of the journal itself, then on starting from rest there will be metal to metal contact or boundary lubrication.

This condition will obtain until such time as the velocity is sufficient to build up a film of sufficient thickness to separate the parts. It is under conditions of metal to metal contact or boundary lubrication that the demands on the bearing materials and the lubricant are greatest.

The material of the shaft or journal is established from considerations of strength and rigidity, and will invariably be steel; such considerations have been discussed in Chapter 3.

Because the conditions under which bearings must operate in service may vary over a wide range, it is necessary that bearing materials be used which have certain desirable properties. Amongst these we must include such factors as mechanical strength; softness and low melting point; low modulus of elasticity; corrosion resistance; high thermal conductivity; and of course, economic considerations. Since these factors cannot all be obtained to a desirable degree in a single material, it is necessary in practice to make a compromise.

The most common bearing materials consist of (a) white metals, (b) copperbase alloys, and (c) aluminium-base alloys.

White metals is a term used to include the tin and lead-base metals, broardly referred to as Babbitts (after Isaac Babbitt, 1839), and since such metals are highly competitive, they are recommended for most applications where the loading is not severe. Babbitt bearings are manufactured with the white metal lined onto steel, cast iron and copper base alloys. Since white metal suffers a reduction in fatigue strength with increase in temperature, and this reduction is a function of thickness, it is usual to limit the thickness to between about 0·100–0·175 mm (0·004–0·007 in), and thicknesses of only 0·025–0·05 mm (about 0.001–0·002 in) are used with copper-lead over the back-up material.

Copper-base alloys including lead-bronze, gun-metal and phosphor-bronze are widely used as bearing materials.

Lead-bronze is the cheapest, and is used for general service bearings. It has a low tendency to seizure, in common with the white metal bearings, and has greater fatigue strength to withstand higher temperatures. Lead bronze bushes are frequently used in the form of single, solid units, i.e. as bushes without the supporting shell surrounding the bearing material, as is required of the Babbitt or white metal bearing materials.

Gun-metal provides a relatively cheap and easy to machine material, having good bearing properties and capable of withstanding somewhat higher loads than the lead-bronze alloys. This alloy also has good resistance to corrosion in sea water.

Phosphor-bronze is used for heavily loaded bearings, where high frictional stresses are likely to occur. Because of the high hardness of this material, it demands the use of a hardened steel journal.

Aluminium-base alloys are available for use as bearing materials, and offer a range of fatigue strengths comparible with that of the lead-bronzes. The corrosion resistance is extremely high, and economically these alloys are highly competitive compared with the copper-base alloys. Because of the relatively high coefficient of thermal expansion, it is suggested by Hunsicker[20] that aluminium alloy bearings should not be used at temperatures above 107–149 °C (225–300 °F), due to the high thermal stresses induced which may lead to cracking on cooling.

As an approximate guide only, Table 4.4 indicates the maximum bearing pressures corresponding to some common bearing materials, as suggested by Neale[21].

Table 4.4

Maximum bearing pressure		*Bearing material*
MN/m²	lbf/in²	
Up to 17·25	2 500	White metals
17·25–27·6	2 500–4 000	Copper-leads
17·25–34·5	2 500–5 000	Aluminium-tins
27·6–41·4	4 000–6 000	Low-tin lead bronzes
41·4–69	6 000–10 000	High-tin lead bronzes
69–103·5	10 000–15 000	Phosphor-bronzes

For applications where bearings are required to operate with little or no lubrication, porous metal bearings impregnated with oil, or graphited bronze bearings have been developed. Plastic materials, such as Teflon, are also used, but only for very light loads, and a backing material is generally required.

For further details regarding materials for plain bearings the reader is referred to other works.[7, 20–24]

4.13. EXTERNALLY PRESSURISED PLAIN BEARINGS

It is frequently desirable to prevent metal to metal contact or boundary lubrication entirely, and one way of achieving this is to supply lubricant to the bearing under pressure. This method is employed when the velocity of the moving surface is insufficient to maintain an oil film, and is necessary for bearing applications involving heavy loads and low speeds. When a bearing operates with full film lubrication due to external pressurisation, this is known as *hydrostatic lubrication*, since the separating film is not dependent upon the velocity of the moving surface. Under these conditions, attention must be given to the adequate supply of lubricant at all times, and in particular to the location of oil supply holes and grooves. A discussion on hydrostatic lubrication is presented by Fuller[7] and details concerning mechanical design and its effects on lubrication can be found in other works[8, 16, 24, 25].

4.14. ROLLING CONTACT BEARINGS

Rolling contact bearings is the name given to bearings which depend upon contact between rolling members to give the desired restraint with a minimum of friction. Such bearings are frequently referred to as anti-friction bearings, but this term is ambiguous, and arises mainly from the fact that for starting conditions and moderate speeds, the friction in a rolling contact bearing is lower

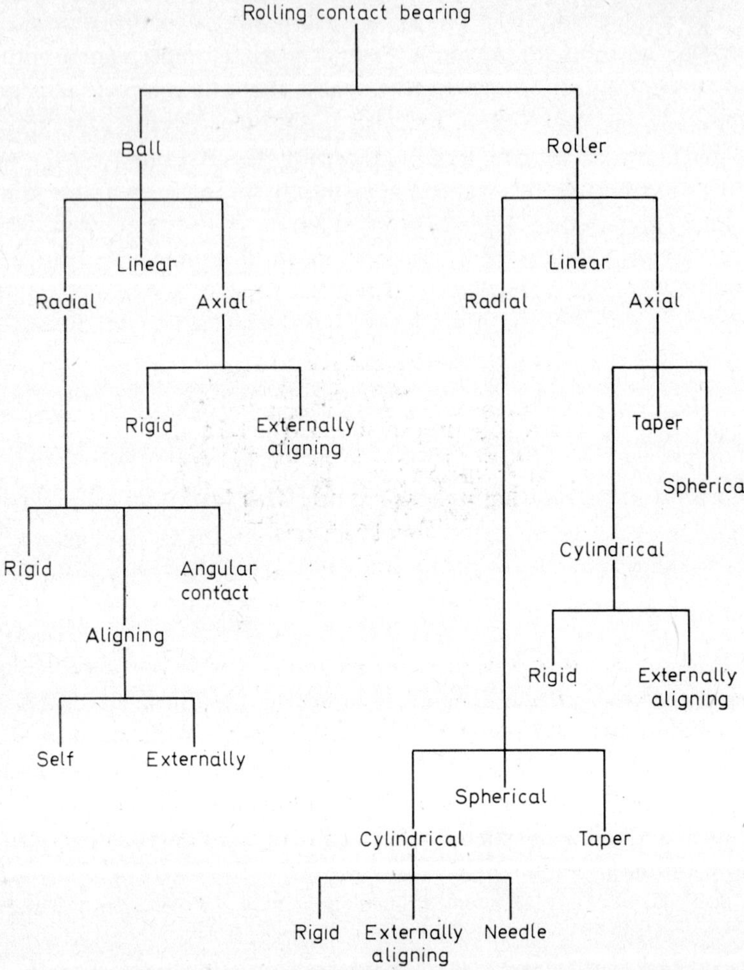

Fig. 4.35. Family tree of main types of rolling contact bearings (Reproduced from Harris[26])

than that obtained in an equivalent journal bearing. However, a journal bearing which is properly designed and operates under full hydrodynamic conditions, will have a coefficient of friction in the same order of magnitude as that of a lubricated rolling contact bearings.

Rolling contact bearings consist essentially of two surfaces in relative motion separated by a row of balls or rollers in contact with each of the two surfaces. Usually, the rolling elements themselves are separated by a cage or separator, which evenly distributes them around the bearing, thus preventing them from rubbing against each other.

The main types of rolling contact bearings can be divided into groups depending upon the basic shape of rolling element, i.e. ball or roller, and further classification depending upon the type

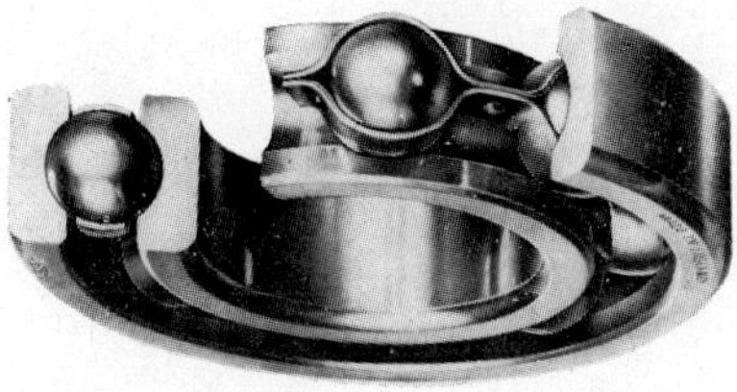

Fig. 4.36. Single-row deep groove ball bearing (Reproduced by kind permission of the Skefko Ball Bearing Company)

of load they are designed to withstand, i.e. radial or axial. The main types of rolling contact bearings can be illustrated by the family tree shown in Fig. 4.35, which is based on that suggested by Harris[26].

Before giving consideration to bearing selection and load carrying capacity, it is desirable to describe briefly some of the more common types of bearings in use. Photographs and other material is reproduced here by kind permission of the Skefko Ball Bearing Company[27, 28].

The most common type of ball bearing is the *single row deep groove ball bearing* of the rigid type, as indicated in Fig. 4.36. These bearings have a deep groove in each ring and therefore have a high radial and axial load carrying capacity, making them suitable for either radial or axial loading, or a combination of both. In certain cases, these bearings may even be more suitable than thrust ball bearings for carrying axial loads, especially at high speeds. This type of bearing has a cylindrical bore, and because it is not self-aligning, accurate parallelism between the shaft and housing is necessary. However, if the outer surface of the bearing is made spherical, the bearing becomes externally aligning, and is then incapable of supporting a moment.

When it is difficult to obtain accurate parallelism between the shaft and housing bore, or where it is likely that there will be considerable deflection of the shaft under load, it may be necessary to consider using a self aligning bearing. Fig. 4.37 shows a *double row self-aligning ball bearing*, having two rows of balls, each running in

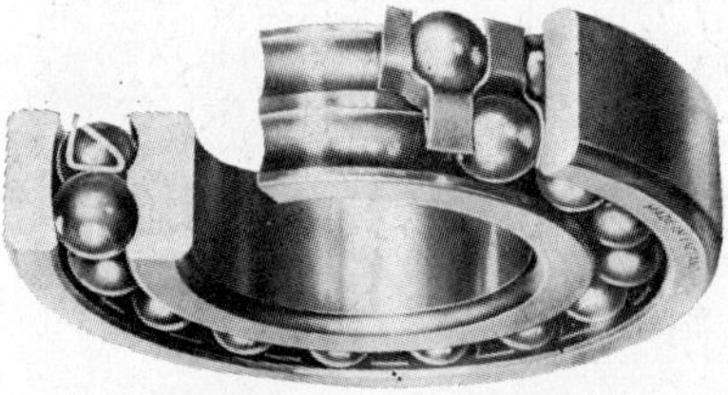

Fig. 4.37. Double-row self aligning ball bearing (Reproduced by kind permission of the Skefko Ball Bearing Company)

its own groove in the inner ring and sharing a common sphered track in the outer ring; this arrangement permits self-alignment of the bearing, and thus prevents them from supporting a moment. Such bearings have a limited axial load carrying capacity. Self-aligning bearings are available with a cylindrical bore, or with a taper bore and adaptor sleeve.

Angular contact ball bearings are designed so that the direction of the load through the balls forms an angle with the radial plane of the bearing, thus permitting them to carry combined axial and radial loads. These bearings are available as either single or double row, as indicated in Fig. 4.38.

The *single row angular contact ball bearing* has a high load carrying capacity, particularly where the axial load predominates. In order to be free from radial clearance, they must be mounted without axial clearance, and this necessitates that a bearing of this type be adjusted towards another bearing of the axially locating type. Care must be taken to see that the bearing is mounted facing in the right direction. Such bearings are usually mounted in pairs.

Double row angular contact ball bearings of standard design may be free from radial and axial clearance, and therefore they are particularly suitable in cases where the bearing is required to locate the shaft with a high degree of accuracy. These bearings are capable

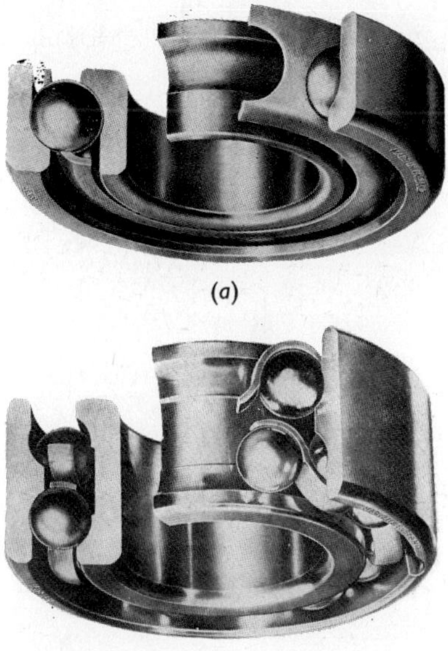

(a)

(b)

Fig. 4.38. Angular contact ball bearings (a) Single-row (b) Double-row (Reproduced by kind permission of the Skefko Ball Bearing Company)

of carrying tilting moments in the axial plane, so that they are suitable for use where rotating components with only one bearing can be incorporated owing to lack of space. They can carry axial loads in either direction or a combination of radial and axial loads.

For applications requiring resistance to axial loading only, *thrust ball bearings* may be used, and these may be either single acting and intended for an axial load in one direction, or double acting for axial loading in either direction. Such bearings are illustrated in Fig. 4.39. These bearings may incorporate a sphered seating and seating ring, thus providing them with some external aligning capacity.

Cylindrical roller bearings, as illustrated in Fig. 4.40, have higher radial load carrying capacity than equivalent ball bearings, due to the greater area of contact between the rolling elements and the

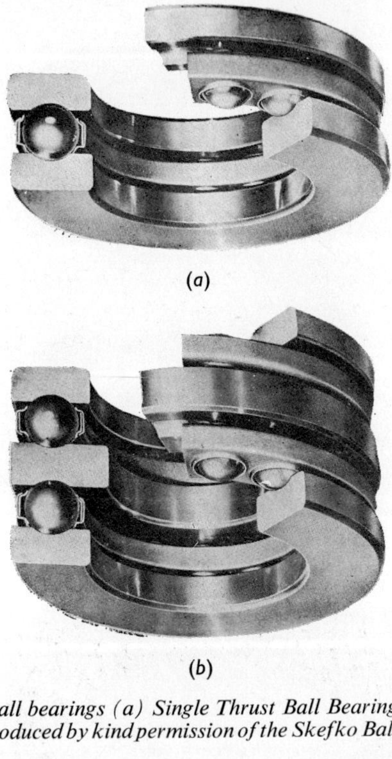

(*a*)

(*b*)

Fig. 4.39. Thrust ball bearings (a) Single Thrust Ball Bearing (b) Double Thrust Ball Bearing (Reproduced by kind permission of the Skefko Ball Bearing Company)

rings. These bearings have very low friction torque charcteristics and are particularly suitable for operation at high speed. The rollers are guided by flanges either on the inner or outer ring, the other ring usually having no flanges. The cylindrical rollers are held together with the flanged ring by means of the cage or separator, even when the free ring is removed. These bearings permit relative axial displacement of the shaft and bearing housing within certain

Fig. 4.40. Cylindrical roller bearing (Reproduced by kind permission of the Skefko Ball Bearing Company)

limits. Like the single row deep groove ball bearing, the cylindrical roller bearing requires accurate parallelism between the shaft and bearing housing. By making the outer surface of the bearing spherical, such bearings can be provided with some external aligning capacity. In their standard form, cylindrical roller bearings are not able to withstand axial loading, although by the provision of a flange on the free rings, such bearings can cope with a limited amount of axial loading.

Needle roller bearings, as illustrated in Fig. 4.41, are used when radial space is seriously limited. The friction associated with this type of bearing is much greater than that obtained with other

Fig. 4.41. Needle roller bearing (Reproduced by kind permission of the Skefko Ball Bearing Company)

cylindrical roller bearings. Such bearings are frequently used for oscillating conditions, and may be provided with or without a cage or separator. When fitted with a cage, they are suitable for relatively heavy loads and high speeds.

Spherical roller bearings (Fig. 4.42) are manufactured in a number of different designs, but a feature common to them all is that the

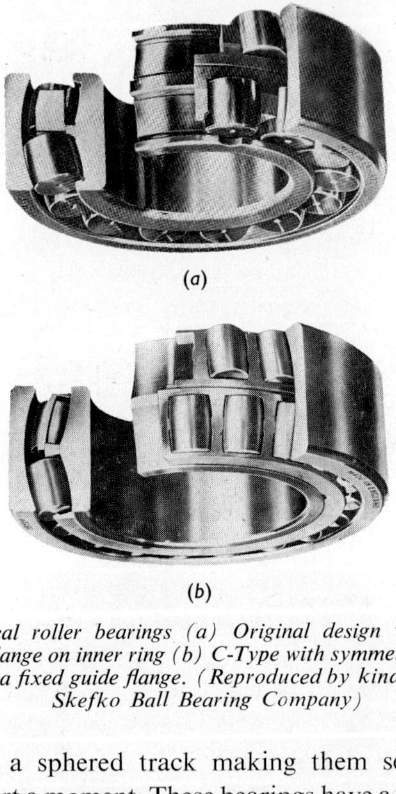

(a)

(b)

Fig. 4.42. Spherical roller bearings (a) Original design with asymmetrical rollers and fixed flange on inner ring (b) C-Type with symmetrical rollers and an inner ring without a fixed guide flange. (Reproduced by kind permission of the Skefko Ball Bearing Company)

outer ring has a sphered track making them self-aligning and unable to support a moment. These bearings have a high radial load carrying capacity and the wide types are also able to carry heavy axial loads. Due to the degree of sliding which occurs, spherical roller bearings have higher friction than cylindrical roller bearings, and they are consequently not suited for high speed operation.

Taper roller bearings, as illustrated in Fig. 4.43, are suitable for heavy combined radial and axial loads, or axial loads only. The tapers of the rollers and the cup (outer ring) and cone (inner ring) tracks have a common apex on the bearing axis. They are available with normal or steep angle execution, the latter having a greater thrust carrying capacity. Since a taper roller bearing locates the shaft in one direction only, it must be paired with another bearing

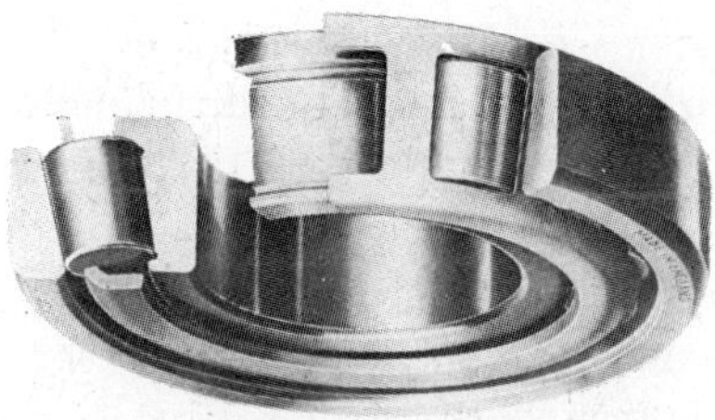

Fig. 4.43. Taper roller bearing (Reproduced by kind permission of the Skefko Ball Bearing Company)

Fig. 4.44. Spherical roller thrust bearing (Reproduced by kind permission of the Skefko Ball Bearing Company)

which locates in the other direction. Depending on the application, the two bearings have to be adjusted against each other so that there is either a small amount of clearance or some pre-load.

Spherical roller thrust bearings, as illustrated in Fig. 4.44, combine very high load carrying capacity with complete self-alignment. In contrast to most other types of thrust bearings, they can also carry radial loads. These bearings are suitable for operation at relatively high speds and heavy loads. Except in certain special circumstances, oil lubrication must be used.

4.15. SELECTION OF ROLLING CONTACT BEARINGS

A methodical procedure for the correct selection of ball and roller bearings is difficult to establish, because it depends so much on the particular application, and many factors are involved. However, if a sufficiently comprehensive knowledge of the characteristics of the different bearing types is obtained, Palmgren[29] suggests that the selection of the bearing type is relatively simple. It is suggested that for small high speed applications ball bearings are generally most suitable, but that for large and heavily loaded applications, roller bearings provide the only satisfactory answer. Where permanent or fluctuating mis-alignment occurs between shaft and bearing housing a self-aligning bearing is desirable, whilst a completely free axial displacement of the shaft necessitates the use of a cylindrical roller bearing. Deep groove ball bearings frequently offer the best solution for applications involving relatively heavy axial or combined loading at high speeds.

The prime consideration with regard to bearing selection concerns the magnitude and direction of the external load, so that the first task will always be that of determining the bearing forces. Since different bearing types have different load carrying capacities, a bearing must be selected which is suitable for carrying the particular type of load, i.e. radial, axial or a combination of the two.

Obviously, the designer must consider the methods and types of lubrication required, since this will have several effects on the bearing itself[26], and also with regard to the design of the bearing housing and protective caps, sealing, etc. Ball bearings are less demanding with regard to lubrication and maintenance than roller bearings, an important consideration if they are to be used on machines serviced by technically unskilled personnel.

A major advantage of rolling contact bearings is that they are readily available as standard units[28, 30], and agreement has been reached by the International Standards Organisation (I.S.O.—Technical Committee 4—Ball and Roller Bearings) with regard to establishing boundary dimensions. The appropriate National Standards, in the United Kingdom the British Standards Institution, and in the United States of America the Anti-Friction Bearing Manufacturers Association (A.F.B.M.A.), have standardised on boundary dimensions based upon the recommendations of I.S.O. 4 plans, and the bearing manufacturers in turn base their range on the relevant national standard to supply customers' needs. Consequently, when selecting bearing types and sizes, preference should be given to bearings whose dimensions conform to the I.S.O. plan. British Standard Specification 292: 1958[31] recommends that metric bearings should be adopted in preference to inch sizes for ball bearings and cylindrical roller bearings, and standard bearings are always preferred to special bearings unless a departure from this is abolutely unavoidable. In such instances, the expert advice of the bearing manufacturer should be sought.

Fig. 4.45 shows the boundary dimensions which have been standardised by international agreement, and also indicates the variation of outside diameter with series for bearings of constant bore, and variation of bore diameter with series for bearings of constant outside diameter.

Bearings are divided into series, the most common being light, medium and heavy, with extra light also being used in many instances. For standard bearings having metric boundary dimensions, these series are represented by a number consisting of two digits, namely light 02, medium 03, and heavy 04. Generally a bearing will be designated by a number sufficient to completely specify its size and series, and will usually be preceded by a letter or number to indicate the type of bearing or special features. Tables 4.5, 4.6 and 4.7 are reproduced by kind permission of the Skefko Ball Bearing Company, from their general publication GB 550 Ball and Roller Bearings[28]; these tables have been selected as typical of the data provided by bearing manufacturers in their catalogues. Table 4.5 represents data applicable to single row deep groove ball bearings for metric dimension series 03 (medium series); Table 4.6 for cylindrical roller bearings–metric dimension series 03 (medium series); and Table 4.7 for taper roller bearings–metric dimension

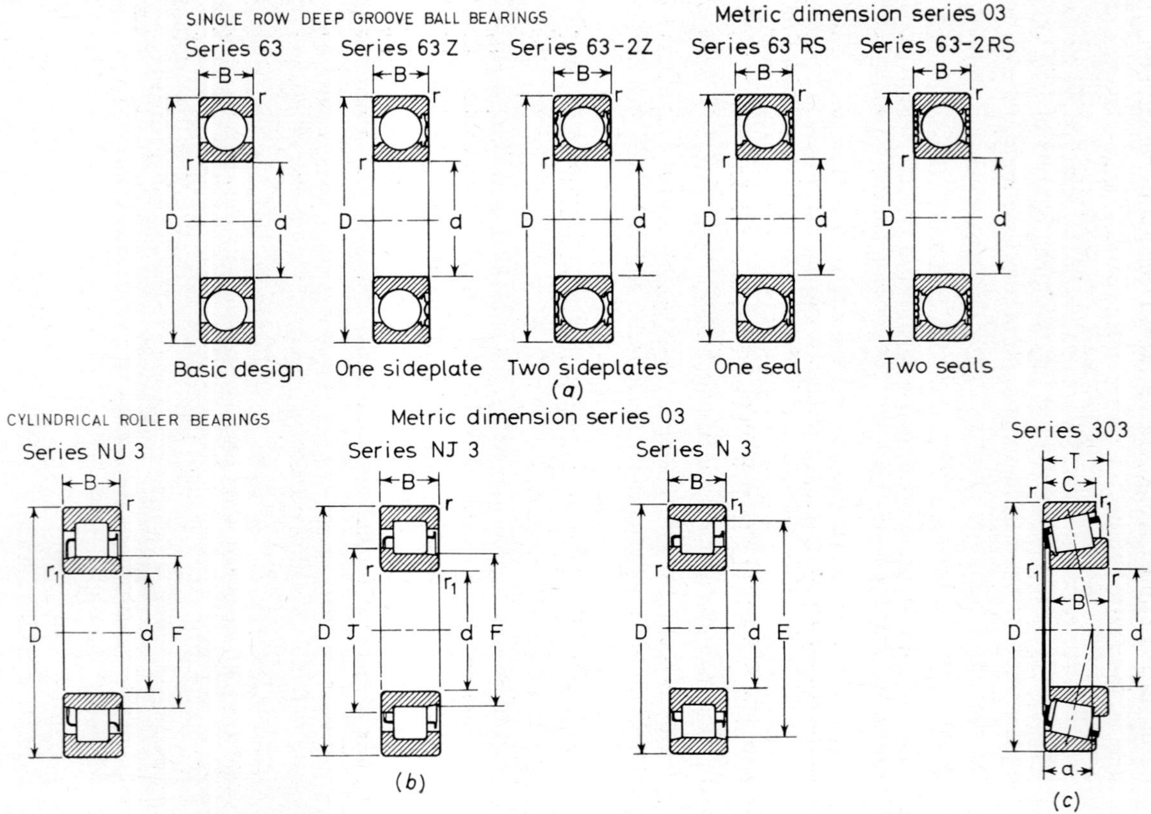

Figures for (a) Table 4.5, (b) Table 4.6 and (c) Table 4.7

Table 4.5 SINGLE ROW DEEP GROOVE BALL BEARINGS
(Repr. from GB 550[28] by kind permission of the SKEFKO Ball Bearing Company)

Bearing No.	*B.S. No.*	*d*		*D*		*B*		*r* ≈	*Basic capacity,* lbf		*Max. permissible speed* r.p.m.
		mm	in.	mm	in.	mm	in.	mm	*Static* C_0	*Dynamic* C	
6300	BRM 010	10	0·3937	35	1·3780	11	0·4331	1	800	1 370	16 000
01	BRM 012	12	0·4724	37	1·4567	12	0·4724	1·5	950	1 700	16 000
02	BRM 015	15	0·5906	42	1·6535	13	0·5118	1·5	1 140	1 930	16 000
6303	BRM 017	17	0·6693	47	1·8504	14	0·5512	1·5	1 370	2 320	13 000
04	BRM 020	20	0·7874	52	2·0472	15	0·5906	2	1 700	2 750	13 000
05	BRM 025	25	0·9843	62	2·4409	17	0·6693	2	2 280	3 600	10 000
6306	BRM 030	30	1·1811	72	2·8346	19	0·7480	2	3 200	4 800	10 000
07	BRM 035	35	1·3780	80	3·1496	21	0·8268	2·5	3 800	5 700	8 000
08	BRM 040	40	1·5748	90	3·5433	23	0·9055	2·5	4 800	7 100	8 000
6309	BRM 045	45	1·7717	100	3·9370	25	0·9843	2·5	6 550	9 150	8 000
10	BRM 050	50	1·9685	110	4·3307	27	1·0630	3	7 800	10 600	6 000
11	BRM 055	55	2·1654	120	4·7244	29	1·1417	3	9 300	12 200	6 000
6312	BRM 060	60	2·3622	130	5·1181	31	1·2205	3·5	10 600	14 000	5 000
13	BRM 065	65	2·5591	140	5·5118	33	1·2992	3·5	12 000	16 000	5 000
14	BRM 070	70	2·7559	150	5·9055	35	1·3780	3·5	13 700	18 000	5 000
6315	BRM 075	75	2·9528	160	6·2992	37	1·4567	3·5	16 000	19 600	4 000
16	BRM 080	80	3·1496	170	6·6929	39	1·5354	3·5	17 600	21 200	4 000
17	BRM 085	85	3·3465	180	7·0866	41	1·6142	4	19 300	22 800	4 000
6318	BRM 090	90	3·5433	190	7·4803	43	1·6929	4	21 600	24 500	3 000
19	BRM 095	95	3·7402	200	7·8740	45	1·7717	4	24 500	26 500	3 000
20	BRM 100	100	3·9370	215	8·4646	47	1·8504	4	29 000	30 500	3 000
6321	BRM 105	105	4·1339	225	8·8583	49	1·9291	4	31 500	31 500	2 500
22	BRM 110	110	4·3307	240	9·4488	50	1·9685	4	36 000	34 500	2 500
24	BRM 120	120	4·7244	260	10·2362	55	2·1654	4	35 600	35 500	2 500
6326	BRM 130	130	5·1181	280	11·0236	58	2·2835	5	43 000	39 000	2 500
28	BRM 140	140	5·5118	300	11·8110	62	2·4409	5	49 000	44 000	2 000
30	—	150	5·9055	320	12·5984	65	2·5591	5	56 000	47 500	2 000

Table 4.6 CYLINDRICAL ROLLER BEARINGS *(Reproduced from GB 550[28] by kind permission of the SKEFKO Ball Bearing Company)*

Bearing No.			B.S.* No.	d		D		B		r ≈	r_1 ≈	Basic capacity, lb		Max. permissible speed r.p.m.
				mm	in.	mm	in.	mm	in.	Millimetres		Static C_0	Dynamic C	
NU 304	NJ 304	N 304	RRM 020	20	0·7874	52	2·0472	15	0·5906	2	1	2 120	3 200	13 000
305	305	305	RRM 025	25	0·9843	62	2·4409	17	0·6693	2	2	3 050	4 500	10 000
306	306	306	RRM 030	30	1·1811	72	2·8346	19	0·7480	2	2	4 250	6 000	10 000
NU 307	NJ 307	N 307	RRM 035	35	1·3780	80	3·1496	21	0·8268	2·5	2	5 200	7 350	8 000
308	308	308	RRM 040	40	1·5748	90	3·5433	23	0·9055	2·5	2·5	6 800	9 150	8 000
309	309	309	RRM 045	45	1·7717	100	3·9370	25	0·9843	2·5	2·5	8 650	11 600	8 000
NU 310	NJ 310	N 310	RRM 050	50	1·9685	110	4·3307	27	1·0630	3	3	10 800	14 000	6 000
311	311	311	RRM 055	55	2·1654	120	4·7244	29	1·1417	3	3	12 700	17 300	6 000
312	312	312	RRM 060	60	2·3622	130	5·1181	31	1·2205	3·5	3·5	16 000	20 400	5 000
NU 313	NJ 313	N 313	RRM 065	65	2·5591	140	5·5118	33	1·2992	3·5	3·5	18 000	22 400	5 000
314	314	314	RRM 070	70	2·7559	150	5·9055	35	1·3780	3·5	3·5	19 600	25 000	5 000
315	315	315	RRM 075	75	2·9528	160	6·2992	37	1·4567	3·5	3·5	24 000	30 500	4 000
NU 316	NJ 316	N 316	RRM 080	80	3·1496	170	6·6929	39	1·5354	3·5	3·5	26 500	32 000	4 000
317	317	317	RRM 085	85	3·3465	180	7·0866	41	1·6142	4	4	29 000	35 500	4 000
318	318	318	RRM 090	90	3·5433	190	7·4803	43	1·6929	4	4	34 000	40 500	3 000
NU 319	NJ 319	N 319	RRM 095	95	3·7402	200	7·8740	45	1·7717	4	4	36 500	45 000	3 000
320	320	320	RRM 100	100	3·9370	215	8·4646	47	1·8504	4	4	43 000	52 000	3 000
321	321	321	RRM 105	105	4·1339	225	8·8583	49	1·9291	4	4	49 000	58 500	2 500
NU 322	NJ 322	N 322	RRM 110	110	4·3307	240	9·4488	50	1·9685	4	4	57 000	65 500	2 500
324	324	324	RRM 120	120	4·7244	260	10·2362	55	2·1654	4	4	65 500	78 000	2 500
NU 326	*NJ 326*	—	—	130	5·1181	280	11·0236	58	2·2835	5	5	86 500	96 500	2 500
328	*328*	—	—	140	5·5118	300	11·8110	62	2·4409	5	5	96 500	108 000	2 000
330	*330*	—	—	150	5·9055	320	12·5984	65	2·5591	5	5	108 000	118 000	2 000
NU 332	*NJ 332*	—	—	160	6·2992	340	13·3858	68	2·6772	5	5	114 000	125 000	2 000
334	*334*	—	—	170	6·6929	360	14·1732	72	2·8346	5	5	129 000	140 000	1 600
336	*336*	—	—	180	7·0866	380	14·9606	75	2·9528	5	5	153 000	160 000	1 600
NU 338	*NJ 338*	—	—	190	7·4803	400	15·7480	78	3·0709	6	6	166 000	173 000	1 600
340	*340*	—	—	200	7·8740	420	16·5354	80	3·1496	6	6	166 000	173 000	1 600

Table 4.7 TAPER ROLLER BEARINGS

(Reproduced from GB 550[28] by kind permission of the SKEFKO Ball Bearing Company)

Bearing No.	*d*		*D*		*T*		*B*		*C*	*r* ≈	r_1 ≈	*a* ≈	*Basic capacity,* lbf		*Max. per-missi-ble speed*
	mm	in.	mm	in.	mm	in.	mm	in.	Millimetres				*Static* C_0	*Dyna-mic C*	r.p.m.
30 302	15	0·5906	42	1·6535	14·25	0·561	13	0·5118	11	1·5	0·5	9	2 160	3 150	8 000
03	17	0·6693	47	1·8504	15·25	0·600	14	0·5512	12	1·5	0·5	10	2 750	4 000	8 000
04	20	0·7874	52	2·0472	16·25	0·640	15	0·5906	13	2	0·8	11	3 450	4 900	8 000
30 305	25	0·9843	62	2·4409	18·25	0·718	17	0·6693	15	2	0·8	13	4 750	6 550	6 000
06	30	1·1811	72	2·8346	20·75	0·817	19	0·7480	16	2	0·8	15	6 300	8 300	6 000
07	35	1·3780	80	3·1496	22·75	0·896	21	0·8268	18	2·5	0·8	16	8 300	10 600	6 000
30 308	40	1·5748	90	3·5433	25·25	0·994	23	0·9055	20	2·5	0·8	19	9 800	12 200	5 000
09	45	1·7717	100	3·9370	27·25	1·073	25	0·9843	22	2·5	0·8	21	12 500	15 000	5 000
10	50	1·9685	110	4·3307	29·25	1·152	27	1·0630	23	3	1	23	14 600	17 300	4 000
30 311	55	2·1654	120	4·7244	31·5	1·240	29	1·1417	25	3	1	24	17 300	19 600	4 000
12	60	2·3622	130	5·1181	33·5	1·319	31	1·2205	26	3·5	1·2	26	20 000	23 600	4 000
13	65	2·5591	140	5·5118	36	1·417	33	1·2992	28	3·5	1·2	28	23 600	27 500	3 000
30 314	70	2·7559	150	5·9055	38	1·496	35	1·3780	30	3·5	1·2	29	27 000	31 000	3 000
15	75	2·9528	160	6·2992	40	1·575	37	1·4567	31	3·5	1·2	31	30 500	34 000	3 000
16	80	3·1496	170	6·6929	42·5	1·673	39	1·5354	33	3·5	1·2	33	33 500	37 500	2 500
30 317	85	3·3465	180	7·0866	44·5	1·752	41	1·6142	34	4	1·5	35	36 500	42 500	2 500
18	90	3·5433	190	7·4803	46·5	1·831	43	1·6929	36	4	1·5	36	41 500	46 500	2 500
30 319	95	3·7402	200	7·8740	49·5	1·949	45	1·7717	38	4	1·5	39	50 000	55 000	2 500
20	100	3·9370	215	8·4646	51·5	2·028	47	1·8504	39	4	1·5	40	56 000	61 000	2 000
21	105	4·1339	225	8·8583	53·5	2·106	49	1·9291	41	4	1·5	41	61 000	65 500	2 000
30 322	110	4·3307	240	9·4488	54·5	2·146	50	1·9685	42	4	1·5	43	65 500	72 000	1 600
24	120	4·7244	260	10·236	59·5	2·342	55	2·1654	46	4	1·5	47	81 500	85 000	1 600

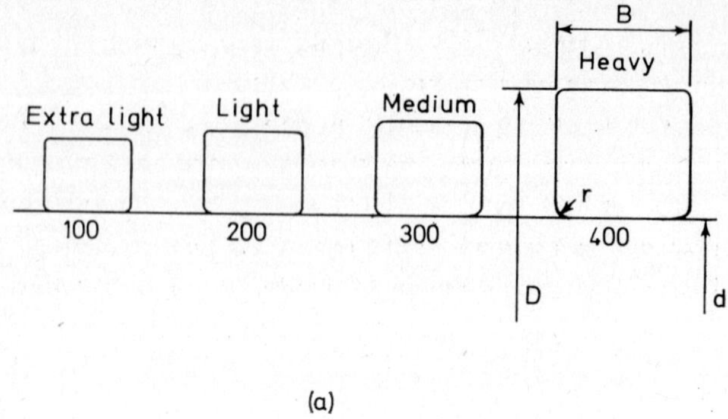

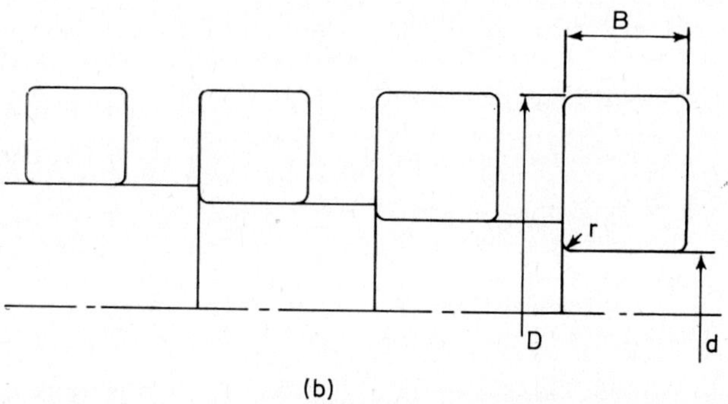

Fig. 4.45. Standard boundary dimensions (a) Variation of outside diameter with series for constant bore (b) Variation of bore with series for constant outside Diameter

series 03. Complete data for many more types and series are included in the reference quoted[28], the tables reproduced here being merely representative.

4.16. THE CARRYING CAPACITY OF ROLLING CONTACT BEARINGS

It is not the purpose here to attempt an authorititive treatise of rolling bearing analysis, and for this purpose the reader should refer to the excellent works of Palmgren[29] and more recently Harris[32]. Nevertheless, in selecting a bearing it is necessary for the designer to have some understanding of the load carrying capacity, and more particularly of the bearing life.

It is worth repeating some of the comments made in the SKF General Catalogue[28], with regard to the terms life and dynamic capacity. In the first instance, the term life refers to the number of loading cycles which may be applied to a bearing before failure by fatigue occurs. This assumes that the bearing is protected from

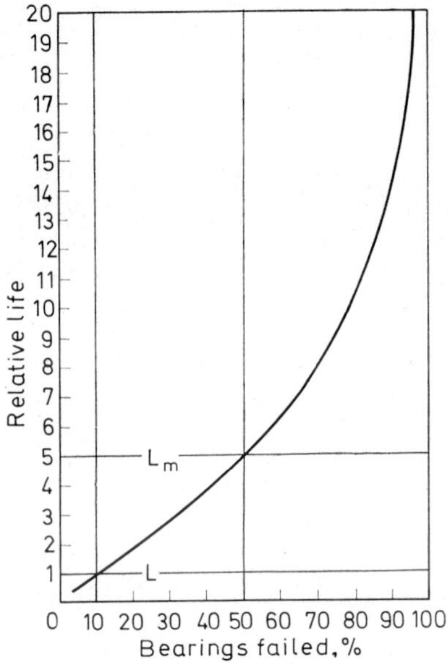

Fig. 4.46. Typical life dispersion curve (Reproduced from GB. 550[28] by kind permission of the Skefko Ball Bearing Company)

dirt and moisture, is properly mounted, well lubricated, properly handled, and so on, so that all causes of failure are eliminated, except fatigue of the material. One must be somewhat more precise, however, when defining life under fatigue conditions, for as previously discussed[33], fatigue life is a stochastic quantity.

In accordance with I.S.O. Recommendation R 15, the nominal life L of a bearing is defined as the life attained or exceeded by 90 per cent of all bearings. Fig. 4.46 shows a typical dispersion curve[28] illustrating the percentage bearings failed for different relative lives, from which it may be observed that 50 per cent of the bearings have a life L_m which is five times greater than the nominal life L on which calculations are based.

Also in accordance with I.S.O. Recommendation R 15 is the definition of dynamic capacity C, as that constant purely radial load which a group of apparently identical bearings can withstand for 10^6 revolutions of the inner ring with the outer ring stationary. Values for dynamic capacity C are published by the bearing manufacturers' and those listed in Tables 4.5, 4.6 and 4.7 give these values in terms of lbf; these values can be converted to Newtons by multiplying by 4·448 22.

Next, we must consider the relationship between load and life and the dynamic capacity. If one group of bearings of a certain type and size are run under a constant load F_1, and another group under a constant load F_2 but otherwise identical conditions, the lives for these two groups will be L_1 and L_2 (say), respectively. Extensive testing indicates that the relationship between these quantities is of the form

$$\frac{L_1}{L_2} = \left(\frac{F_2}{F_1}\right)^p$$

or

$$F_1(L_1)^{1/p} = F_2(L_2)^{1/p} = \text{Constant} \qquad (4.35)$$

In the special case where one of the groups is run under a purely radial load which gives a life of 10^6 cycles, then if this load is C (by definition of dynamic capacity), then

$$L = (C/P) \qquad (4.36)$$

where P is the equivalent radial load (which will presently be discussed), and L represents the nominal life in millions of revolutions. Equation 4.36 is termed the life equation.

The exponent p is dependent upon the type of contact or Hertzian stress condition[33] which applies, and considerable research has been necessary to establish its value. The values which have been accepted internationally for practical use are

$$p = 3 \text{ for ball bearings, and}$$
$$p = 10/3 \text{ for roller bearings.}$$

The value (C/P) is termed the loading ratio, and appropriate values corresponding to different lives, expressed in hours at different speeds, are reproduced in Tables 4.8 and 4.9 for ball and roller bearings respectively, reproduced by kind permission of The Skefko Ball Bearing Company.

In the above discussion it has been assumed that the load acting on the bearing is a purely radial one, and that the outer ring remains stationary whilst the inner ring rotates. The more general situation might be that of combined radial and axial loading, possibly with the outer ring rotating. Under these circumstances, it is necessary to determine the equivalent radial load P, which has the same damaging effect as the combined radial and axial loading, and taking account of the possibility of rotation of the outer ring. For this purpose, the equivalent radial load may be calculated from the equation[28, 32],

$$P = XVF_r + YF_a \tag{4.37}$$

where P = equivalent radial bearing load,
F_r = actual radial bearing load,
F_a = actual axial bearing load,
X = a radial factor,
Y = a thrust factor,
V = a rotation factor.

The rotation factor V depends upon whether the inner ring or outer ring rotates, and has the values indicated in Table 4.10[28].

The factors X and Y depend upon the type of bearing and the ratio of (F_a/VF_r). Values for these factors are listed in the manufacturers' catalogues, and are reproduced by permission from the Skefko catalogue[28] in Tables 4.11 and 4.12 for ball and roller bearings respectively. In the special case where a purely radial load acts on a bearing, i.e. $F_a = 0$, the value for P is simply equal to VF_r, reference

Table 4.8 LOADING RATIO (C/P) FOR DIFFERENT LIVES, EXPRESSED IN WORKING HOURS, AT DIFFERENT SPEEDS (BALL BEARINGS)

(Reproduced from GB 550[28] by kind permission of the SKEFKO Ball Bearing Company)

Life in hours L_h	*Revolutions per minute*													
	10	*16*	*25*	*40*	*63*	*100*	*125*	*160*	*200*	*250*	*320*	*400*	*500*	*630*
100									1·06	1·15	1·24	1·34	1·45	1·56
500				1·06	1·24	1·45	1·56	1·68	1·82	1·96	2·12	2·29	2·47	2·67
1 000			1·15	1·34	1·56	1·82	1·96	2·12	2·29	2·47	2·67	2·88	3·11	3·36
1 250		1·06	1·24	1·45	1·68	1·96	2·12	2·29	2·47	2·67	2·88	3·11	3·36	3·63
1 600		1·15	1·34	1·56	1·82	2·12	2·29	2·47	2·67	2·88	3·11	3·36	3·63	3·91
2 000	1·06	1·24	1·45	1·68	1·96	2·29	2·47	2·67	2·88	3·11	3·36	3·63	3·91	4·23
2 500	1·15	1·34	1·56	1·82	2·12	2·47	2·67	2·88	3·11	3·36	3·63	3·91	4·23	4·56
3 200	1·24	1·45	1·68	1·96	2·29	2·67	2·88	3·11	3·36	3·63	3·91	4·23	4·56	4·93
4 000	1·34	1·56	1·82	2·12	2·47	2·88	3·11	3·36	3·63	3·91	4·23	4·56	4·93	5·32
5 000	1·45	1·68	1·96	2·29	2·67	3·11	3·36	3·63	3·91	4·23	4·56	4·93	5·32	5·75
6 300	1·56	1·82	2·12	2·47	2·88	3·36	3·63	3·91	4·23	4·56	4·93	5·32	5·75	6·20
8 000	1·68	1·96	2·29	2·67	3·11	3·63	3·91	4·23	4·56	4·93	5·32	5·75	6·20	6·70
10 000	1·82	2·12	2·47	2·88	3·36	3·91	4·23	4·56	4·93	5·32	5·75	6·20	6·70	7·23
12 500	1·96	2·29	2·67	3·11	3·63	4·23	4·56	4·93	5·32	5·75	6·20	6·70	7·23	7·81
16 000	2·12	2·47	2·88	3·36	3·91	4·56	4·93	5·32	5·75	6·20	6·70	7·23	7·81	8·43
20 000	2·29	2·67	3·11	3·63	4·23	4·93	5·32	5·75	6·20	6·70	7·23	7·81	8·43	9·11
25 000	2·47	2·88	3·36	3·91	4·56	5·32	5·75	6·20	6·70	7·23	7·81	8·43	9·11	9·83
32 000	2·67	3·11	3·63	4·23	4·93	5·75	6·20	6·70	7·23	7·81	8·43	9·11	9·83	10·6
40 000	2·88	3·36	3·91	4·56	5·32	6·20	6·70	7·23	7·81	8·43	9·11	9·83	10·6	11·5
50 000	3·11	3·63	4·23	4·93	5·75	6·70	7·23	7·81	8·43	9·11	9·83	10·6	11·5	12·4
63 000	3·36	3·91	4·56	5·32	6·20	7·23	7·81	8·43	9·11	9·83	10·6	11·5	12·4	13·4
80 000	3·63	4·23	4·93	5·75	6·70	7·81	8·43	9·11	9·83	10·6	11·5	12·4	13·4	14·5
100 000	3·91	4·56	5·32	6·20	7·23	8·43	9·11	9·83	10·6	11·5	12·4	13·4	14·5	15·6
200 000	4·93	5·75	6·70	7·81	9·11	10·6	11·5	12·4	13·4	14·5	15·6	16·8	18·2	19·6

Life in hours L_h	Revolutions per minute													
	800	*1 000*	*1 250*	*1 600*	*2 000*	*2 500*	*3 200*	*4 000*	*5 000*	*6 300*	*8 000*	*10 000*	*12 500*	*16 000*
100	1·68	1·82	1·96	2·12	2·29	2·47	2·67	2·88	3·11	3·36	3·63	3·91	4·23	4·56
500	2·88	3·11	3·36	3·63	3·91	4·23	4·56	4·93	5·32	5·75	6·20	6·70	7·23	7·81
1 000	3·63	3·91	4·23	4·56	4·93	5·32	5·75	6·20	6·70	7·23	7·81	8·43	9·11	9·83
1 250	3·91	4·23	4·56	4·93	5·32	5·75	6·20	6·70	7·23	7·81	8·43	9·11	9·93	10·6
1 600	4·23	4·56	4·93	5·32	5·75	6·20	6·70	7·23	7·81	8·43	9·11	9·93	10·6	11·5
2 000	4·56	4·93	5·32	5·75	6·20	6·70	7·23	7·81	8·43	9·11	9·83	10·6	11·5	12·4
2 500	4·93	5·32	5·75	6·20	6·70	7·23	7·81	8·43	9·11	9·83	10·6	11·5	12·4	13·4
3 200	5·32	5·75	6·20	6·70	7·23	7·81	8·43	9·11	9·83	10·6	11·5	12·4	13·4	14·5
4 000	5·75	6·20	6·70	7·23	7·81	8·43	9·11	9·83	10·6	11·5	12·4	13·4	14·5	15·6
5 000	6·20	6·70	7·23	7·81	8·43	9·11	9·83	10·6	11·5	12·4	13·4	14·5	15·6	16·8
6 300	6·70	7·23	7·81	8·43	9·11	9·83	10·6	11·5	12·4	13·4	14·5	15·6	16·8	18·2
8 000	7·23	7·81	8·43	9·11	9·83	10·6	11·5	12·4	13·4	14·5	15·6	16·8	18·2	19·6
10 000	7·81	8·43	9·11	9·83	10·6	11·5	12·4	13·4	14·5	15·6	16·8	18·2	19·6	21·2
12 500	8·43	9·11	9·83	10·6	11·5	12·4	13·4	14·5	15·6	16·8	18·2	19·6	21·2	22·9
16 000	9·11	9·83	10·6	11·5	12·4	13·4	14·5	15·6	16·8	18·2	19·6	21·2	22·9	24·7
20 000	9·83	10·6	11·5	12·4	13·4	14·5	15·6	16·8	18·2	19·6	21·2	22·9	24·7	26·7
25 000	10·6	11·5	12·4	13·4	14·5	15·6	16·8	18·2	19·6	21·2	22·9	24·7	26·7	28·8
32 000	11·5	12·4	13·4	14·5	15·6	16·8	18·2	19·6	21·2	22·9	24·7	26·7	28·8	31·1
40 000	12·4	13·4	14·5	15·6	16·8	18·2	19·6	21·2	22·9	24·7	26·7	28·8	31·1	
50 000	13·4	14·5	15·6	16·8	18·2	19·6	21·2	22·9	24·7	26·7	28·8	31·1		
63 000	14·5	15·6	16·8	18·2	19·6	21·2	22·9	24·7	26·7	28·8	31·1			
80 000	15·6	16·8	18·2	19·6	21·2	22·9	24·7	26·7	28·8	31·1				
100 000	16·8	18·2	19·6	21·2	22·9	24·7	26·7	28·8	31·1					
200 000	21·2	22·9	24·7	26·7	28·8	31·1								

Table 4.9 LOADING RATIO (C/P) FOR DIFFERENT LIVES, EXPRESSED IN WORKING HOURS, AT DIFFERENT SPEEDS (ROLLER BEARINGS)

(Reproduced from GB 550[28] by kind permission of the SKEFKO Ball Bearing Company)

Life in hours L_h	Revolutions per minute													
	10	16	25	40	63	100	125	160	200	250	320	400	500	630
100									1·05	1·13	1·21	1·30	1·39	1·49
500				1·05	1·21	1·39	1·49	1·60	1·71	1·83	1·97	2·11	2·26	2·42
1 000			1·13	1·30	1·49	1·71	1·83	1·97	2·11	2·26	2·42	2·59	2·78	2·97
1 250		1·05	1·21	1·39	1·60	1·83	1·97	2·11	2·26	2·42	2·59	2·78	2·97	3·19
1 600		1·13	1·30	1·49	1·71	1·97	2·11	2·26	2·42	2·59	2·78	2·97	3·19	3·42
2 000	1·05	1·21	1·39	1·60	1·83	2·11	2·26	2·42	2·59	2·78	2·97	3·19	3·42	3·66
2 500	1·13	1·30	1·49	1·71	1·97	2·26	2·42	2·59	2·78	2·97	3·19	3·42	3·66	3·92
3 200	1·21	1·39	1·60	1·83	2·11	2·42	2·59	2·78	2·97	3·19	3·42	3·66	3·92	4·20
4 000	1·30	1·49	1·71	1·97	2·26	2·59	2·78	2·97	3·19	3·42	3·66	3·92	4·20	4·50
5 000	1·39	1·60	1·83	2·11	2·42	2·78	2·97	3·19	3·42	3·66	3·92	4·20	4·50	4·82
6 300	1·49	1·71	1·97	2·26	2·59	2·97	3·19	3·42	3·66	3·92	4·20	4·50	4·82	5·17
8 000	1·60	1·83	2·11	2·42	2·78	3·19	3·42	3·66	3·92	4·20	4·50	4·82	5·17	5·54
10 000	1·71	1·97	2·26	2·59	2·97	3·42	3·66	3·92	4·20	4·50	4·82	5·17	5·54	5·94
12 500	1·83	2·11	2·42	2·78	3·19	3·66	3·92	4·20	4·50	4·82	5·17	5·54	5·94	6·36
16 000	1·97	2·26	2·59	2·97	3·42	3·92	4·20	4·50	4·82	5·17	5·54	5·94	6·36	6·81
20 000	2·11	2·42	2·78	3·19	3·66	4·20	4·50	4·82	5·17	5·54	5·94	6·36	6·81	7·30
25 000	2·26	2·59	2·97	3·42	3·92	4·50	4·82	5·17	5·54	5·94	6·36	6·81	7·30	7·82
32 000	2·42	2·78	3·19	3·66	4·20	4·82	5·17	5·54	5·94	6·36	6·81	7·30	7·82	8·38
40 000	2·59	2·97	3·42	3·92	4·50	5·17	5·54	5·94	6·36	6·81	7·30	7·82	8·38	8·98
50 000	2·78	3·19	3·66	4·20	4·82	5·54	5·94	6·36	6·81	7·30	7·82	8·38	8·98	9·62
63 000	2·97	3·42	3·92	4·50	5·17	5·94	6·36	6·81	7·30	7·82	8·38	8·98	9·62	10·3
80 000	3·19	3·66	4·20	4·82	5·54	6·36	6·81	7·30	7·82	8·38	8·98	9·62	10·3	11·0
100 000	3·42	3·92	4·50	5·17	5·94	6·81	7·30	7·82	8·38	8·98	9·62	10·3	11·0	11·8
200 000	4·20	4·82	5·54	6·36	7·30	8·38	8·98	9·62	10·3	11·0	11·8	12·7	13·6	14·6

Life in hours L_h	Revolutions per minute													
	800	1 000	1 250	1 600	2 000	2 500	3 200	4 000	5 000	6 300	8 000	10 000	12 500	16 000
100	1·60	1·71	1·83	1·97	2·11	2·26	2·42	2·59	2·78	2·97	3·19	3·42	3·66	3·92
500	2·59	2·78	2·97	3·19	3·42	3·66	3·92	4·20	4·50	4·82	5·17	5·54	5·94	6·36
1 000	3·19	3·42	3·66	3·92	4·20	4·50	4·82	5·17	5·54	5·94	6·36	6·81	7·30	7·82
1 250	3·42	3·66	3·92	4·20	4·50	4·82	5·17	5·54	5·94	6·36	6·81	7·30	7·82	8·38
1 600	3·66	3·92	4·20	4·50	4·82	5·17	5·54	5·94	6·36	6·81	7·30	7·82	8·38	8·98
2 000	3·92	4·20	4·50	4·82	5·17	5·54	5·94	6·36	6·81	7·30	7·82	8·38	8·98	9·62
2 500	4·20	4·50	4·82	5·17	5·54	5·94	6·36	6·81	7·30	7·82	8·38	9·98	9·62	10·3
3 200	4·50	4·82	5·17	5·54	5·94	6·36	6·81	7·30	7·82	8·38	8·98	9·62	10·3	11·0
4 000	4·82	5·17	5·54	5·94	6·36	6·81	7·30	7·82	8·38	8·98	9·62	10·3	11·0	11·8
5 000	5·17	5·54	5·94	6·36	6·81	7·30	7·82	8·38	8·98	9·62	10·3	11·0	11·8	12·7
6 300	5·54	5·94	6·36	6·81	7·30	7·82	8·38	8·98	9·62	10·3	11·0	11·8	12·7	13·6
8 000	5·94	6·36	6·81	7·30	7·82	8·38	8·98	9·62	10·3	11·0	11·8	12·7	13·6	14·6
10 000	6·36	6·81	7·30	7·82	8·38	8·98	9·62	10·3	11·0	11·8	12·7	13·6	14·6	15·6
12 500	6·81	7·30	7·82	8·38	8·98	9·62	10·3	11·0	11·8	12·7	13·6	14·6	15·6	16·7
16 000	7·30	7·82	8·38	8·98	9·62	10·3	11·0	11·8	12·7	13·6	14·6	15·6	16·7	17·9
20 000	7·82	8·38	8·98	9·62	10·3	11·0	11·8	12·7	13·6	14·6	15·6	16·7	17·9	19·2
25 000	8·38	8·98	9·62	10·3	11·0	11·8	12·7	13·6	14·6	15·6	16·7	17·9	19·2	20·6
32 000	8·98	9·62	10·3	11·0	11·8	12·7	13·6	14·6	15·6	16·7	17·9	19·2	20·6	
40 000	9·62	10·3	11·0	11·8	12·7	13·6	14·6	15·6	16·7	17·9	19·2	20·6		
50 000	10·3	11·0	11·8	12·7	13·6	14·6	15·6	16·7	17·9	19·2	20·6			
63 000	11·0	11·8	12·7	13·6	14·6	15·6	16·7	17·9	19·2	20·6				
80 000	11·8	12·7	13·6	14·6	15·6	16·7	17·9	19·2	20·6					
100 000	12·7	13·6	14·6	15·6	16·7	17·9	19·2	20·6						
200 000	15·6	16·7	17·9	19·2	20·6									

Table 4.10 ROTATIONAL FACTOR, V.

Bearing Type	*Rotating Inner Ring*	*Rotating Outer Ring*
Self-aligning ball bearings	1	1
All other radial bearings	1	1·2

to the tables indicating that $X = 1$ for this instance. Further, the effect of an axial load does not significantly influence the equivalent load P until its magnitude is such that the ratio (F_a/VF_r) exceeds a specified value; this value is designated e in Tables 4.11 and 4.12. Note that for a single row deep groove ball bearing the value for e is a function of the ratio F_a/C_0, where C_0 is the static capacity, defined by I.S.O. and listed in Tables 4.5, 4.6 and 4.7 as that load which will give a maximum permanent deformation of about 0.000 1 of the rolling element diameter at the most heavily stressed contact. In these tables, the values are given in lbf, but may readily be converted to units of Newtons. When the ratio of (F_a/VF_r) exceeds the limiting value of e corresponding to the particular bearing under consideration, the appropriate values for X and Y can be obtained from the tables, and the equivalent radial load P can be calculated from equation 4.37.

If the life (hours) corresponding to a constant speed of rotation (rev/min) is known, then the desired value for C/P may be readily obtained from Table 4.8 or 4.9. Since P has previously been calculated from equation 4.37, the required value for dynamic capacity C can be obtained from

$$C = (C/P)P$$

and a suitable bearing with the necessary C value can be selected.

A similar procedure may be applied for the selection of spherical roller thrust bearings; in this instance, however, the load P is the equivalent purely axial load considered centrally acting, determined[28] from the equation

$$P = F_a + 1{\cdot}2F_r \tag{4.38}$$

The appropriate tables for these bearings are not included in this chapter, but full details will be found in manufacturers' catalogues.

In the case of angular contact ball bearings and taper roller bearings, an imposed radial load will give rise to an induced axial or thrust load, in addition to any external axial load, and this must frequently be included in the calculations. As already mentioned, these types of bearings are usually arranged so that they can be

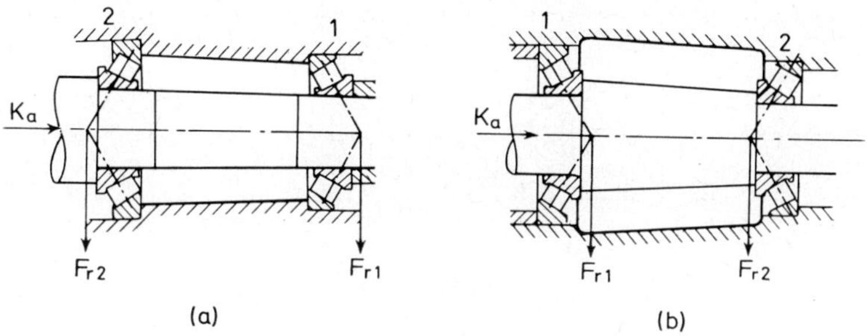

Fig. 4.47. Adjustment of taper roller bearings (Reproduced from GB 550[28] by kind permission of the Skefko Ball Bearing Company)

adjusted against each other, as indicated in Fig. 4.47. Using the notation indicated in this figure, the axial thrust load induced can be determined from Table 4.13. Bearing 1 is subjected to a radial load F_{r1}, bearing 2 to a radial load F_{r2} and an axial load K_a. If the axial load K_a acts in the opposite direction to that indicated in Fig. 4.47, the formulae listed in Table 4.13 are still valid provided that the bearing position numbers are reversed, i.e. bearing 1 becomes bearing 2, and vice versa.

The factors Y_1 and Y_2 represent the thrust factors for bearings 1 and 2 respectively, for the ratio $F_a/VF_r > e$, and can be obtained from Table 4.12.

As a final point, brief mention should be made of the fact that a bearing may sometimes be subjected to a load which fluctuates in value. Suppose that a constant load F_1 acts for N_1 revolutions, F_2 for N_2 revolutions, F_3 for N_3 revolutions, and so on. The equivalent mean load having the same influence on bearing life can be approx-

Table 4.11. FACTORS X AND Y FOR BALL BEARINGS

(Reproduced from GB 550[28] by kind permission of the Skefko Ball Bearing Company)

Bearing type		$\frac{F_a}{VF_r} \leq e^*$		$\frac{F_a}{VF_r} > e$		*e*
		X	*Y*	*X*	*Y*	
Single row						
deep groove	$\frac{F_a}{C_o}$ = 0·025				2·0	0·22
ball bearings,	0·04				1·8	0·24
Series:	0·07				1·6	0·27
EL, R, 60,	0·13	1	0	0·56	1·4	0·31
62, 63, 64,	0·18				1·3	0·34
EE, RLS,	0·24				1·2	0·37
RMS and	0·33				1·1	0·40
FL	0·56				1·0	0·44
Magneto bearings						
Series: EN, L		1	0	0·50	2·50	0·20
Double row self-aligning ball bearings						
135, 126, 127, 108, 129		1	1·8	0·65	2·8	0·34
1 200 —1 203			2·0		3·1	0·31
1 204 —1 205			2·3		3·6	0·27
1 206 —1 207			2·7		4·2	0·23
1 208 —1 209		1	2·9	0·65	4·5	0·21
1 210 —1 212			3·4		5·2	0·19
1 213 —1 222			3·6		5·6	0·17
2 200 —2 204			1·3		2·0	0·50
2 205 —2 207			1·7		2·6	0·37
2 208 —2 209			2·0		3·1	0·31
2 210 —2 213		1	2·3	0·65	3·5	0·28
2 214 —2 220			2·4		3·8	0·26
2 221 —2 222			2·3		3·5	0·28
1 300 —1 303			1·8		2·8	0·34
1 304 —1 305			2·2		3·4	0·29
1 306 —1 309		1	2·5	0·65	3·9	0·25
1 310 —1 322			2·8		4·3	0·23
2 301			1·0		1·6	0·63
2 302 —2 304			1·2		1·9	0·52
2 305 —2 310		1	1·5	0·65	2·3	0·43
2 311 —2 318			1·6		2·5	0·39

Bearing type	$\frac{F_a}{VF_r} \leq e^*$		$\frac{F_a}{VF_r} > e$		e
	X	Y	X	Y	
Double row self-aligning ball bearings					
RL 4—RL 6		2·1		3·3	0·29
RL 7—RL 8		2·3		3·6	0·27
RL 9—RL11		2·7		4·2	0·23
RL 12—RL14	1	2·9	0·65	4·5	0·21
RL 15—RL18		3·4		5·2	0·19
RL 20—RL36		3·6		5·6	0·17
RL 38—RL48		4·2		6·5	0·15
RM 3—RM 6		1·8		2·8	0·34
RM 7—RM10		2·1		3·3	0·29
RM11—RM14	1	2·4	0·65	3·8	0·26
RM15—RM18		2·7		4·2	0·23
RM20—RM48		2·9		4·5	0·21
Single row angular contact ball bearings Series:					
72B, 73B (α = 40°)	1	0	0·35	0·57	1·14
ALS, AMS (α = 30°)			0·39	0·76	0·80
Double row angular contact ball bearings Series: 32, 33	1	0·73	0·62	1·17	0·86

imated[28] by the equation

$$F_m = \left[\frac{F_1^3 N_1 + F_2^3 N_2 + F_3^3 N_3 + \ \dots \ F_n^3 N_n}{\sum_{i=1}^{n} N_i} \right]^{1/3} \qquad (4.39)$$

4.17. AXIAL LOCATION OF BEARINGS

It is essential in a bearing arrangement design to ensure that the correct fits between bearings and seatings are chosen. To a large extent the choice of fit will depend upon the working conditions,

Table 4.12. FACTORS X AND Y FOR ROLLER BEARINGS
(Reproduced from GB 550[28] by kind permission of the Skefko Ball Bearing Company)

Bearing type	$\frac{F_a}{VF_r} \leq e$		$\frac{F_a}{VF_r} > e$		*e*
	X	*Y*	*X*	*Y*	
Spherical roller bearings					
23 024 —23 084		2·7		4·0	0·25
23 088CA—230/500CA	1	3·3	0·67	4·9	0·21
24 024C —24 080CA		2·3		3·5	0·29
24 084CA—240/500CA	1	2·4	0·67	3·6	0·28
23 120C —23 128C		2·4		3·6	0·28
23 130C —23 132C		2·3		3·5	0·29
23 134 —23 188	1	2·1	0·67	3·1	0·32
23 192CA—231/500CA		2·3		3·5	0·29
24 122C —24 128C		1·9		2·9	0·35
24 130C —24 172CA	1	1·8	0·67	2·7	0·37
24 176CA—24 192CA		1·9		2·9	0·35
22 205 —22 207		2·5		3·7	0·27
22 208 —22 209		2·9		4·4	0·23
22 210 —22 215		3·4		5·0	0·20
22 216	1	2·6	0·67	3·9	0·26
22 217C —22 220C		2·9		4·4	0·23
22 222C —22 232C		2·6		3·9	0·26
22 234 —22 264		2·4		3·6	0·28
23 218C —23 220C	1	2·2	0·67	3·3	0·31
23 222C —23 264CA		2·0		3·0	0·34
21 304 —21 305		2·8		4·2	0·24
21 306 —21 310		3·2		4·8	0·21
21 311 —21 319	1	3·4	0·67	5·0	0·20
21 320 —21 322		3·7		5·5	0·18
22 308 —22 311		1·6		2·4	0·41
22 312C —22 315C		1·9		2·9	0·35
22 316C —22 324C		2·0		3·0	0·34
22 326 —22 340	1	1·8	0·67	2·7	0·37
22 344 —22 356		1·9		2·9	0·35
Taper roller bearings					
30 203 —30 204				1·75	0·34
30 205 —30 208				1·60	0·37
30 209 —30 222	1	0	0·4	1·45	0·41
30 224 —30 230				1·35	0·44
32 206 —32 208				1·60	0·37
32 209 —32 222	1	0	0·4	1·45	0·41
32 224 —32 230				1·35	0·44
30 302 —30 303				2·10	0·28
30 304 —30 307	1	0	0·4	1·95	0·31
30 308 —30 324				1·75	0·34
31 305 —31 318	1	0	0·4	0·73	0·82
32 303				2·10	0·28
32 304 —32 307	1	0	0·4	1·95	0·31
32 308 —32 324				1·75	0·34

Table 4.13.

	Conditions	*Axial Bearing Load*
1	$\frac{F_{r1}}{Y_1} \geqslant \frac{F_{r2}}{Y_2}$ $K_a \geqslant 0$	$F_{a1} = 0{\cdot}5\,\frac{F_{r1}}{Y_1}$ $F_{a2} = F_{a2} + K_a$
2	$\frac{F_{r1}}{Y_1} < \frac{F_{r2}}{Y_2}$ $K_a \geqslant 0{\cdot}5\left(\frac{F_{r2}}{Y_2} - \frac{F_{r1}}{Y_1}\right)$	
3	$\frac{F_{r1}}{Y_1} < \frac{F_{r2}}{Y_2}$ $K_a \leqslant 0{\cdot}5\left(\frac{F_{r2}}{Y_2} - \frac{F_{r1}}{Y_1}\right)$	$F_{a1} = F_{a2} - K_a$ $F_{a2} = 0{\cdot}5\,\frac{F_{r2}}{Y_2}$

but other factors such as the size and type of bearing, the method of lubrication, the bearing clearance, and so on, also require careful consideration. The limits of both the bore and the outside diameter of bearings are standardised, and the desired fits are obtained by adopting suitable limits for the shaft and bearing housing. A thorough discussion regarding shaft and housing fits can be found in the Skefko catalogue[28], together with comprehensive tables from which the correct limits may be selected.

A bearing with a cylindrical bore inner ring is usually mounted on the shaft with a tight fit, and axial location is provided by means of a flat shoulder on the shaft, as illustrated in Fig. 4.48. Table 4.14

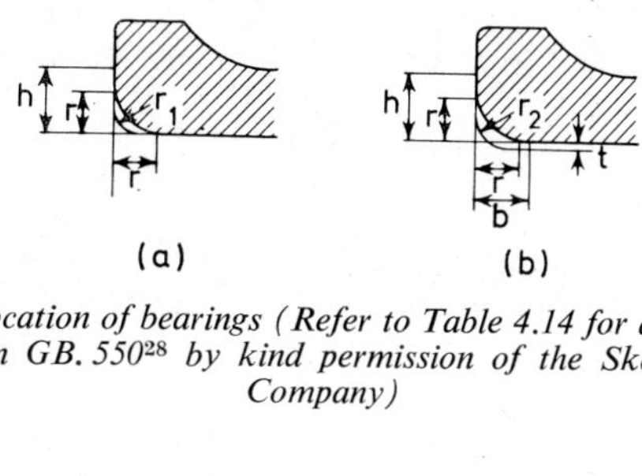

Fig. 4.48. Axial location of bearings (Refer to Table 4.14 for actual dimensions) (Reproduced from GB. 550[28] by kind permission of the Skefko Ball Bearing Company)

Table 4.14. FILLETS AND SHOULDER HEIGHTS FOR BALL, CYLINDRICAL AND SPHERICAL ROLLER BEARINGS (REFER TO FIG. 4.48)

(Reproducedfrom GB 550[28] by kind permission of the Skefko Ball Bearing Company)

For bearings with metric dimensions **Dimensions in millimetres**

Bearing nominal corner radius r	Shoulder height*		Fillet r_1 max	Fillet with relief		
	h	h_{min}		t	r_2	b
0·5	1	1	0·3	—	—	—
1	2·5	2	0·6	—	—	—
1·5	3	2·5	1	0·2	1·3	2
2	3·5	3	1	0·3	1·5	2·4
2·5	4·5	3·5	1·5	0·4	2	3·2
3	5	4	2	0·5	2·5	4
3·5	6	4·5	2	0·5	2·5	4
4	7	5·5	2·5	0·5	3	4·7
5	9	7·5	3	0·5	4	5·9
6	11	9	4	0·6	5	7·4
8	14	11	5	0·6	6	8·6
10	18	14	6	0·6	7	10

For bearings with inch dimensions **Dimensions in inches**

Bearing nominal corner radius r	Shoulder height*		Fillet r_1 max	Fillet with relief		
	h	h_{min}		t	r_2	b
0·02	$\frac{3}{64}$	$\frac{3}{64}$	0·016	—	—	—
$\frac{1}{32}$	$\frac{5}{64}$	$\frac{5}{64}$	0·025	—	—	—
0·04	$\frac{3}{32}$	$\frac{5}{64}$	0·03	—	—	—
$\frac{1}{16}$	$\frac{1}{8}$	$\frac{7}{64}$	0·04	0·008	0·05	0·075
$\frac{3}{32}$	$\frac{3}{16}$	$\frac{5}{32}$	0·06	0·015	0·08	0·125
$\frac{1}{8}$	$\frac{7}{32}$	$\frac{3}{16}$	0·08	0·02	0·10	0·16
$\frac{5}{32}$	$\frac{9}{32}$	$\frac{7}{32}$	0·10	0·02	0·12	0·185
$\frac{3}{16}$	$\frac{11}{32}$	$\frac{9}{32}$	0·12	0·02	0·15	0·225

* All the shoulder heights given are for radial bearings only. Shoulder heights h_{min} are intended only for radial bearings on which the axial load is negligible; they are suitable for deep groove ball bearings with sideplates or seals; they are unsuitable for taper roller bearings and single row angular contact ball bearings.

These details apply equally to shafts and housings.

indicates the shoulder height required for ball, cylindrical and spherical roller bearings, in order that sufficient abutment may be provided. The values given for h are suitable for normal thrust loads, h_{min} being used for shielded and sealed ball bearings. Values given correspond to different bearing corner radii, and the details are equally applicable to shafts and housings. Table 4.14 is not applicable to taper roller bearings, but details may be obtained from the Skefko catalogue[28].

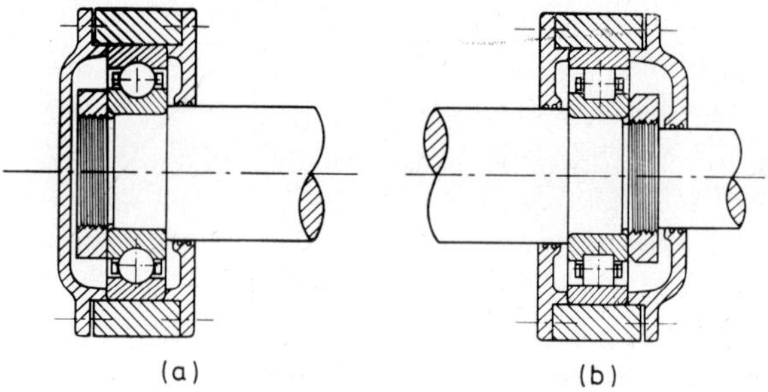

Fig. 4.49. Bearing arrangement using single-row deep groove ball bearing and cylindrical roller bearing

Various bearing arrangement designs may be adopted depending upon the particular application, and Fig. 4.49 illustrates the common combination of a deep groove ball bearing and a cylindrical roller bearing. This arrangement is very suitable for many electric motors, where the bearings operate with relatively long centre distances, and the shaft may be subjected to considerable temperature variation. With this arrangement, the inner rings of both bearings are made an interference fit on the shaft, and the outer rings, for convenience of assembly, are generally made a transition or sliding fit in the housings. Both inner and outer rings of both bearings are located axially, but since relative movement of the inner and outer ring is possible for the cylindrical roller bearing, any axial thrust will be taken by the single row deep groove ball bearing.

4.18. LUBRICATION OF ROLLING CONTACT BEARINGS

Rolling contact bearings may be lubricated by grease or mineral oil of the correct quality, depending upon the application. Generally speaking, at low and intermediate speeds, grease is to be preferred, since it has the advantage of considerably improving the efficiency of sealing, and also the maintenance required is usually reduced[34].

With grease lubrication, if the conditions are such that very infrequent lubrication is required, as might be the case for low speeds, replenishing or renewing the grease may be performed by removing the bearing cap. On the other hand, if lubrication is required more frequently, it will be necessary to incorporate lubricating ducts and nipples in the design, so that the bearing may be supplied with grease by means of a grease gun.

When a bearing is required to operate at high temperatures, either because of high speeds or because the bearing is heated due to external heating, and the lubrication interval becomes so short as to be impractical, it is necessary to employ oil lubrication[35]. As mentioned previously, the design of spherical roller thrust bearings is such that they must almost always be oil lubricated.

The simplest type of oil lubricating system is that which utilises an oil bath, the level of which should not exceed much above the lowest point of the outer ring. This method is quite suitable for moderate speeds, but for high speed operation, or where the temperature rise of the oil is likely to be excessive, it may be necessary to provide some means of external cooling. These, and other methods of oil lubrication, are discussed in the references previously mentioned.

Associated with the problem of lubrication is that of ensuring adequate sealing and protection of the bearing from the ingress of dust and moisture. Where grease lubrication is used, the sealing arrangement will usually be relatively simple, although the conditions under which the bearing is expected to operate is of obvious importance. On the other hand, when oil is the lubricant, the seal must perform the dual purpose of protecting the bearing from the ingress of dirt, and at the same time keep the oil inside the housing. The type and design of seal depends very much upon the situation, and may range from a simple felt ring, to somewhat complex labyrinth arrangements with some form of oil flinger fitted.

4.19. EXAMPLE 3

By way of simple illustration of the calculation of load carrying capacity for bearings, consider the example of the forage harvester discussed in Section 3.8. For this application, since there will be effectively no axial loading on the bearings, any type of radial bearing would be satisfactory. If accurate parallelism between the shaft and housings can be obtained, a suitable choice of bearing would be the single row deep groove ball bearing.

For the shaft carrying the pulley A and chain sprocket B (refer to Section 3.8), the loading on the bearings has been previously calculated and found to be

$$R_1 = 3\,885 \text{ N } (868 \text{ lbf})$$
$$R_2 = 4\,585 \text{ N } (1\,025 \text{ lbf})$$

and these are purely radial loads.

Corresponding to a life of 5 000 h at a speed of 450 rev/min, the required value for (C/P) may be obtained from Table 4.8 for ball bearings. It is found that

$$\frac{C}{P} = 5{\cdot}12$$

For bearing R_1, and using English units of lbf since the C values in the bearing manufacturers' catalogues are in these units, C corresponding to the value for $P = R_1 = 868$ lbf (3 885 N) is found from

$$C = 5{\cdot}12 \times 868 = 4\,440 \text{ lbf } (19\,900 \text{ N})$$

Referring to Table 4.5 for a medium series bearing, indicates that a suitable choice would be that having a bore of 30 mm, i.e. BRM

030, or using the SKF Bearing No. 6 306. Similarly, if tables for extra light, light and heavy series are referred to the Skefko catalogue[28], it will be found that appropriate bearings may be selected, as listed in Section 3.8. An identical procedure is used for bearing R_2.

Based on strength requirements, it was illustrated in Section 3.8 that the most suitable bearing was a BRL 035. Reference to the manufacturers' catalogues indicates that the bearing corner radius for this bearing is 2 mm. Now referring to Table 4.14, it is seen that

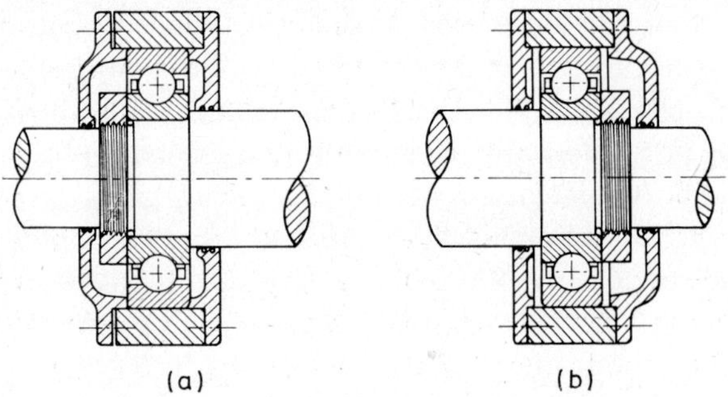

Fig. 4.50. Bearing arrangement using two single-row deep groove ball bearings.

the maximum fillet radius for the shaft is 1 mm, and the required shoulder height for axial location is 3.5 mm. The mounting of the bearings supporting the shaft might typically be as illustrated in Fig. 4.50, which assumes that single row deep groove ball bearings are to be used at each end. The inner ring of both bearings would be made an interference fit on the shaft. The bearing at the left hand support is axially located on both its inner and outer ring, but the right hand bearing has its outer ring free axially, this outer ring being a good sliding fit in the housing, enabling the bearing to freely position itself. With this arrangement, the left hand bearing is referred to as a locating bearing and the right hand bearing as a non-locating bearing. If the facility for the outer ring of the non-locating bearing to move axially was not made, there would be a danger of nipping the two bearings axially one against the other, which could result in premature failure.

4.20. CONCLUDING REMARKS

Sliding bearings are used extensively where full hydrodynamic lubrication can be obtained, especially at high speeds where they are less noisy than rolling contact bearings. Externally pressurised sliding bearings overcome the disadvantages associated with metal to metal contact during starting or at low speeds, but require external supply systems such as pumps, filters, etc.

If high starting torques are anticipated, rolling contact bearings may be preferable because of the inherently lower starting friction associated with the rolling elements. Further, such bearings are readily available as standard components, and it is necessary only for the designer to select the appropriate type and size from a bearing catalogue. Rolling contact bearings generally require less attention to lubrication than sliding bearings, particularly if prepacked grease lubrication is used, and failure of the lubricant is not likely to be as catastrophic as it might be for a sliding or journal bearing. Rolling contact bearings usually give some warning of failure by becoming noisier, whereas sliding bearings may fail suddenly and without warning. Many types of rolling contact bearings are capable of withstanding combinations of radial and axial loads, whereas sliding bearings are usually designed to withstand one type of loading only. Since the bearing clearance in rolling contact bearings is much less than that required in journal bearings, it is possible to obtain more accurate alignment and rigidity, a factor of particular importance for applications such as geared drives and machine tools. The ratio of length to diameter for rolling contact bearings makes them suitable where axial length is at a premium.

The decision to use sliding or rolling contact bearings depends upon the circumstances, factors such as reliability, cost, maintenance, convenience, etc., playing an important part.

REFERENCES

1. LIPSON, C., *Wear Considerations in Design*, Prentice Hall Inc., Englewood Cliffs, N. J. (1967)
2. TOWER, B., 'First Report of the Research Committee on Friction: on friction experiments (friction of lubricated bearings)', *Proc. Inst Mech. Engrs.*, 632 (1883); 29 (1884)
3. TOWER, B., 'Second Report of the Research Committee on Friction: Experiments on the oil pressure in a bearing', *Proc. Inst. Mech. Engrs.*, 58 (1885)
4. REYNOLDS, O., 'On the Theory of Lubrication and its Application to Mr. Beauchamp Tower's Experiments', *Phil. Trans. Roy. Soc.* (London), **177** (1886)
5. MICHELL, A. G. M., *Lubrication*, Blackie and Son, London (1950)
6. KINGSBURY, A., 'Development of the Kingsbury Thrust Bearing', *Mech. Eng.*, **72,** 957–967 (1950)
7. FULLER, D. D., *Theory and Practice of Lubrication for Engineers*, John Wiley and Sons, Inc., New York (1956)
8. PHELAN, R. M., *Fundamentals of Mechanical Design*, McGraw-Hill Book Company, Inc., New York, 2nd edn. (1962)
9. PETROFF, N., 'Friction in Machines and the effect of the Lubricant', (a) In Russian: *Engineering Journal*, St. Petersburg, 71–140, 228–279, 377–436, 535–564 (1883); (b) German translation by L. Wurzel, Hamburg, L. Voss (1887)
10. SOMMERFELD, A., 'Zur Hydrodynamischen Theorie der Schmiermittel —Reibung', (On the Hydrodynamic Theory of Lubrication), *Z. Math. Physik*, **50,** 97–155 (1904)
11. HERSEY, M. D., *Theory and Research in Lubrication*, John Wiley and Sons, Inc., New York (1966)
12. OCVIRK, F. W., 'Short Bearing Approximations for Full Journal Bearings', *Natl. Aeronautics and Space Adminstn., Techn. Note*. 2808 (1952)
13. DUBOIS, G. B., OCVIRK, F. W. and WEHE, R. L., 'Experimental Investigation of Eccentricity Ratio, Friction, and Oil Flow of Long and Short Bearings with Load Number Charts', *Natl. Aeronautics and Space Administn., Tech. Note* 3491 (1955)
14. MCKEE, S. A. and MCKEE. R. T., 'Friction of Journal Bearings as Influenced by Clearance and Length'. *Trans. Am. Soc. Mech. Engrs*, **51,** 161–171 (1929)
15. KARELITZ, G. B., 'Performance of Oil-Ring Bearings', *Trans. Am. Soc. Mech. Engrs.*, **62,** 57–70 (1930)
16. SHIGLEY, J., *Machine Design*, McGraw-Hill Book Company, Inc., New York (1956)
17. LASCHE, O., 'Bearings for High Speed', *Traction and Transmission*, **6,** 33–64 (1903)
(Translated from: 'Die Reibungsverhältnisse in hagern mit höher Umfangs-geschwinddigkeit, *ZVDI*, **46,** 1881–90 (1902))
18. HERSCHEL, W. H., 'Standardization of the Saybolt Universal Viscometer', *Bur. Standards Technical Paper* 112, (1919)
19. RAIMONDI, A. A. and BOYD, J., 'A Solution for the Finite Journal Bearing and its Application to Analysis and Design', Parts 1, 2 and 3, *A. S. L. E. Trans.*, **1,** no. 1, 159–209 (April, 1958) Pergamon Press, New York (1958)
20. HUNSICKER, H. Y., 'Aluminium Alloy Bearings—Metallurgy, Design and Service Characteristics', *Sleeve Bearing Materials*, 82–118, Amer. Soc.. for Metals (1949)

21. NEALE, M. J., 'Selection of Bearings', Lubrication and Wear, 547, *Conference of I. Mech. E.*, London (1967)
22. HOLLIGAN, P. T., 'Materials for Plain Bearings', *Tribology*, **1**, no. 1, 16–20 (January, 1968)
23. S. A. E. HANDBOOK, 'Bearing and Bushing Alloys', SAE J459c, 168–170, Society of Automotive Engineers, New York (1968)
24. GREENWOOD, D. C., *Mechanical Power Transmission*, McGraw-Hill Book Company Inc., New York (1962)
25. O'CONNOR, J. J., BOYD, J. and AVALLONE, E. A., *Standard Handbook of Lubrication Engineering*, McGraw-Hill Book Company Inc., New York (1968)
26. HARRIS, J. H., *The Lubrication of Rolling Bearings*, Shell-Mex and B.P. Ltd., London (1967)
27. '*SKF Bearings—Mounting and Maintenance*", *GB 562*, The Skefko Ball Bearing Company Ltd., Luton, Bedfordshire (1965)
28. '*SKF Ball and Roller Bearings*', *General Catalogue GB 550*, The Skefko Ball Bearing Company Ltd., Luton, Bedfordshire (1964)
29. PALMGREN, A., 'Ball and Roller Bearing Engineering', *SKF Industries, Inc.*, Philadelphia, Pa. 3rd edn. (1959)
30. '*New Departure Ball Bearings – General Catalogue*', New Departure Division of General Motors Corporation, Bristol, Connecticut, 25th edn. (1961) (Sole Distributors in the United Kindgom Ac-Delco Division of General Motors Ltd., 23, Buckingham Gate, London, S.W.1)
31. B. S. 292: 1958 'Dimensions of Ball Bearings and Parallel-Roller Bearings', British Standards Institution, 2 Park Street, London, W.1
32. HARRIS, T. A., *Rolling Bearing Analysis*, John Wiley and Sons Inc., New York (1966)
33. DUGGAN, T. V., *Applied Engineering Design and Analysis*, Design Technology Series, Iliffe Books Ltd., London (1970)
34. '*SKF—Grease Lubrication of Rolling Bearings*', *TSP 5296*, *Reg. 741*, The Skefko Ball Bearing Company Ltd., Luton, Bedfordshire (1966)
36. '*SKF—Oil Lubrication of Rolling Bearings*', *TSP 5297*, *Reg. 742*, The Skefko Ball Bearing Company Ltd., Luton, Bedfordshire (1958)

Chapter 5

Flexible Power Transmission Elements

A brief discussion concerning some of the advantages associated with flexible power transmission elements has been presented in Chapter 1, and in the present chapter the relevent theory and practice will be discussed.

Various types of belts are used for the transmission of power and, with the exception of toothed belts, they rely upon friction between the belt and pulley for the success of the drive. Because of the slip and creep which may occur in flat and V-belts, the angular velocity ratio between the two shafts is not constant, nor is it exactly equal to the ratio of the pulley diameters. Toothed belts are a comparatively recent development, and like chain and sprocket drives, they incorporate a positive drive with a constant velocity ratio.

The choice of a suitable flexible drive will depend upon a number of factors, including the speed, velocity ratio, centre distance, positive-drive requirements, power to be transmitted, shaft relationship, resistance to shock loading, maintenance requirements, and so on; these, and other factors are discussed by Greenwood[1].

5.1. FLAT BELTS

The basic theory of belt driving may be understood by considering an endless flat belt stretched over two pulleys, as illustrated in Fig. 5.1. When the pulleys are at rest, the initial tension throughout the belt is T, as shown in Fig. 5.1a. When the driving pulley rotates,

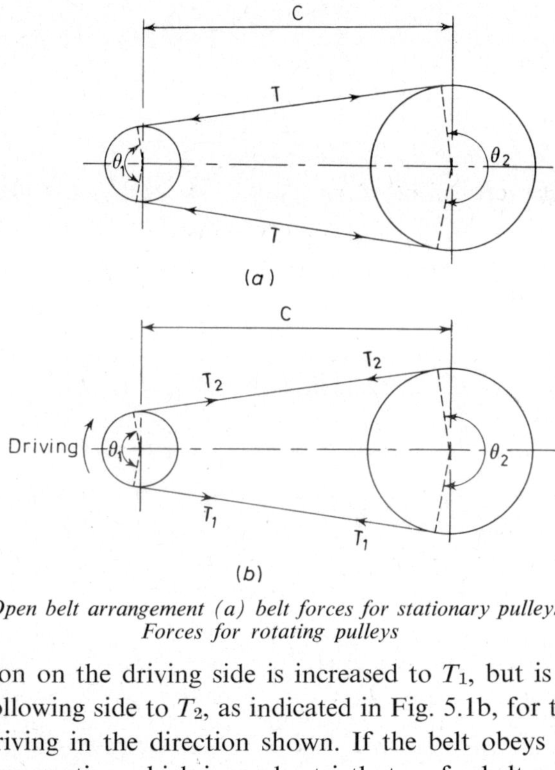

Fig. 5.1. Open belt arrangement (a) belt forces for stationary pulleys (b) Belt Forces for rotating pulleys

the tension on the driving side is increased to T_1, but is reduced on the following side to T_2, as indicated in Fig. 5.1b, for the small pulley driving in the direction shown. If the belt obeys Hooke's law (an assumption which is rarely strictly true for belt materials), the increase in length in the tight side and the reduction in length in the slack side must be equal, and the average tensions remain unaltered, i.e.

$$T = \tfrac{1}{2}(T_1+T_2) \tag{5.1}$$

which is sufficiently accurate for most purposes.

If the difference in the pulls on the tight and slack sides of the belt is great enough to overcome frictional resistance between the contacting surface of the belt and pulley, that portion of the belt in contact with the pulley begins to slide, and so-called *belt slip* occurs. The limiting value for the difference between T_1 and T_2 to ensure that belt slip does not occur, can be shown to be obtained from the equation[2]

$$\frac{T_1 - T_c}{T_2 - T_c} = \exp(f\theta) \tag{5.2}$$

where T_1 = tension on tight side of belt,
T_2 = tension on slack side of belt,
T_c = centrifugal tension in the belts,
f = coefficient of friction between belt and pulley,
θ = angle of lap, i.e. angle subtended by active arc of belt on pulley, radians.

The centrifugal tension in the belt is due to the peripheral velocity, and acts in addition to whatever static tension may already exist. The value for T_c is obtained from the equation for centrifugal force, namely

$$T_c = mV^2 \quad \text{N (or lbf)} \tag{5.3}$$

where m = mass of belt per unit length, kg/m (or lb/ft),
= γbt
γ = mass density of belt material, kg/m^3 (or lb/in^3),
b = width of belt, m (or in)
t = thickness of belt, m (or in),
V = velocity of belt, m/s (or ft/s).

If the centrifugal effects are negligible, i.e. the value for T_c is small compared to the tensions, equation 5.2 may be simplified to

$$\frac{T_1}{T_2} = \exp(f\theta) \tag{5.4}$$

The maximum permissible value for the ratio of T_1/T_2 to ensure that a flat belt will not slip round the pulley without driving, is frequently taken to be 3 to 1 for an arc of contact of 180 degrees.

The effective pull in the belt, from the point of view of transmitting power, is $T_1 - T_2$, and the relationship between this effective pull and the power transmitted can be obtained from the usual equations, i.e.

$$W = \frac{\pi DN(T_1 - T_2)}{60} \quad \text{J/s or W} \tag{5.5}$$

or

$$\text{hp} = \frac{\pi DN(T_1 - T_2)}{33\,000} \tag{5.6}$$

where D = pulley diameter (m in equation 5.5, ft in equation 5.6),
N = speed of rotation of pulley, rev/min,
$(T_1 - T_2)$ = effective belt pull (N in equation 5.5, lbf in equation 5.6).

The required cross-sectional area of the belt may be determined by equating the tensile stress to the working strength of the belt material, i.e.

$$S_w = \frac{T_1}{A} = \frac{T_1}{bt} \tag{5.7}$$

where S_w = working strength of the belt material,
A = cross-sectional area of belt.

Since belt failure is due mainly to fatigue, the working strength of the belt material is dependent upon speed, pulley diameters, belt tension and total belt length. The pulley diameter should not be less than about 100 times the belt thickness. Specific data is provided by the belt manufacturer.

Reference to equations 5.5 and 5.6 indicates that for maximum power transmission, the belt should be operated at as high a velocity as possible, but equation 5.3 indicates that the centrifugal force is proportional to the square of the velocity. Since centrifugal force has the effect of pulling the belt away from the pulley, this has the effect of increasing the possibility of slip. Further, to obtain a high velocity large pulleys are required, which necessitates increased centre distances. To satisfy these conflicting requirements, a satisfactory compromise is generally obtained by operating the belt at speeds between 10 m/s and 25 m/s (about 2 000 ft/min to 5 000 ft/min), although belt speeds in excess of this are possible, dependent upon the material. In the absence of more specific requirement, a belt speed in the region of 20 m/s (about 4 000 ft/min) is suggested as a reasonable value.

By the nature of transmitting power, a given section of the belt is stretched more on the tight side than on the slack side. Consequently, the driving pulley always receives a greater length of belt than it gives out, the reverse being true for the driven pulley. This creates relative motion between the belt and pulley, known as *belt creep*, not to be confused with belt slip mentioned earlier.

The power transmitted by flat belts may be limited by stretching of the belt, i.e. by belt creep, or by slipping of the belt on the pulley, i.e. by belt slip. Belt creep may be kept to acceptable values by the use of sufficiently low allowable stresses in the belt, and belt slip is dependent upon the angle of contact, θ, and the coefficient of friction between the belt and the pulley surface, f. The value for the coefficient of friction depends upon the belt material and the pulley material, as well as the belt speed. An average value for leather belts on cast-iron pulleys is $f = 0{\cdot}45$. The angle of contact is dependent upon the geometry of the drive, and whether the belt is open or crossed, as indicated in Fig. 5.2.

From the geometry of the open drive shown in Fig. 5.2a, the angle of contact for the small and large pulley respectively, is seen to be given by

$$\theta_1 = \pi - 2 \sin^{-1}\left(\frac{D-d}{2C}\right) \text{ rad} \tag{5.8}$$

$$\theta_2 = \pi + 2 \sin^{-1}\left(\frac{D-d}{2C}\right) \text{ rad} \tag{5.9}$$

and the total belt length L is

$$L = [4C^2 - (D-d)^2]^{1/2} + \tfrac{1}{2}(D\theta_2 + d\theta_1) \tag{5.10}$$

One point which is worth mentioning is that, for a horizontal open belt arrangement, a larger angle of lap or contact angle will be obtained on both pulleys if the rotation of the driving pulley can be arranged so that the slack side of the belt is on top.

For the crossed-drive indicated in Fig. 5.2b, the angle of contact for the small and large pulleys are the same, and is given by the equation

$$\theta = \pi + 2\sin^{-1}\left(\frac{D+d}{2C}\right) \quad \text{rad} \tag{5.11}$$

The total belt length is then obtained from the expression

$$L = [4C^2 - (D-d)^2]^{1/2} + \frac{\theta}{2}(D+d) \tag{5.12}$$

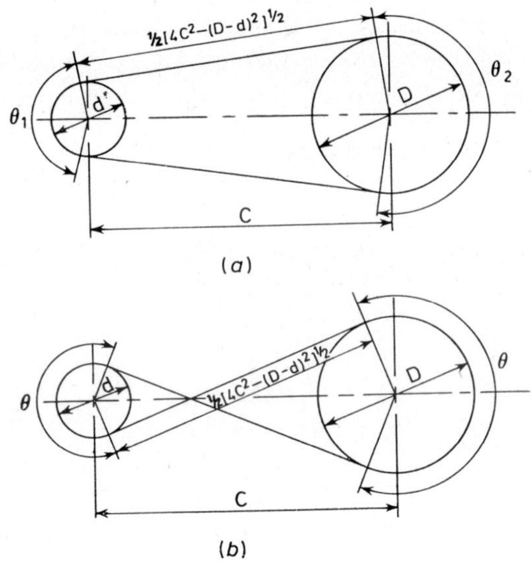

Fig. 5.2. Angles of contact and belt lengths (a) Open belt arrangement (b) Crossed belt arrangement

The main requirements for belt materials are high strength and durability, high flexibility and high coefficient of friction, coupled with low cost and good resistance against environmental conditions. Various materials are used for flat belts, including leather, rubber, and fabrics. Leather belts are the most common, and are specified as single, double or triple-ply, depending upon the number of layers which are used in the construction.

Flat belt pulleys are usually manufactured from cast iron or fabricated steel, from wood or wooden rims attached to cast iron spiders, or from compressed paper. The pulley rim may be flat or crowned, and connected to the hub by means of a web or a number of radial arms, the rim being either integral with the remainder of the pulley, or a separate attachment. To facilitate assembly, split-type pulleys are sometimes used.

It is important to mention that pulleys should be balanced to operate at the appropriate speed, this being particularly important for high speed operation. The stresses in the rim, due to rotation, must be kept to acceptable levels by limiting the peripheral velocity of the rim, and Table 5.1 indicates suitable rim speeds for various types of pulley construction, based on suggestions by Greenwood[1].

Table 5.1. NORMAL FLAT BELT PULLEY SPEEDS

Type of Pulley	*Rim Speed*	
	m/s	ft/min
Cast iron	18–23	3 500–4 500
Fabricated Steel	20–28	4 000–5 500
All Wood	25–30	5 000–6 000
Wood rim	40–50	8 000–10 000
Compressed paper or fibre	40–50	8 000–10 000

Where possible, standard pulleys are recommended, and manufacturers' catalogues give complete information regarding sizes and types, including standard diameters, face widths and bores.

Flat belts are used in applications involving large centre distances, but their use has decreased considerably in recent times, due to the fact that most modern machines have a built-in drive. Fig. 5.3, provided by J. H. Fenner and Co. Ltd., and reproduced by kind permission, illustrates an application of flat belts.

Fig. 5.3. Flat belts driving heavy duty ball mill mixers of oxide and abrasive materials (Reproduced by kind permission of J. H. Fenner & Co.)

5.2. V-BELTS

Although the initial cost of flat belt drives is generally less than that for V-belt drives, the latter provides a more efficient and compact arrangement.

Fig. 5.4a indicates a V-belt seated in its pulley groove, and demonstrates the basic principle of V-belt operation. In this case, a locking or wedge grip is developed as the belt enters and seats itself in the pulley groove. The top portion is placed under tension and the bottom portion is forced into compression. The sides of the belt are consequently induced to bulge, but the restraint offered by the pulley groove prevents this and causes the belt to grip the side walls of the groove. The groove itself is cut so that the V-belt rides sufficiently high to ensure that there is always an adequate air space between the base of the belt and the bottom of the groove, in order that the wedge grip is not lost, which would lead to an appreciable loss in power transmitting capacity.

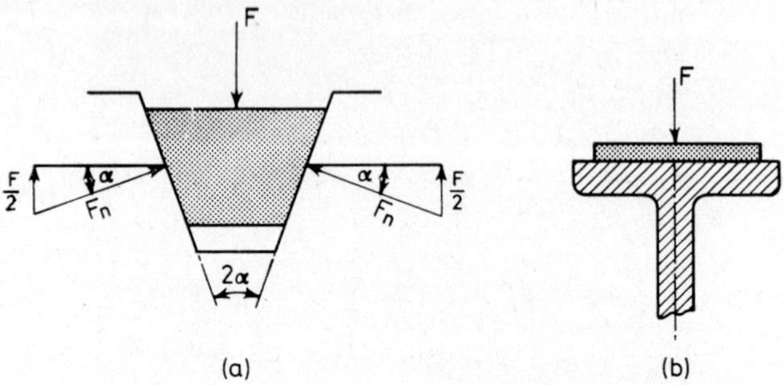

Fig. 5.4. Forces acting on (a) V-Belts and (b) Flat Belt

Referring to Fig. 5.4, which shows a V-belt and a flat belt each acted upon by a radial force F, it will be of interest to consider the frictional force between the belt and pulley for each case.

For the V-belt, the radial force F is reacted by the side walls of the pulley, the normal forces F_n being given by

$$F_n = \frac{F/2}{\sin\alpha} \tag{5.13}$$

where α = half pulley groove angle.
The frictional grip of the belt is equal to the normal forces multiplied by the coefficient of friction, i.e.

$$\text{Frictional force} = 2\,F_n f$$

$$= F\frac{f}{\sin\alpha}$$

$$= F/f' \tag{5.14}$$

where $f' = f/\sin\alpha$ = virtual coefficient of friction.

For the flat belt the frictional grip is equal to the radial force multiplied by the coefficient of friction, i.e.

$$\text{Frictional force} = Ff \tag{5.15}$$

Comparing equations 5.14 and 5.15, it is seen that to take account of the wedging action obtained with the V-groove, it is only necessary to replace f by the virtual coefficient of friction f', in the belt formulae previously presented.

The groove angle is of obvious importance in as much as it affects the gripping action of the belt. A small groove angle will give rise to a high gripping force between the belt and the pulley walls, but the force required to pull the belt out of the groove will also be high, which may result in loss of power and belt wear. Consequently, angles are chosen which give the best all round performance, and this is obtained using a groove angle of about 38 degrees; the actual groove angle varies slightly with decrease in pulley diameter. Fig.5.5 indicates the dimensions of a V-belt cross-section.

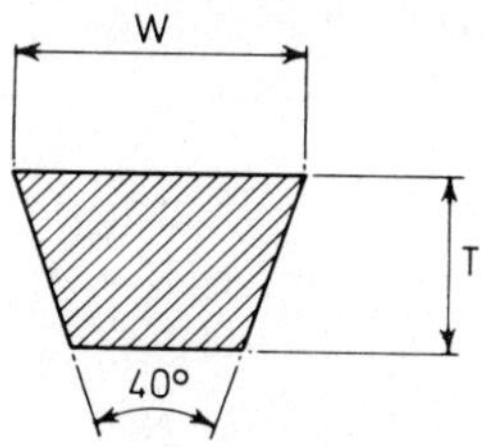

Fig. 5.5. Dimensions of V-Belt cross-section

The nominal groove angle is specifield as 40 degrees in the relevent British Standard Specifications[3], but angles of 38, 36, 34 and 32 degrees are specified, according to the pulley diameter and dimensions of the groove cross-section.

Standard V-belt drives enable a much higher tension ratio to be utilised than can be obtained with flat belts, usually about at least three times as high. This directly affects the effective belt pull $(T_1 - T_2)$, which also affects the torque which can be transmitted. Further, since the total force on the shaft is $(T_1 + T_2)$, for a specified torque the higher ratio of tensions enables this force to be reduced. Thus, the advantages of V-belts over flat belts are quite significant. The higher loading permissible before belt slip occurs enables very short centre distances and high speed ratios to be used, without increasing the loading on the bearings.

In the interest of standardisation, standard sizes of belts have been adopted, divided into five categories, and designated *A*, *B*, *C*, *D* and *E*, depending upon the dimensions and cross-section. Table 5.2 indicates the nominal cross-sectional dimensions of standard *V*-belts, adopted in the United Kingdom and the United States of America. The usual metric nominal sizes are also indicated.

Table 5.2. APPROXIMATE CROSS-SECTION DIMENSIONS OF STANDARD V-BELTS

Cross-section Symbol	*Nominal top width (W)*		*Nominal Thickness (T)*	
	(in)	(mm)	(in)	(mm)
A	0·500	13	0·315	8
B	0·655	17	0·405	11
C	0·875	22	0·530	14
D	1·250	32	0·750	19
E	1·500	38	0·905	23

It should be mentioned that individual manufacturers' belts may deviate slightly from these dimensions[4].

Extra high power belts, known as Premium V-belts, are also available in the same range of sizes as standard industrial belts[4]. Both standard and premium belts produced by J. H. Fenner and Co. (and other manufacturers), have resistance to abrasion, heat and oil and electrical conductivity, this latter reducing the risk of static electricity and a build-up of voltage.

To provide an alternative form of power transmission which results in more compact drives, so-called narrow or wedge-belts have been developed. These belts have a higher ratio of depth to top width, and are termed *Spacesaver Wedge-Belts* by Fenner and Co.[4]. Dimensions for narrow V-belts and their corresponding pulleys for industrial drives have been standardised, and three belt sections are provided[5], designated 3V, 5V and 8V; these belts have the nominal cross-sectional dimensions indicated in Table 5.3, which also indicated the approximate metric equivalents. The designations used by J. H. Fenner and Co. Ltd[4] for their *Spacesaver Wedge-Belts* are Alpha, Beta and Delta.

Table 5.3. APPROXIMATE CROSS-SECTION DIMENSIONS OF STANDARD NARROW V-BELTS

Cross-section	*Nominal top width* (*W*)		*Nominal thickness* (*T*)	
Symbol	(in)	(mm)	(in)	(mm)
3V	0·375	9·5	0·315	8
5V	0·625	15	0·530	14
8V	1·000	25	0·905	23

The increase in compactness provided by the use of narrow or *Spacesaver* wedge-belt drives is well illustrated by Fig. 5.6 and 5.7, reproduced by kind permission of J. H. Fenner and Co.

In designing a V-belt drive, the essential information required is

(a) driving speed,
(b) driven speed,
(c) power to be transmitted,
(d) type of driven machine.

In addition, the following information is also desirable:

(a) centre distance required,
(b) hours of operation per day,
(c) diameters and lengths of shafts,
(d) operating conditions,
(e) any restrictions on pulley diameter or width.

The main factors influencing the design are availability, cost, adequate belt speed, adequate pulley diameter for section of belt being used, suitability of pulley proportions for the installation, and compactness.

It is convenient to divide V-belt drives into three categories, namely; those which use a standard motor speed decreasing to a lower driven speed; those which involve any other driving speed decreasing to a lower driven speed; and those which involve any driving speed increasing to a higher driven speed. Manufacturers catalogues are usually very comprehensive[4], and detailed tables are provided for the selection of drives which use standard motor speeds. The procedure for the design of a drive not included in the

Fig. 5.6. Standard V-belt drive for Broom and Wade compressor.

Motor: 135 hp *(101* kW*) 975* rev/min
Motor Pulley: 14 in *(356* mm*) × 13D*
Driven Pulley: $41\frac{7}{8}$ in *(1065* mm*) × 13D*
Belts: Thirteen D180
Centre Bistance: 46·3 in *(1/175* mm*)*
Widht of Drive: $17\frac{5}{8}$ in *(448* mm*)*

(Reproduced by kind permission of J. H. Fenner and Co. Hull)

standard drive tables is somewhat different, but clear and complete guidance is also included for this situation by the manufacturers.

Of prime importance for the correct operation of V-belt drives is the initial tension in the belts. Also, since some stretching of the belts will occur in service, it is necessary to make some allowances available in the centre distances for adjustment. In assembling a drive, it is recommended[4] that the motor or prime mover should be moved towards the driven unit so that the V-belts may be placed in their respective grooves by hand. When the belts have been placed in the pulley grooves, the motor or prime mover should be moved away from the driven unit to apply uniform tension to the V-belts.

Fig. 5.7. Spacesaver Wedge-Belt Drive for Broom and Wade Compressor

Motor: 135 hp *(101* kW*) 975* rev/min
Motor Pulley: 13·2 in *(335* mm*) × 12 Beta*
Driven Pulley: 41·25 in *(1050* mm*) × 12 Beta*
Belts: Twelve Beta 1800
Cente Distance: 44·9 in *(1140* mm*)*
Width of Drive: 8 $\frac{9}{16}$ in *(217* mm*)*

(Reproduced by kind permission of J.H. Fenner and Co. Ltd., Hull)

If it is not possible to arrange for the centre distance to be adjusted, it may be necessary to use a jockey pulley in order to obtain the required tension. The best arrangement for a jockey pulley, from the point of view of flexing of the belts, it to use a grooved pulley in the usual way, and position this to run on the inside of the belts on the slack side. It is further suggested that such a jockey pulley should be mounted as near the larger pulley as possible.

It is frequently more convenient to use a flat faced jockey pulley on the outside of the belts, and this should preferably be positioned on the slack side within about one-third of the centre distance from the driving pulley.

In order to reduce the initial cost of the drive, it is sometimes possible to drive from a V-grooved pulley to a flat pulley. This arrangement might be particularly desirable if the large flat pulley is already in existence, but is should be mentioned that such a drive can never be as efficient as a V-V drive because there is no wedge-grip on the flat pulley. Further, the belts must be operated at higher tensions, which reduces the V-belt life and increases the loading on the bearings.

For satisfactory operation of V-flat drives, the speed ratio should be 3 to 1 or more, and the centre distance must not be greater than the difference in pulley diameters in order that adequate arc of contact is provided.

For further details relating to the calculation of belt section and the design of drives, the normal V-belt drive procedure should be followed, as illustrated in manufacturers catalogues[4].

5.3. TOOTHED OR TIMING BELTS

Unlike flat and V-belts, which depend upon friction between the contacting surfaces of the belts and pulleys for the transmission of power, toothed belts (frequently referred to as timing belts) obtain

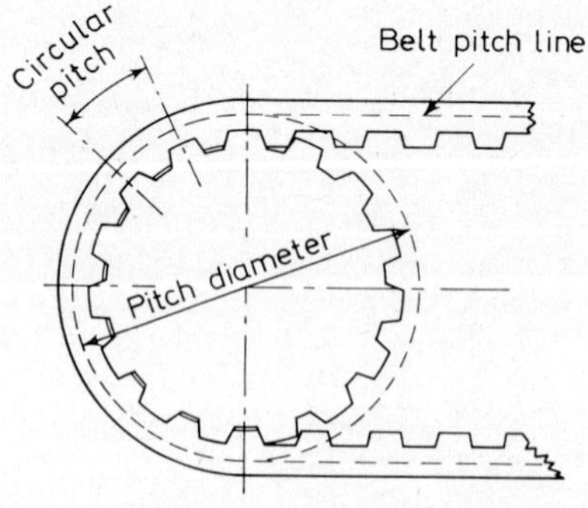

Fig. 5.8. Toothed belt drive

their grip by virtue of moulded teeth in the belt meshing with the teeth in both driving and driven pulleys. Fig. 5.8 illustrates a toothed belt meshing with its corresponding pulley, and indicates some of the nomenclature used.

Toothed belts do not stretch or slip, and consequently have the advantage over flat and V-belts in that they provide a positive

Fig. 5.9. Toothed belt driving a conveyor belt (Reproduced by kind permission of J. H. Fenner and Co. Ltd., Hull)

drive at a constant velocity ratio. After obtaining correct initial tension in the belt, no re-tensioning whatever is required, so that they can operate at fixed centres. Further, since the drive does not rely on friction, the static and running tensions are low, thus reducing the load on the shafts and bearings. Toothed belts have a very thin section, ensuring that the heat generated due to flexing around the pulleys is kept to a minimum. Also, since these belts have a high strength to mass (weight) ratio, on correctly designed drives belt speeds up to 80 m/s (16 000 ft/min) are possible.

Toothed or timing belts have many of the advantages associated with gears and chain drives, but without the necessity for lubrication or oil tight casings, a convenience which should be given careful consideration.

Toothed belt drives are quite versatile in that they may be used for ratios up to about 8·5 to 1, with power transmission capabilities ranging from very low to about 380 kW (500 hp). They find appli-

Fig. 5.10. Machine tool drive incorporating toothed belt drive (Reproduced by kind permission of J. H. Fenner and Co. Ltd., Hull)

cation in many fields, including machine tools, printing machinery, textile machinery, fans and blowers, compressors, household appliances, and miscellaneous light drives. For comprehensive details of available commercial units, the reader is referred to manufacturers catalogues, such as that of Fenner[4]. Figs 5.9 and 5.10 illustrate two applications of toothed belt drives, provided by J. H. Fenner and Co. Ltd., and reproduced by permission.

5.4. CHAIN DRIVES

Chain drives may be considered intermediate between belt drives and geared drives. They are capable of providing a more compact and positive method of transmitting power than belts, and are well adapted to either long or short centre distances. Further, a single chain may be arranged to drive more than one unit[6].

The manufacture of power transmission chains and sprockets is a specialised field, and the mechanical designer will usually make his selection from manufacturers catalogues, rather than carrying out a design from first principles. In the United Kingdom, the size and duty of power transmission roller chains, chain wheels and cutters, have been standardised in accordance with the specification of the British Standards Institution[7], which covers steel roller chains of simple, duplex and triplex construction. In other countries, the appropriate National Standards should be consulted.

An important advantage associated with chain drives when compared to gears is that the flexibility of the chain enables such a drive to absorb relatively higher shock loads than a rigid gear system. If properly selected and installed, chain drives will operate with a very high efficiency, a figure in the region of 98 percent being achieved under ideal conditions. Compared with belts, chain drives require relatively good shaft alignment, particularly for wide drives.

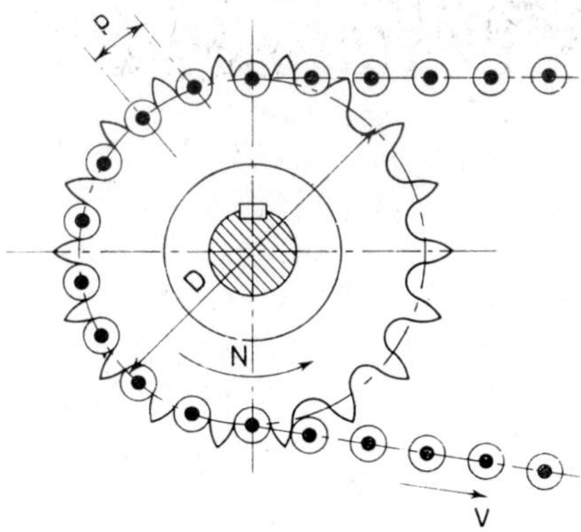

Fig. 5.11. Chain and sprocket

Fig. 5.11 illustrates the action of a chain in its sprocket, with some common chain nomenclature also being indicated.

The three principal types of chain in common use are the roller chain, the block chain and the inverted tooth or silent chain, as indicated in Fig. 5.12. Of these, the roller chain is the most common, and such chains are suitable for low and intermediate service speeds, i.e. up to about 13 m/s (2 500 ft/min). For higher speeds, i.e. up to about 20 m/s (4 000 ft/min), the so-called silent chains are recommended. Block chains are used mainly for supporting weights

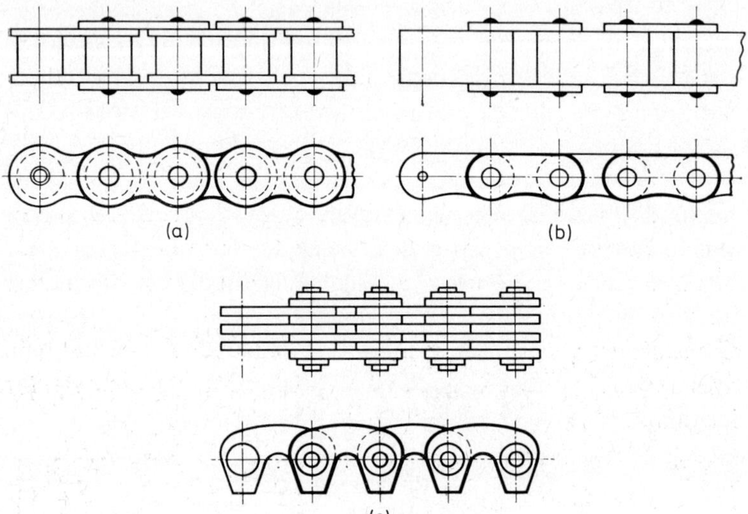

Fig. 5.12. Construction of (a) simple roller chain, (b) block chain, and (c) inverted tooth or silent chain

and for conveying purposes, although such chains have been used for light power transmission, but they are extremely noisy and are not generally recommended for such application.

The linear velocity of a chain is given by the equation

$$V = \frac{pTN}{60} \text{ m/s (in/s)} \tag{5.16}$$

where p = pitch of chain, m (or in),
T = number of sprocket teeth,
N = sprocket speed of rotation, rev/min,
V = chain peripheral velocity, m/s (or in/s).

Although the chain peripheral velocity is usually limited to the figures previously given, speeds up to 33 m/s (6 500 ft/min) can be obtained, but for such high speeds the chain manufacturer should be consulted.

An empirical relationship which has been widely used in practice[8] in estimating a reasonable pitch, is represented by the following equations:

$$p \leq \left(\frac{115\,000}{N}\right)^{2/3} \text{ mm} \tag{5.17}$$

or

$$p \leq \left(\frac{900}{N}\right)^{2/3} \text{ in} \tag{5.18}$$

In order that wear on the chain pin might be reduced, it is desirable to use a sprocket with no less than 17 teeth. Further, the larger the sprocket, the less will be the load on the chain for a given

Fig. 5.13. Application of single roller chains (Reproduced by kind permission of J. H. Fenner and Co. Ltd., Hull)

Fig. 5.14. Conveyor system incorporating chain and flat belt drive (Reproduced by kind permission of J. H. Fenner and Co. Ltd., Hull)

Fig. 5.15. Application of triple-chain drive (Reproduced by kind permission of J. H. Fenner and Co. Ltd., Hull)

power to be transmitted. Consequently, these factors indicate that from an operational point of view, large sprockets with many teeth are desirable, but important factors such as cost, space limitations, lubrication requirements, etc., favour smaller sprockets with a small number of teeth. Again, we have a situation typical of design, in that a compromise is necessary. In reaching a decision, the designer should be guided by the recommendations given in the manufacturers catalogues, which are based upon considerable experience[4,6]. Figs 5.13, 5.14 and 5.15 illustrate some applications of chain drives, and are reproduced by kind permission of J. H. Fenner and Co.

5.5. SHAFT LOADING DUE TO BELT AND CHAIN DRIVES

The effective pull in a belt or chain, from the point of view of transmitting power, is $(T_1 - T_2)$, as indicated by equations 5.5 and 5.6. However, as far as shaft loading is concerned, for the purposes of

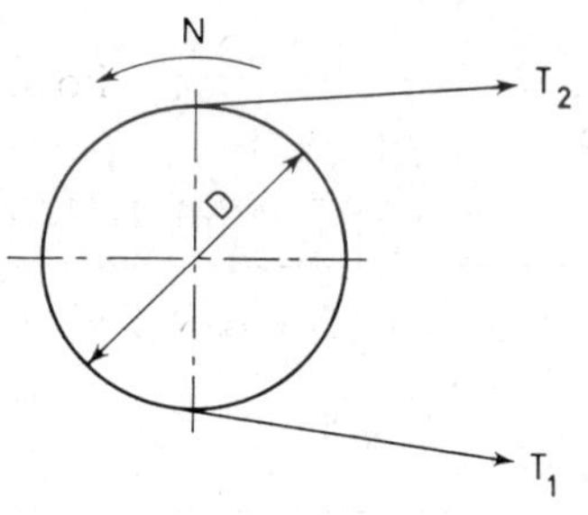

Fig. 5.16. Forces due to belt and chain tensions

calculating bending stresses and determining bearing loads, it is the total pull $(T_1 + T_2)$ which is important, as illustrated in Fig. 5.16. Repeating equations (5.5) and (5.6), the general formulae for calculating the effective pull are

$$T_1 - T_2 = \frac{60W}{\pi DN} \text{ (N)}$$

or

$$T_1 - T_2 = \frac{33\,000\,hp}{\pi DN} \text{ (lbf)}$$

where the symbols and units are as previously defined.

When calculating the total belt or chain pull (T_1+T_2), it is usual to introduce a service factor, designated F_x, such that the total belt pull is determined from

$$P = T_1 + T_2 = \frac{60W}{\pi DN} F_x$$

$$= \frac{19\,100\,kW}{DN} F_x \tag{5.19}$$

In equation 5.19, P is in Newtons; the power transmitted is kW (kilowatts); the effective diameter of the pulley or chain sprocket D in metres; and the speed of rotation is N rev/min.

In English units,

$$P = T_1 + T_2 = \frac{126\,000\,hp}{DN} F_x \tag{5.20}$$

where P is in pounds-force (lbf); hp is the horsepower transmitted;

Table 5.4 SERVICE FACTORS FOR BELTS AND CHAINS

Type of Drive	*Service Factor,* F_x
Single Ply Flat Belts	2·0
Double Ply Flat Belts	2·5
Triple Ply Flat Belts	3·0
V-Belts	1·5
Single Chains	1·0
Double Chains	1·25

D the effective pulley or sprocket diameter in inches; and the speed of rotation is N rev/min.

The service factor F_x in equations 5.19 and 5.20 depends upon the type of drive, and values based on considerable operating experience[9, 10] are given in Table 5.4.

5.6. CONCLUDING REMARKS

In this chapter, flexible power transmission elements have been discussed in sufficient detail to enable the reader to appreciate the principles involved. No attempt has been made to repeat the detailed and comprehensive tables relating to basic rating and selection, which can be found in the various manufacturers catalogues.

The decision to use a particular method of flexible drive depends upon the particular circumstances which apply, and factors requiring consideration include speed of operation and power transmission requirements; velocity ratio and centre distances; positive drive requirements; resistance to shock loading; maintenance requirements, lubrication and reliablity, cost, etc.

REFERENCES

1. GREENWOOD, D. C., *Mechanical Power Transmission* McGraw Hill Book Company, New York (1962)
2. GREEN, W. G., *Theory of Machines*, Blackie and Son Ltd., Glasgow (1955)
3. B. S. 1440: 1968, 'Endless V-Belt Drives—Sections, *A*, *B*, *C*, *D* and *E* for Industrial Purposes', British Standards Institution, London (1968)
4. *Fenner Mechanical Power Transmission*, Catalogue 100/68, J. H. Fenner and Co. Ltd., Hull (1968)
5. B. S. 3790: 1964, 'Endless Narrow *V*-Belt Drives for Industrial Purposes—Sections 3*V*, 5*V* and 8*V*', British Standards Institution, London (1964)
6. *Renold Power Transmission Diagrams and Technique*, Renold Chains Limited, Manchester (1953)
7. B. S. 228: 1962, 'Transmission Roller Chains, Chain Wheels and Cutters', British Standards Institution, London (1962)
8. BLACK, P. H. and ADAMS, O. E., *Machine Design*, McGraw-Hill Book Company, New York, 3rd edn. (1968)
9. *Engineering Handbook*, Hy-Eng. 1, Hyatt Bearing Division, General Motors Corporation, New Jersey, 13th Edn. (1951)
10. *The Timken Engineering Journal—Section 1*, The British Timken Division of the Timken Roller Bearing Company, Northampton (1965)

CHAPTER 6

Design of Gears

Gears are one of the most important groups of machine elements encountered in mechanical engineering. These elements, which may be defined as toothed wheels used to transmit power and motion from one shaft to another by means of successively engaging teeth, enable a positive action producing a definite velocity ratio to be obtained. Further, they are capable of being used at high loads and speeds to give a compact means of transmitting power, usually with high efficiency.

Essentially, the teeth on a gearwheel may be considered as a series of cams, and the motion imparted by one tooth to another in engagement with it depends upon the respective profiles of the mating surfaces. Of obvious importance, then, is the shape of teeth or the *tooth form*, the essential characteristic being that of providing smooth and uniform angular motion to be transmitted by the connecting teeth through all operating positions of the two profiles; such action is termed *conjugate*.

Although theoretically there are several tooth forms which fulfil the requirements of conjugate action, practical considerations limit the choice to two, namely the *cycloid* and the *involute*. Because of the flexibility in design and the practical advantages associated with the involute profile, this curve is now used almost universally in spur, helical, bevel and worm gears. The reader will know from courses in geometry[1] that an involute is the curve formed by the end of a straight line unwrapped from a circle called the *base circle*.

For continuity of action, it is important to ensure that before a pair of engaging teeth reach the point of disengagement, a succeeding pair of teeth are brought into engagement.

Properly designed gears require careful attention to a number of factors if reliability is to be obtained. These include the proper selection of materials and geometry to satisfy strength and wear requirements commensurate with satisfactory life; suitable methods of lubrication and the maintenance of a satisfactory operating temperature by heat dissipation; and certain practical limitations which are not always evident, and which may be expressed in terms of limit equations[2,3].

In this chapter an attempt will be made to present sufficient coverage of the more common types of gears to enable the engineer to design such units. It must be emphasised, however, that the many and varied theoretical aspects of gearing are quite complex, and no attempt will be made here to fully discuss some of these considerations. There are several excellent treatises dealing with the kinematics and special problems associated with gears, amongst those requiring particular mention are the works of Merritt[4], Tuplin[5], Buckingham[6] and Dudley[7]. In addition, gears and their associated problems have been the subject of various symposia[8,9]. It is suggested that the reader who requires a more detailed study than that presented in this chapter should consult these and the other references indicated.

6.1. SPUR AND HELICAL GEARS

Spur and helical gears are used to connect parallel shafts which rotate in opposite directions.

The most common type of gears encountered are those known as *straight spur gears*, such as the pair indicated in Fig. 6.1. These gears provide an efficient drive for slow or moderate speeds, but become noisy at high speeds unless manufactured to a high degree of accuracy and surface finish; they are also somewhat limited in their load transmitting capability. Being of simple form, they can be produced relatively cheaply. Since the teeth of spur gears are straight and parallel to the shaft axes on which they are mounted, the forces acting at the teeth do not give rise to any axial forces.

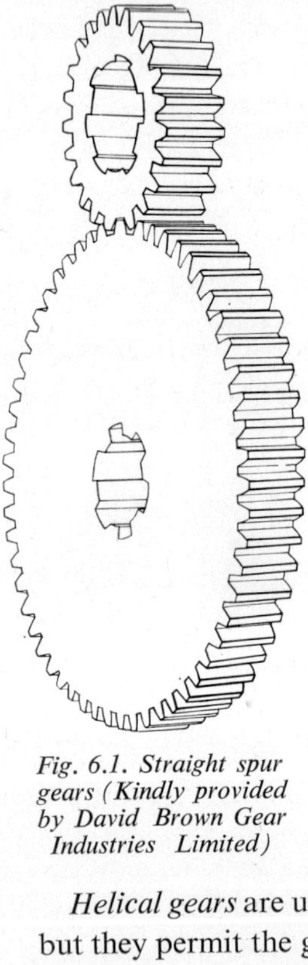

Fig. 6.1. Straight spur gears (Kindly provided by David Brown Gear Industries Limited)

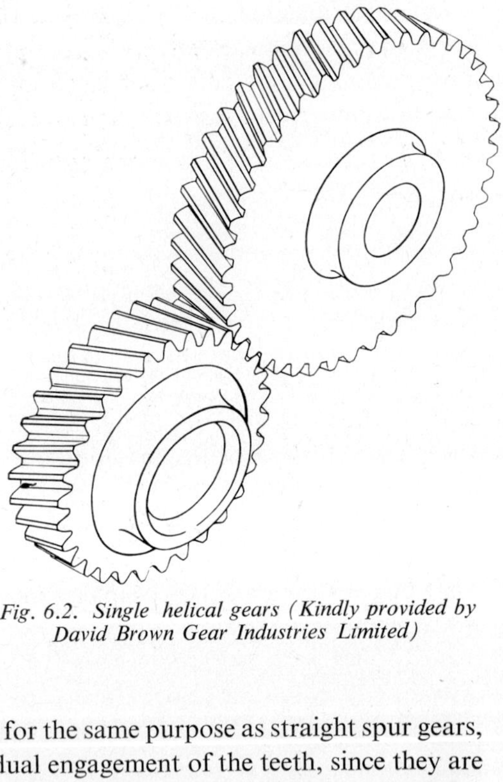

Fig. 6.2. Single helical gears (Kindly provided by David Brown Gear Industries Limited)

Helical gears are used for the same purpose as straight spur gears, but they permit the gradual engagement of the teeth, since they are cut with the teeth inclined at an angle to the axis of the gear. Thus, for the same standard of accuracy, helical gears run more quietly and are capable of greater load carrying capacity than spur gears. Helical gears may be divided into two groups, namely *single helical gears* and *double helical gears*, as illustrated in Fig. 6.2 and 6.3. Since the teeth of helical gears are inclined to the axis of the shaft, the forces acting at the teeth give rise to an end thrust which, in the case of single helical gears, might in many instances be a limiting factor to the load carrying capability.

One way of eliminating the axial thrust from a single helical gear is by mounting a pair of such gears of opposite hand on the shaft, or by the use of *double helical gears* or so-called *herringbone*

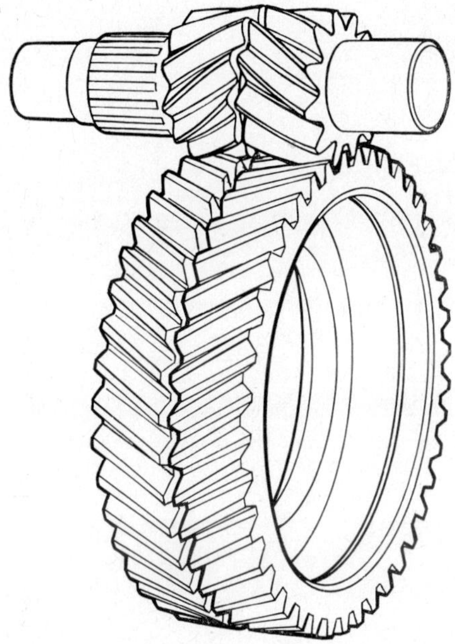

Fig. 6.3. Double helical gears (Kindly provided by David Brown Gear Industries Limited)

gears, as indicated in Fig. 6.4. The axial loads caused by the tooth forces are consequently balanced, and this enables a larger helix angle to be used than is possible for single helical gears. Consequently, double helical gears have the gradual engagement necessary for high speed operation, and since there is no limitation due to axial thrust, they may be used for heavier loads than can be accommodated by single helical gears.

Spur and helical gears are manufactured from *gear blanks* (the name given to the disc or cylinder before the teeth have been machined), by means of milling, shaping or hobbing, Fig. 6.5. For reasons of economy and convenience of manufacture, gear cutters have been standardised, and unless very special requirements justify the use of special cutters, the designer should ensure that

Fig. 6.4. Helical gears used in ball mill drive for cement plant (Reproduced by kind permission of R. W. Transmissions Limited)

Fig. 6.5. Hobbing a double helical pinion on a vertical hobbing machnine (Repro duced by kind permission of David Brown Gear Industries Ltd.)

the specification for the gears enables them to be produced using standard available cutters.

The engineer must carry the design to the stage where he can specify sufficient details for the gears to be manufactured, which will involve making decisions and calculations necessary to enable the materials to be selected and the geometry to be defined.

It is necessary at this stage to define some of the terms used in connection with gears, and it should be noted that spur gears may be considered a special case of helical gears where the helix angle is zero. Fig. 6.6 indicates the principal features associated with involute teeth.

The symbols used here relating to geometrical data is in accordance with the International Organisation for Standardization[10].

Pitch diameter (d_1' or d_2') is the diameter of the pitch cylinder (or circle).

Base circle (d_{b1} or d_{b2}) is the circle from which the involute curve is generated.

Circular pitch (p) is the length of arc that corresponds to one tooth interval, i.e. the distance measured along the pitch circle between corresponding profiles of adjacent teeth. Thus,

$$p = \frac{\pi d_1'}{z_1} = \frac{\pi d_2'}{z_2} \tag{6.1}$$

where d_1' = pitch diameter of smaller gear (termed a pinion),
d_2' = pitch diameter of larger gear (termed a wheel or gear),
z_1 = number of teeth on pinion,
z_2 = number of teeth on gear.

Normal pitch (p_n) of a helical gear is the distance between similar faces of successive teeth measured along a helix lying on the pitch cylinder normal to the teeth, i.e. the circular pitch referred to the normal plane. The relationship between the circular pitch (p) and the normal pitch (p_n) is evident from Fig. 6.6, i.e.

$$p_n = p \cos \beta \tag{6.2}$$

where β = helix angle of the teeth.

Base pitch (p_b) is the distance measured along the base circle between corresponding profiles of adjacent teeth.

Axial pitch (p_x) of a helical gear is the distance measured parallel to the axis between similar faces of successive teeth, and from

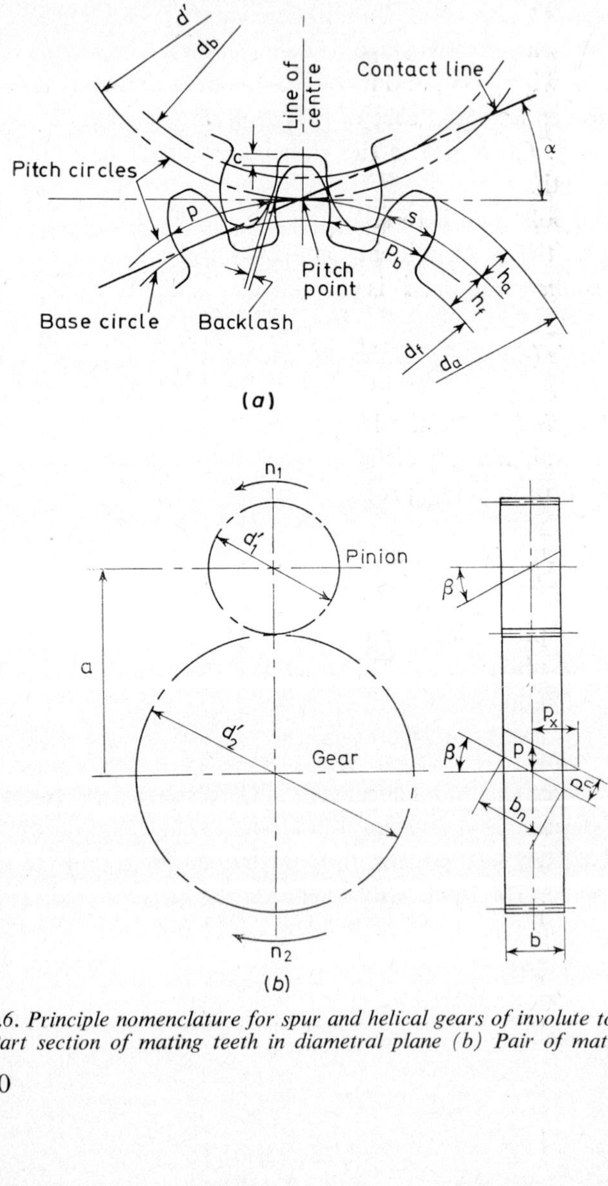

Fig. 6.6. Principle nomenclature for spur and helical gears of involute tooth form (a) Part section of mating teeth in diametral plane (b) Pair of mating gears

Fig. 6.6 and equation 6.2

$$p_x = p_n/\sin\beta = p\cot\beta \qquad (6.3)$$

Diametral pitch (P) is the number of teeth divided by the pitch diameter, i.e.

$$P = \frac{z_1}{d_1'} = \frac{z_2}{d_2'} \qquad (6.4)$$

From equations 6.1 and 6.4 the important relationship between circular pitch (p) and diametral pitch (P) is obtained, namely,

$$Pp = \pi \qquad (6.5)$$

Module (m) is the reciprocal of the diametral pitch.

Normal diametral pitch (P_n) is used in helical gears to indicate the diameteral pitch referred to the normal plane.

The relationship between the normal pitch (p_n) and the normal diametral pitch (P_n) is similar to equation 6.5, i.e.

$$P_n p_n = \pi \qquad (6.6)$$

so that the relationship between the normal diametral pitch (P_n) and the diametral pitch (P) is given by

$$P_n = \frac{P}{\cos\beta} \qquad (6.7)$$

Combining equations 6.4 and 6.7 gives the result

$$P_n = \frac{z_1}{d_1'\cos\beta} = \frac{z_2}{d_2'\cos\beta} \qquad (6.8)$$

which indicates that the helix angle β is dependent upon the pitch diameter.

Contact line or *line of action* is the common tangent to the two base circles.

Pitch point is that point at which the contact line cuts the line of centres.

Pressure angle (α) is the acute angle formed between the contact line and the common tangent to the pitch circles. For the involute tooth form, the pressure angle is constant along the path of contact, but for all other tooth forms it is a variable.

Addendum (h_{a1} or h_{a2}) is the height by which a tooth projects beyond the pitch circle. The diameter of the circle coinciding with the tooth tips is the *tip diameter* (d_{a1} for the pinion and d_{a2} for the gear).

Dedendum (h_{f1} or h_{f2}) is the depth of a tooth space below the pitch circle.

The diameter of the circle coinciding with the tooth roots is the *root diameter* (d_{f1} for the pinion and d_{f2} for the gear).

Working depth is the depth in the tooth space to which the mating gear extends and is the sum of the addenda of the two gears, i.e. $h_{a1}+h_{a2}$.

Bottom clearance (c) is the shortest distance between the tip of the tooth and the bottom of its mating space.

Tooth thickness (s) is the length of arc of the pitch circle between opposite faces of the same tooth..

Backlash is the free movement between gear teeth when the gears are set at their working centre distance, i.e. the amount by which the width of the tooth space exceeds the thickness of the engaging tooth on the pitch circles.

Facewidth (b) is the width of a gear tooth measured in the direction of the axis of the gear.

Gear ratio (u) is the ratio of the number of teeth in the gear wheel to the number of teeth in the pinion, which is equal to the ratio of the angular velocity of the pinion to the angular velocity of the wheel, i.e.

$$u = \frac{z_2}{z_1} = \frac{n_1}{n_2} = \frac{d'_2}{d'_1} \tag{6.9}$$

Centre distance (a) of a pair of gears is the perpendicular distance between the axes of the gears. Thus, for gears working at standard centre distance,

$$a = \tfrac{1}{2}(d'_1+d'_2) \tag{6.10}$$

Other terms will be introduced where necessary.

6.2. STANDARD TOOTH FORMS FOR SPUR AND HELICAL GEARS

The actual shape of the involute is determined by the diameter of the base circle, and the working pressure angle is dependent upon the working centre distance, since the contact line remains tangential to the base circles. Consequently, the effect of increasing the centre distance is to increase the pressure angle, but for the involute tooth form a certain amount of tolerance can be accepted in the centre distance without ill effects on the mode of engagement of the teeth; this is one of the outstanding practical advantages of the involute tooth form.

It is of obvious advantage to standardise on tooth form, and this requires not only a decision regarding the form of the curve (namely the involute), but also a decision in respect to pressure angle. Since low pressure angles provide for quiet running but are weaker than tooth forms utilising higher pressure angles, a compromise must be made. Tooth forms having a pressure angle of 14.5 degrees have been in wide use in the past, but modern practice in the United Kingdom is to promote the use of the full depth involute with a pressure angle of 20 degrees. In the United States, the two tooth forms which have been standardised are full depth teeth utilising pressure angles of either 20 or 25 degrees; previous standard tooth forms using 14·5 or 20 degrees, and the 20 degree stub are now obsolete.

6.3. GEAR FORCES AND STRENGTH AND WEAR REQUIREMENTS FOR SPUR GEARS

Before proceeding to a consideration of the strength and wear requirements for spur gears, let us first of all examine the gear forces which are created when power is transmitted. Fig. 6.7 indicates the forces acting on a spur gear when transmitting power. The resultant force F_n which acts normal to the tooth profile at the point of contact, can be resolved into two mutually perpendicular components, namely a tangential tooth force F_t, which is responsible for transmitting the torque M_t, and a radial force F_t, which is often referred to as the separating force, since it is always directed towards the centre of the gears. The tangential tooth force as a

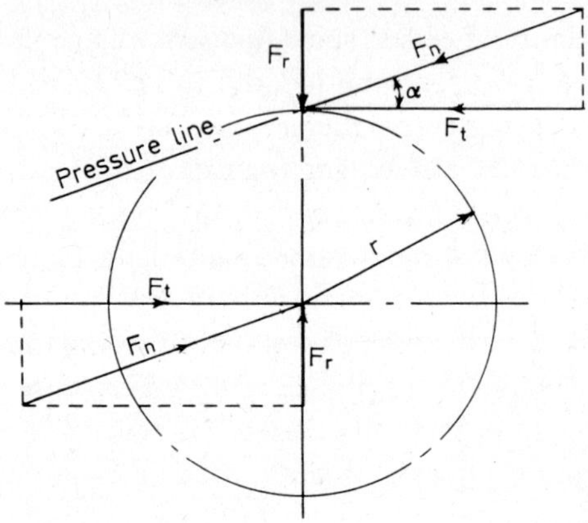

Fig. 6.7. Forces acting on spur gear when transmitting power

result of transmitting the torque is readily obtained from the definition of torque, i.e.

$$F_t = \frac{M_t}{r} \tag{6.11}$$

where r = radius from shaft axis to the pitch circle diameter, = $d_1'/2$ for the pinion and $d_2'/2$ for the gear.

The torque transmitted, M_t, can be obtained from the usual power transmission equations (equations 3.1 and 3.2), remembering to use the correct speed of rotation depending upon whether the pinion or the gear is being considered.

From simple geometry (Fig. 6.7), the radial or separating force may be obtained from the equation

$$F_r = F_t \tan \alpha \tag{6.12}$$

where α = pressure angle.

As far as loading on the shaft is concerened, for shaft strength and bearing reaction considerations, the resulting normal force is of importance, and this is simply the vector sum of F_r and F_t, i.e.

$$F_n = \sqrt{(F_r^2+F_t^2)} \tag{6.13}$$

This is very much a simplified approach, but nevertheless usually quite adequate from the point of view of determining shaft stresses and bearing reactions. In reality, however, the forces acting between pairs of successively engaging teeth are of a dynamic nature, and this factor is important when considering the strength of gear teeth.

A simplified approach to assessing the *strength of gear teeth* was proposed in 1892 by Lewis[11], and although some of the assumptions which he made are far from true, nevertheless the so-called *Lewis equation* has been extensively used in practice for many years (particularly in the United States), with a good deal of success. Lewis assumed that the worst condition for loading of the teeth occurs when engaging teeth are just coming into contact, and that the entire load is carried by a single pair of teeth, Fig. 6.8. The

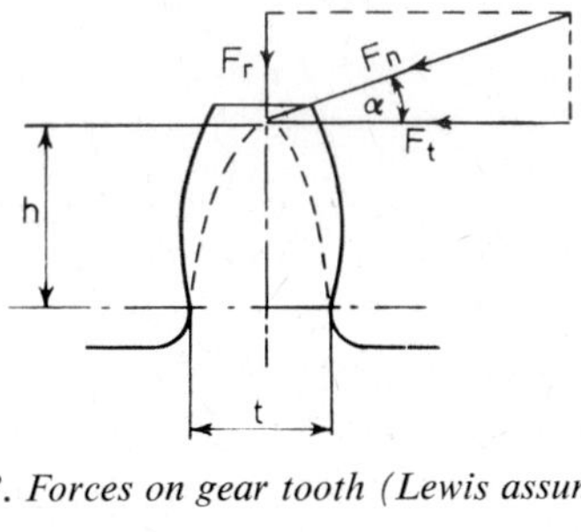

Fig. 6.8. Forces on gear tooth (Lewis assumption)

resultant or normal tooth force F_n can be resolved into two components in the usual way, the radial component F_r tending to compress the tooth and the tangential component causing bending of the tooth. Lewis assumed that the radial component could be neglected, and that the tooth could be treated as a cantilever of parabolic (constant bending strength) profile, inscribed within the actual tooth and tangential to it at the fillets. The stresses in the tooth may then be calculated using the bending theory, i.e.

$$\sigma = \frac{M}{Z}$$

From the geometry indicated in Fig. 6.8, and letting the face width of the tooth be b, we have

$$M = F_t h$$

$$Z = \frac{bt^2}{6}$$

so that

$$\sigma = \frac{6F_t h}{bt^2} \tag{6.14}$$

To obtain the equation in its more usual form, the circular pitch p is introduced, and

$$\sigma = \frac{6F_t hp}{bt^2 p}$$

$$\sigma = \frac{F_t}{bpy} \tag{6.15}$$

where $y = t^2/6\,hp$.

The factor y is known as the Lewis form factor, and is a dimensionless quantity which depends only on the tooth system used and the number of teeth in the gear. Values for the form factor are well documented in the American literature for standard tooth systems, and a useful graphical method for determining y has been presented by Black and Adams[12].

It is important to mention that, because of the assumptions made in the derivation of the Lewis equation, caution must be exercised in estimating a value for the allowable bending stress. This will obviously be dependent upon the gear material, and such factors as the accuracy of manufacture, fillet radius, dynamic loading and fatigue considerations, are of great importance. It is therefore preferable to use values based on actual operating experience, and apply these values to the design of similar gears. Letting the allowable design stress be denoted S_d, and replacing the circular pitch

p by the diametral pitch P ($p = \pi/P$), equation 6.15 may be re-written in the following form:

$$F_b = \frac{F_t}{b} = \frac{S_d y \pi}{P} \tag{6.16}$$

where F_b = allowable tangential tooth load per unit face width based on bending strength.

If the gears are to be used very infrequently, the design stress may be taken as the yield strength divided by an appropriate reserve factor, but this is generally not the case, and this approach is unrealistic. A more logical approach is to base the design stress on the fatigue strength, and to include for the effect of fatigue strength reduction due to the fillet radius, surface finish, etc., as discussed in the first volume in this series[2]. Thus, corresponding to any particular life requirements, the design strength would be more realistically obtained from the equation

$$S_d = \frac{S'_a}{RF} \tag{6.17}$$

where S'_a = fatigue strength corresponding to life required, allowing for the factors present in the gear (see equation 3.23), obtained from a corrected fatigue curve as discussed in Chapter 4 of the first book in this series[2].

Suggested values for the basic design strengths, S_d, are listed in the literature[12], and to allow for the dynamic effects due to the velocity of the tooth action, it is usual to multiply the design strength by a *velocity factor*, C_v, where C_v is less than unity, and is dependent on the pitch line velocity. When the velocity is measured in ft/min, i.e. working in the English system, the value for C_v may be determined from one of the following equations:

$$C_v = \frac{600}{600 + V} \text{ (for } V \text{ less than 2 000 ft/min)} \tag{6.18}$$

$$C_v = \frac{1\,200}{1\,200 + V} \text{ (for } V \text{ 2 000–4 000 ft/min)} \tag{6.19}$$

$$C_v = \frac{78}{78 + \sqrt{V}} \text{ (for } V \text{ greater than 4 000 ft/min)} \tag{6.20}$$

In terms of SI metric units, measuring the pitch line velocity in m/s,

$$C_v = \frac{3{\cdot}05}{3.05+V} \text{ (for } V \text{ less than 10 m/s)} \tag{6.21}$$

$$C_v = \frac{6{\cdot}1}{6{\cdot}1+V} \text{ (for } V \text{ 10–20 m/s)} \tag{6.22}$$

$$C_v = \frac{5{\cdot}6}{5{\cdot}6+\sqrt{V}} \text{ (for } V \text{ greater than 20 m/s)} \tag{6.23}$$

Thus, equation 6.16 may be written as

$$F_b = \frac{F_t}{b} = \frac{S_d(\pi y)C_v}{P} \tag{6.24}$$

The velocity factors indicated above are only approximate at the best, since they are assumed to be independent of the material from which the gears are made, and the accuracy to which the gears are cut is not accounted for. This latter factor in particular has a pronounced effect on the dynamic loading on gear teeth.

Probably the best known method for calculating the dynamic load on gear teeth is that due to Buckingham[6,13], which expresses the dynamic load by the equation

$$F_d = F_t + \sqrt{[F_a(2F_2 - F_a)]} \tag{6.25}$$

where F_d = dynamic load,
F_t = tangential (steady) load due to transmitted torque,
F_a = acceleration force,
F_2 = force required to deform the teeth the amount of the effective tooth error.

A very nice discussion on the Buckingham equation and its application has been presented by Shigley[14], to which the reader is referred.

Before leaving the question of strength of gear teeth, it is appropriate to indicate the method which has been used extensively in the United Kingdom, based on the recommendations of the British

Standards Institution[15]. This specification, which has been prepared by a committee consisting of representatives of Government Departments, leading British Gear Manufacturers and Scientific Institutions, embodies the standard practice for the design of machine cut gears.

Based on satisfying strength requirements, the following equation is recommended for determining the allowable tangential load per unit length of face width,

$$F_b = \frac{F_t}{b} = \frac{S_b Y X_b}{P} \qquad (6.26)$$

where S_b = bending strength factor for the material,
Y = strength factor,
X_b = speed factor for strength,
P = diametral pitch.

The similarity between equations 6.24 and 6.26 will at once be evident. Charts for the determination of the strength factor (Y) and the speed factor for strength (X_b) are reproduced in Fig. B.1 and B.2 respectively of Appendix B, by kind permission of the British Standards Institution. Values for the bending strength factor, S_b, for some commonly used gear materials are listed in Table B.1 of Appendix B.

In addition to satisfying the bending strength requirements, gears must also be designed so that they do not fail by wear, i.e. by abrasive wear or pitting. This involves a consideration of the contact pressure or Hertzian stress[2] which occurs at the centre of the contact surfaces, and on the surface fatigue properties of the material. Tooth failure by wear, i.e. pitting, is generally far more significant than tooth breakage in the sense that it occurs far more frequently.

A commonly used method for assessing wear as a criterion for straight spur gears is the application of the Buckingham wear equation[16], which may be written in the form

$$F_c = \frac{F_t}{b} = d'KQ \qquad (6.27)$$

where F_c = allowable tangential load per unit face width based on wear,
d' = pitch diameter,
K = material form and material factor,
Q = ratio factor.

The material and form factor K is obtained from the equation

$$K = \frac{S_c^2 \sin \alpha}{1{\cdot}4}\left(\frac{1}{E_1}+\frac{1}{E_2}\right) \tag{6.28}$$

where S_c = surface fatigue strength (determined experimentally),
α = pressure angle,
E_1 = modulus of elasticity for the pinion,
E_2 = modulus of elasticity for the gear.

Values for K recommended by Buckingham can be found in the literature[12, 14, 16], and it is suggested by Phelan[17] that values for S_c are approximated to by the equation

$$S_c = 180\,(BHN_1+BHN_2) \tag{6.29}$$

where S_c = surface fatigue strength (10^3 lbf/in^2)
and BHN_1 and BHN_2 are the Brinell Hardness numbers of the pinion and gear. This approximation is a very useful one.
The ratio factor is obtained from the equation

$$Q = \frac{2u}{u+1} \tag{6.30}$$

Based on satisfying wear requirements, the British Standards Institution [15] recommends the following equation for determining the allowable tangential load per unit length of face width,

$$F_c = \frac{F_t}{b} = \frac{S_c Z X_c}{P^{0{\cdot}8}} \tag{6.31}$$

where S_c = surface strength factor for the material,
Z = zone factor,
X_c = speed factor for wear,
P = diametral pitch.

Charts for the determination of the zone factor (Z) and the speed factor for wear (X_c) are reproduced in Figs B.3 and B.4 respectively of Appendix B, by kind permission of the British Standard Institution. Values for the surface strength factor, S_c, for some commonly used gear materials are listed in Table B.1 of Appendix B.

6.4. GEAR FORCES AND STRENGTH AND WEAR REQUIREMENTS FOR HELICAL GEARS

In the case of a single helical gear, in addition to the tangential tooth force F_t and the radial or separating force F_r, there will also exist an axial force or thrust F_x, as illustrated in Fig. 6.9. The tangen-

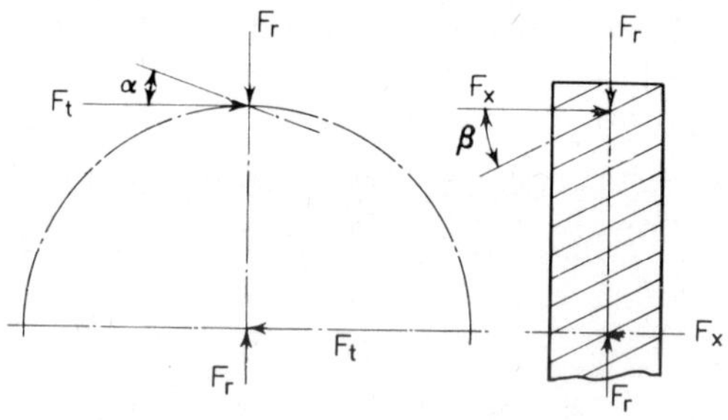

Fig. 6.9. Forces acting on helical gear

tial force F_r is obtained from the basic relationship between torque transmitted, and the product of tangential force and the radius at which it acts, as for spur gears, i.e. equation 6.11 applies, and

$$F_t = \frac{M_t}{r}$$

The radial separating force and the axial force can be obtained from the geometry indicated in Fig. 6.9. Thus,

$$F_r = F_t \tan \alpha = F_t \frac{\tan \alpha_n}{\cos \beta} \qquad (6.32)$$

and

$$F_x = F_t \tan \beta \tag{6.33}$$

where α = pressure angle measured in a plane perpendicular to the axis of the gear,

α_n = pressure angle measured in a plane perpendicular to the tooth,

β = helix angle measured from the axis of the gear.

Of obvious importance in the overall design of a gear shaft and its supporting bearings is the direction and magnitude of the axial thrust, since the supporting bearings must be capable of withstanding this thrust (in addition to the radial forces which act). Consequently, a limiting factor to the load carrying capability of a single helical gear might be due to this axial thrust, and this is usually limited by limiting the helix angle β to within the range 15–30 degrees, a good average being about 20 degrees[18].

The direction of the axial thrust depends upon the hand of the helix, the direction of rotation, and whether the gear under consideration is driving or is being driven.

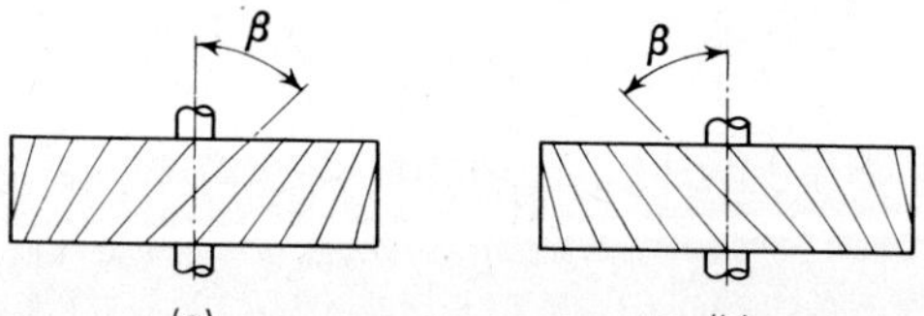

Fig. 6.10. Handing of helical gears (a) Right-hand helix (b) Left-hand helix

The handing of the helix is determined by viewing the teeth looking on the end face in the direction of the axis (i.e. along the shaft); if the teeth recede in a clockwise direction the helix is right-hand, and if the teeth recede in an anti-clockwise direction the helix is left hand. Fig. 6.10 illustrates right hand and left hand helical gears, and it will be obvious that mating gears must be of opposite hand.

Fig. 6.11 illustrates the directions of the axial thrusts F_x (which would act on the bearings) for a pair of mating helical gears. The right hand helix is the driving gear and the left hand helix the driven gear; the directions of rotation are as indicated. For any particular combination of helix angle, direction of rotation and driving and driven gear, the actual direction of the axial thrust can be reasoned

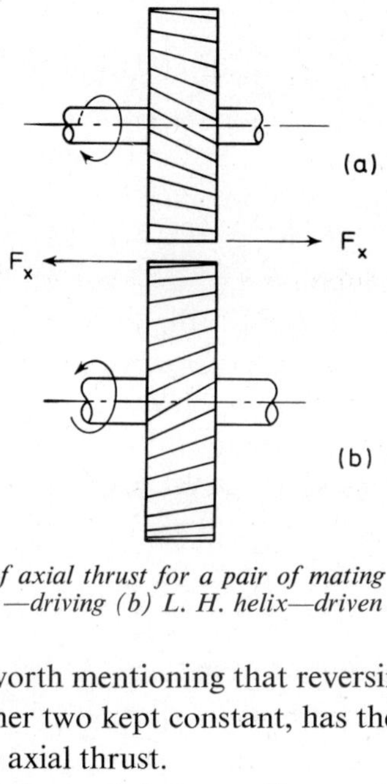

Fig. 6.11. Directions of axial thrust for a pair of mating gears (a) R. H. helix —driving (b) L. H. helix—driven

but it is probably worth mentioning that reversing any one of these factors with the other two kept constant, has the effect of reversing the direction of the axial thrust.

It should also be observed that as well as applying an axial thrust to the bearings of the supporting shaft, the axial force acting at the gear tooth also subjects the shaft to a point moment, acting at the position of the gear. This imposes bending stresses on the shaft and will also influence the shaft deflection. By mounting two single helical gears of opposite hand on the same shaft side by side, the axial thrust may be eliminated; the same result is obtained by the use of double-helical (so-called herringbone) gears.

In assessing the *strength of helical gear teeth*, one approach is to use a modified form of the Lewis equation used for spur gears.

The derivation of the modified equation is similar to that demonstrated for spur gears, but the cross-section of the tooth considered must be a normal section. Thus, making the assumption that one pair of teeth carry the entire load, the modified Lewis equation for helical gears is given by:

$$F_b = \frac{F_t}{b} = \frac{S_d(\pi y_n) C_v}{P_n} \tag{6.34}$$

where y_n = Lewis form factor for standard tooth system, and the numbers of teeth in the *formative gear set* (as discussed below),
P_n = normal diametral pitch,

and other symbols are as previously defined. The values for the Lewis form factor are obtained from standard tables[3, 12, 14], or by the graphical solution referred to previously[12], the only difference between y and y_n being that in the latter case for helical gears, the virtual number of teeth are used.

By considering the radii of curvature of the pitch cylinder surfaces at the pitch point in the normal plane, it may be shown[4] that

$$d'_{1n} = \frac{d'_1}{\cos^2 \beta} \tag{6.35}$$

and

$$d'_{2n} = \frac{d'_2}{\cos^2 \beta} \tag{6.36}$$

where d'_{1n} = diameter of pinion referred to the normal plane,
d'_{2n} = diameter of the gear referred to the normal plane.

It must be emphasised that d'_{1n} and d'_{2n} are not diameters of true circles, but merely twice the radii of curvature of the pitch points of the ellipses formed by the intersections of the normal plane and the pitch cylinders[4]; as such, they represent the diameters of the hypothetical gearset in the normal plane, sometimes referred to as the formative gearset.

Substituting for d'_1 and d'_2 from equations 6.35 and 6.36 into equation 6.8, we obtain the relationships,

$$P_n = \frac{z_1}{d'_{1n} \cos^3 \beta} = \frac{z_2}{d'_{2n} \cos^3 \beta}$$

from which

$$z_{1n} = P_n d'_{2n} = \frac{z_1}{\cos^3 \beta} \tag{6.37}$$

and

$$z_{2n} = P_n d'_{2n} = \frac{z_2}{\cos^3 \beta} \tag{6.38}$$

Equations 6.37 and 6.38 give the relationships between the number of teeth for the formative gearset in the normal plane and for the actual gearset in the diametral plane, where

z_{1n} = number of teeth in the pinion referred to the formative gearset, and

z_{2n} = number of teeth in the gear referred to the formative gearset.

Thus, for purposes of determining y_n for use in equation 6.34, the hypothetical or virtual number of teeth in the formative gearset must be used.

The assumption that one pair of teeth take the entire load is not true for accurately machined helical gears, the load actually being shared by more than a single pair. To account for this in the modified Lewis equation, it is suggested by Dudley[7] that the profile contact ratio, designated ε_γ, be introduced into equation 6.34 to give

$$F_b = \frac{F_t}{b} = \frac{S_d(\pi y_n) C_v \varepsilon_\gamma}{P_n} \tag{6.39}$$

The profile contact ratio will be discussed in a little more detail presently, but for the time being it is sufficient to state that it is an index of the average number of teeth in contact in any fixed diametral plane. It is a function of a number of factors, and can be obtained either by a simple layout of the gears or by calculation. As a first approximation, Shigley[14] suggests an average value of about 1.4, but it must be mentioned that a value lower than this is frequently obtained. Johnson[3] has tabulated values for ε_γ as a function of helix angle, number of pinion teeth, and gear ratio.

The British Standard method for satisfying strength requirements is to use equation 6.26, but with the following modification. The strength factor Y, obtained from Fig. B.1, applies specifically for straight spur gears and helical gears where the helix angle β is equal to 30 degrees. For helix angles other than 30 degrees, the strength factor obtained from Fig. B.1 must be multiplied by $1{\cdot}33 \cos^2 \beta$, i.e.

$$F_b = \frac{F_t}{b} = \frac{S_b Y_1 X_b}{P} \tag{6.40}$$

where

$$Y_1 = (1{\cdot}33 \cos^2 \beta)\, Y \tag{6.41}$$

and the other symbols are as previously defined (see equation 6.26).

In assessing the *wear resistance of helical gears*, one approach is to use the Buckingham equation, which for a helical gear is

$$F_c = \frac{F_t}{b} = \frac{d' K Q \varepsilon_\gamma}{\cos^2 \beta} \tag{6.42}$$

where the symbols are as previously defined (see equation 6.27), except that in equation 6.28, the pressure angle should be taken as the pressure angle of the tooth form in the normal plane.

The British Standard method for assessing wear requirements is to use equation 6.31, but the zone factor Z_1 (say), is obtained from Fig. B.5, which is reproduced in Appendix B by kind permission of the British Standards Institution. The zone factor obtained from Fig. B.5 is correct for helix angles of 30 degrees, but for other helix angles it must be multiplied by $0{\cdot}75 \sec^2 \beta$, i.e.

$$Z = (0{\cdot}75 \sec^2 \beta)\, Z_1 \tag{6.43}$$

6.5. INVOLUTE INTERFERENCE AND UNDERCUTTING OF SPUR AND HELICAL GEARS

When generating a gear, if the cutting tool extends beyond the point of tangency of the contact line with the base circle, involute interference occurs which results in *undercutting* of the teeth and a loss in the proper involute tooth form; this has the effect of weakening the teeth. Further, if an attempt is made to mesh a gearset whose teeth have the condition of involute interference, binding of the teeth will result for certain angular positions of the gearset.

Involute interference should not occur in properly designed gears since it can be readily avoided. It is most likely to occur with gears of standard proportions having a small number of teeth. For *spur gears* having full depth involute form, the number of teeth at which undercutting or involute interference occurs, may be shown[19] to be given by the equation

$$z_{1\min} = \frac{2}{\sin^2 \alpha} \tag{6.44}$$

In the case of *helical gears*, equation 6.44 gives the minimum number of teeth in the formative gearset to avoid undercutting. Thus, replacing $z_{1\ \min}$ by $z_{1n\ \min}$ in equation 6.44 and then substituting for z_{1n} from equation 6.37, the *actual* minimum number of teeth to avoid undercutting or involute interference in a helical gear is given by the equation

$$z_1 = \frac{2}{\sin^2 \alpha} \cos^3 \beta \tag{6.45}$$

Corresponding to a pressure angle of 20 degrees, equations 6.44 and 6.45 reduce to

$$z_1 = 17{\cdot}1 \quad \text{(for spur gears)} \tag{6.44a}$$

and

$$z_1 = 17{\cdot}1 \cos^3 \beta \quad \text{(for helical gears)} \tag{6.45a}$$

It should be emphasised that these results indicate the theoretical minimum number of teeth to avoid undercutting without modification, but in practice a satisfactory tooth shape for straight spur gears is obtained when the minimum number of teeth is about 30. Note that the above results indicate that helical gears can have smaller number of teeth than straight spur gears, without danger of undercutting.

If the number of teeth on the pinion is less than that calculated above, or if it is desirable to improve the surface contact, undercutting may be avoided by increasing the pressure angle of the cutter; reducing the tooth depth (stub teeth); or by so-called *modification*. Since increasing the pressure angle or reducing the tooth depth requires special cutters, these methods are undesirable, and a more satisfactory approach is effected by modification. Modification simply involves cutting the teeth to the standard depth on an oversize or undersize blank.

The two ways of applying modification are[18]:

1. Increasing the diameter of the pinion and leaving the mating wheel standard. This results in the centre distance and running pressure angle being increased
2. Increasing the diameter of the pinion and reducing the diameter of the wheel by an equal amount. The gears then work at standard centres and pressure angle but have unequal addenda

The British Standard Specification[15] states that in order to obtain full involute action and avoid undercutting when a pinion of small number of teeth is in contact with a 20 degree basic rack, and to obtain better zone and strength factors, the addendum varies with the pitch and number of teeth in the mating gears. The recommended values for spur gears[15] are given by the equations

$$h_{a1} = \frac{p}{\pi}(1+k_1) \tag{6.46}$$

$$h_{a2} = \frac{p}{\pi}(1+k_2) \tag{6.47}$$

where h_{a1} = addendum of the pinion,
h_{a2} = addendum of the gear,
k_1 = correction factor for pinion,
k_2 = correction factor for gear.

When the tooth sum (z_1+z_2) exceeds 60, then

$$k_1 = 0.4\left(1-\frac{z_1}{z_2}\right) \tag{6.48}$$

or

$$k_1 = 0.02(30-z_1) \tag{6.49}$$

whichever is the greatest, and

$$k_1 = -k_2 \tag{6.50}$$

The teeth will then work at standard centre distance.
When the tooth sum (z_1+z_2) is less than 60, then

$$k_1 = 0.02(30-z_1) \tag{6.51}$$

and

$$k_2 = 0.02(30-z_2) \tag{6.52}$$

and the centre distance will be extended.

It is desirable at this stage to introduce a term which figures prominently in gear calculation, namely so-called *involute functions.* An involute function is defined by the equation

$$\text{inv}\ \theta = \tan\theta - \theta \tag{6.53}$$

where $(\tan\theta - \theta)$ is the involute of the angle θ. Standard tables listing values of the involute function corresponding to various angles can be found in various texts[19] and handbooks[20].

Using involute functions, the running pressure angle, α_r at which modified gears having extended centres will operate can be shown[19] to be given by the equation

$$\text{inv}\ \alpha_r = \frac{2(k_1+k_2)\tan\alpha}{z_1+z_2} + \text{inv}\ \alpha \tag{6.54}$$

which may also be used for helical gears, provided that the formative gearset is used.

The relationship between the standard centre distance a and the modified or extended centre distance a_r, at which modified gears run, can be shown[19] to be given by the equation

$$a_r = a\frac{\cos\alpha}{\cos\alpha_r} \tag{6.55}$$

where

$$a = \frac{z_1+z_2}{2P} \tag{6.56}$$

In the case of helical gears, the modification factors can be obtained by considering the normal plane. This means that p is replaced by p_n in equations 6.46 and 6.47, and that the virtual number of teeth (z_{1n} and z_{2n}) in the formative gearset be used in place of z_1 and z_2 in equations 6.48 to 6.56.

At this stage it is pertinent to mention that when designing gears, it is necessary to bear in mind the necessity for economic design, and in this respect standard cutters should be used whenever possible. Cutters are characterised by the diametral pitch P (or the module m) and the cutting pressure angle α; in the case of cutters for helical gears, these factors may refer to either the plane of motion or the plane normal to the tooth, depending upon whether hobbing or gear shaping is used. If helical gears are manufactured by hobbing, the standard cutters are usually based on the normal plane, i.e. P_n (or m_n) and α_n are the standard characteristics. When gear shaping is the method used for producing helical gears, the cutters or shapers are based on the diametral plane, i.e. P (or m) and α are the standard characteristics.

The British Standard Specification[15] lists in tabular form the equivalents of diametral, circular and metric pitches (module). Tables 6.1 and 6.2 indicate typical (standard) values for a range of standard available cutters. This does not, of course, infer that all manufacturers of spur and helical gears will have available all the standard cutters listed. In Table 6.2, the metric module $m = 1/P = = d'/z$ is used to define the cutter, and this is a good time to mention that all the equations involving P or P_n throughout this chapter, may be written in terms of the module m or normal module m_n, as the

Table 6.1. DIAMETRAL PITCH AND EQUIVALENT CIRCULAR PITCH

P	0·5	0·75	1·0	1·25	1·5	1·75	2·0	2·25
p (in)	6·283 2	4·188 8	3·141 6	2·513 3	2·094 4	1·795 2	1·570 8	1·396 3
p (mm)	159·59	106·39	79·79	63·83	53·20	45·60	39·90	35·46
P	2·5	2·75	3·0	3·5	4·0	5·0	6·0	7·0
p (in)	1·256 0	1·142 4	1·047 2	0·897 6	0·785 4	0·628 3	0·523 6	0·448 8
p (mm)	31·90	28·99	26·59	22·79	19·95	15·95	13·29	11·39
P	8·0	9·0	10	11	12	13	14	15
p (in)	0·392 7	0·349 1	0·314 2	0·285 6	0·261 8	0·241 7	0·224 4	0·209 4
p (mm)	9·974	8·86	7·97	7·25	6·64	6·13	5·69	5·32
P	16	17	18	19	20	22	24	25
p (in)	0·196 3	0·184 8	0·174 5	0·165 3	0·157 1	0·142 8	0·130 9	0·125 7
p (mm)	4·98	4·69	4·43	4·19	3·98	3·62	3·32	3·19
P	26	28	30	32	34	36	38	40
p (in)	0·120 8	0·112 2	0·104 7	0·098 2	0·092 4	0·087 3	0·082 7	0·078 5
p (mm)	3·07	2·85	2·65	2·49	2·34	2·21	2·09	1·99
P	42	44	46	48	50	56	60	
p (in)	0·074 8	0·071 4	0·068 3	0·065 4	0·062 8	0·056 1	0·052 4	
p (mm)	1·90	1·81	1·73	1·66	1·59	1·42	1·33	

Table 6.2. METRIC MODULE AND EQUIVALENT CIRCULAR PITCH

m (mm)	50	45	40	35	30	25	20	18
p (mm)	157·08	141·37	125·66	109·95	94·25	78·54	62·83	56·54
p (in)	6·184 4	5·565 9	4·947 5	4·239 0	3·710 6	3·092 2	2·473 7	2·226 4
m (mm)	16	15	14	13	12	11	10	9
p (mm)	50·26	47·12	43·98	40·84	37·70	34·56	31·41	28·27
p (in)	1·979 0	1·855 3	1·731 6	1·607 9	1·484 2	1·360 6	1·236 9	1·113 2
m (mm)	8	7	6·5	6	5·5	5	4·5	4
p (mm)	25·13	21·99	20·42	18·85	17·27	15·70	14·13	12·56
p (in)	0·989 5	0·865 8	0·8040	0·742 1	0·680 3	0·618 4	0·556 6	0·494 7
m (mm)	3·5	3	2·5	2	1·5	1	0·5	
p (mm)	10·99	9·42	7·85	6·28	4·71	3·14	1·57	
p (in)	0·432 9	0·371 1	0·309 2	0·247 4	0·185 5	0·123 7	0·061 8	

case may be. For purposes of comparison, the equivalent metric dimensions for circular pitch are listed in Table 6.1, and similarly, the equivalent inch dimensions for circular pitch are listed in Table 6.2.

6.6. FURTHER DESIGN CONSIDERATIONS FOR SPUR AND HELICAL GEARS

Apart from strength and wear requirements and the prevention of undercutting or involute interference, there are many more factors which the engineer concerned with gear design must consider. One of the most important of these factors relates to the face width, which as well as being sufficient to satisfy strength and wear requirements, must also satisfy rigidity and manufacturing considerations.

For *spur gears*, it is generally accepted that there is a practical limitation on the face width which can be allowed in order to be assured of proper tooth contact. This limitation is taken to be a function of the circular pitch, and Johnson[3] suggests the use of a face width factor, such that

$$b = C_f p \qquad (6.57)$$

where b = face width,
C_f = face width factor,
p = circular pitch.

The upper limit on C_f depends upon the degree of precision to which the gears are made, and Johnson suggests that a feasible value for C_f is 4. One manufacturer[18] suggests that the value for C_f should not generally exceed 5 to 6, except for profile ground and shaved gears when it should not exceed about 8.

In the case of *helical gears*, it is convenient to refer to the formative gear set, and equation 6.57 may be written as

$$b_n = C_{fn} p_n \qquad (6.58)$$

where b_n = face width in the normal plane (tooth length),
C_{fn} = face width factor in the normal plane,
p_n = normal circular pitch.

Johnson[3] suggests that the values associated with C_{fn} should be similar to those suggested in connection with spur gears.

The face width for helical gears must be sufficient to give overlap, which means that it must be greater than the axial circular pitch. This is a function of the helix angle, and may be calculated using equation 6.3. Dean[21] recommends that the face width should be at least twice the axial pitch to obtain reasonable benefit from the helical action.

Based on rigidity requirements[18] the face width should not be less than about one-tenth of the pitch diameter of the wheel (the larger gear), but should not exceed about twice the pitch diameter of the pinion.

These limitations on face width may be expressed by the following limit equations:

For Spur Gears:

$$d_2'/10 \leq b \leq C_f p \tag{6.59}$$

$$b \leq 2d_1' \tag{6.60}$$

For Helical Gears:

$$b_n \leq C_{fn} p_n \tag{6.61}$$

$$2p_x \leq b \leq 2d_1' \tag{6.62}$$

$$d_2'/10 \leq b \tag{6.63}$$

The maximum bore of the pinion is usually based on the radial distance from the keyway to the root circle not being less than half the circular pitch. Assuming standard keyway proportions (Appendix A), this maximum dimension may be approximated by the equations

$$\text{Max. pinion bore} = \frac{d_1' - 1{\cdot}8p}{1{\cdot}2} \quad \text{(for spur gears)} \tag{6.64}$$

or

$$\text{Max. pinion bore} = \frac{d_1' - 1{\cdot}8p_n}{1{\cdot}2} \quad \text{(for helical gears)} \tag{6.65}$$

If the shaft is larger than this, it will be necessary to make the pinion integral with the shaft, or to increase the pitch diameter of the pinion.

Another important consideration for the smooth and quiet running and transmission of power, is that the average number of teeth in contact for mating spur gears should be greater than unity. This may be calculated by the equation for the so-called *profile contact ratio* (often referred to as the *contact ratio*), which is given[18,19] by the equation

$$\varepsilon_\gamma = \frac{0{\cdot}5[(d_{a1}^2 - d_{b1}^2)^{1/2} + (d_{a2}^2 - d_{b2}^2)^{1/2}] - a \sin \alpha}{p \cos \alpha} \tag{6.66}$$

where ε_γ = profile contact ratio,
d_{a1} = tip circle diameter of the pinion,
d_{a2} = tip circle diameter of the gear,
d_{b1} = base circle diameter of the pinion,
d_{b2} = base circle diameter of the gear,
a = centre distance,
p = circular pitch,
α = pressure angle.

The tip diameters may be calculated from the equations

$$d_{a1} = d_1' + 2h_{a1} \tag{6.67}$$

$$d_{a2} = d_2' + 2h_{a2} \tag{6.68}$$

where h_{a1} and h_{a2} may be determined from equations 6.46 and 6.47. The base circle diameters may be calculated from the equations

$$d_{b1} = \frac{z_1 p \cos \alpha}{\pi} \tag{6.69}$$

$$d_{b2} = \frac{z_2 p \cos \alpha}{\pi} \tag{6.70}$$

For smooth and quiet running the contact ratio should preferably be not less than about 1.4.

Another consideration for helical gears, which has already been touched upon, is the *face contact ratio*, designated ε_β, and this is given by the equation

$$\varepsilon_\beta = \frac{b \tan \beta}{p} = \frac{b}{p_x} \tag{6.71}$$

where ε_β = face contact ratio (or overlap ratio),
b = face width,
β = helix angle,
p = circular pitch,
p_x = axial circular pitch = $p \cot \beta$ (by equation 6.3).

As mentioned previously in connection with face width, it is suggested that the face contact ratio be greater than 2 in order to obtain reasonable benefit from the helical action.

It is better not to attempt a too big reduction in a single step, limiting reductions being about 8 to 1 for spur gears and 10 to 1 for helical gears. Where reduction ratios in excess of these are required, multiple steps should be used, which will result in better proportioning of the gears.

6.7. CROSSED-HELICAL GEARS

Straight spur and helical gears are used for connecting parallel shafts which rotate in opposite directions. On occasions, however, the necessity may arise to connect skewed shafts, i.e. shafts whose axes lie at an angle to one another, but do not meet, as indicated in Figs 6.12 and 6.13. This requirement may be satisfied by the use of *crossed-helical gears* (formerly known as *spiral gears*), which have appropriately selected helix angles to enable correct meshing when connecting skewed shafts.

Crossed-helical gears are only suitable for transmitting light power at moderate speeds, due to the fact that kinematically point contact and high sliding velocity are produced, which results in poor wear resisting properties. Such gears are best proportioned based on previous experience with similar applications, since the

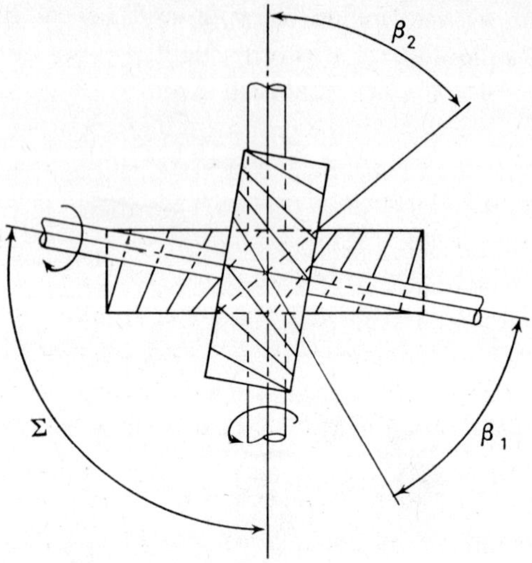

Fig. 6.12. Crossed helical gears indicating shaft angle

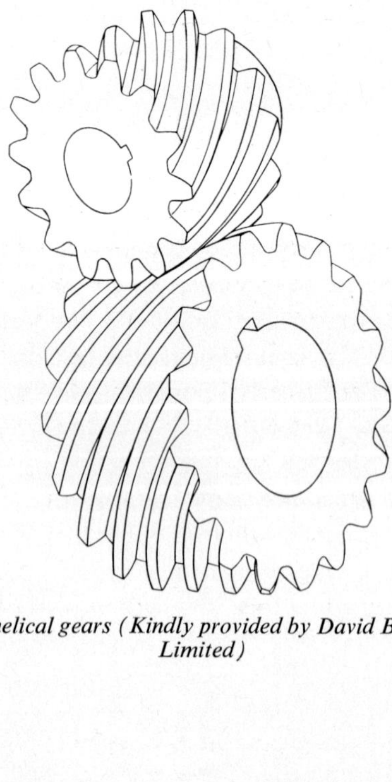

Fig. 6.13. Crossed helical gears (Kindly provided by David Brown Gear Industries Limited)

strength and wear formulae for spur and helical gears are not applicable. In this respect, it is of particular importance to select a combination of materials that will avoid scuffing of the teeth. Because of the elasticity of the gear teeth materials, the hypothetical point contact will, in fact, be increased to a small area, but even so the consequential Hertzian or contact stresses will nevertheless be a limiting condition. Such gears should be run-in gradually, when the performance from them will be somewhat increased.

The design of crossed-helical gears usually involves satisfying a specified shaft angle, centre distance and angular velocity ratio. This needs to be achieved economically using standard cutters. For the teeth of crossed-helical gears to mesh correctly, it is necessary that the two gears have the same pressure angle and pitch in the normal plane. The helix angles for mating crossed helical gears need not, however, be the same, and may be of either the same or opposite hand depending upon the shaft skew.

The relationship between the shaft angle and the helix angles of the mating gears is given by the equation

$$\Sigma = |\beta_1 \pm \beta_2| \tag{6.72}$$

where Σ = shaft angle,
β_1 = helix angle of the pinion,
β_2 = helix angle of the gear.

The plus sign is used when both helix angles are of the same hand, and the minus sign is used when they are of opposite hand. When the shaft angle is greater than the helix angles of both gears, the gears must be of the same hand in order that the sum of the helix angles add up to the shaft angle. Conversely, when either or both of the helix angles are greater than the shaft angle, the gears must be of opposite hand so that the difference between the two angles equals the shaft angle. The driver should have the larger of the two helix angles, and an attempt should be made to make the helix angles of both gears approximately equal. Further, it is suggested[18] that the helix angle of either gear should not exceed 60 degrees in order to avoid difficulty in manufacture and inefficiency in driven gears.

The pitch circle diameter of the pinion and the gear is related to the normal diametral pitch, the number of teeth and the helix angles by equation 6.8 previously presented. By transposition,

$$d_1' = \frac{z_1}{P_n \cos\beta_1} \tag{6.73}$$

$$d_2' = \frac{z_2}{P_n \cos\beta_2} \tag{6.74}$$

As a good starting point, Tuplin[15] suggests tentatively taking the pinion pitch diameter as

$$d_2' = \frac{2a}{1+u} \tag{6.75}$$

or

$$d_1' = a/3 \tag{6.76}$$

whichever is the greatest. This approximation may be used to indicate a good first approximation for situations where the gear ratio does not exceed about 5.

It may be shown[17] that the relationship between the angular velocity ratio, pitch circle diameters and helix angles is given by the equation

$$u = \frac{z_2}{z_1} = \frac{n_1}{n_2} = \frac{d_2' \cos\beta_2}{d_1' \cos\beta_1} \tag{6.77}$$

Further, the shaft centre distance is determined[18] from the equation

$$a = \frac{z_1(\sec\beta_1 + u\sec\beta_2)}{2P_n} \tag{6.78}$$

There is no direct solution to the problem of satisfying fixed centres, angular velocity ratio and shaft angle, and it is necessary to satisfy equations 6.72, 6.77 and 6.78 by trial and error. Tentative decisions need to be taken regarding the normal pitch, tooth numbers and helix angles, and then the helix angles adjusted until the relationships are satisfied. The minimum number of teeth to avoid undercutting or involute interference is given by equation 6.45a, as used for helical gears generally.

There is no reason why the face width of the pinion and the gear should be the same for crossed-helical gears, but in each case the face width should not be less than about one-twelfth of the pitch diameter nor less than 3 p_n sin β. In terms of limit equations:

$$b \geq d'/12 \tag{6.79}$$

$$b \geq 3p_n \sin \beta \tag{6.80}$$

A final point requiring comment is that relating to the shaft supporting bearings. Because of the angularity of the teeth, axial thrusts are produced by the tooth reactions in much the same way as occurs with helical gears connecting parallel shafts. It is, therefore, important to establish the directions of these thrusts, and to provide supporting bearings accordingly. The direction of the axial thrust depends upon the factors mentioned previously, namely the hand of the helix, the direction of rotation, and whether the gear is the driver or the driven.

6.8. BEVEL GEARS

Bevel gears are used when it is required to transmit power and motion between intersecting shafts, probably the most common application being when the shafts are at 90 degrees to each other, Fig. 6.14. In the special case of right angle bevel gears when equal gears are used, i.e. the angular velocity ratio is unity, such gears are known as mitre gears. However, almost any shaft angle is permissible with properly designed bevel gears, as illustrated in Figs 6.15–6.18, but the internal bevel gear (Fig. 6.18) should be avoided.

Bevel gears may be classified according to the shape of the teeth, the simplest type having teeth which are radial towards the point of intersection of the shaft axes; such gears are used for applictions requiring moderate speeds, and are known as *straight bevel gears*, Fig. 6.19. For higher speeds and load carrying capacity, a more gradual tooth engagement is required, and this is obtained by the use of *spiral bevel gears*, Fig. 6.20. Spiral bevel gears bear the same general relationship to straight bevel gears that helical gears do to straight spur gears. The teeth of spiral bevel gears may have any

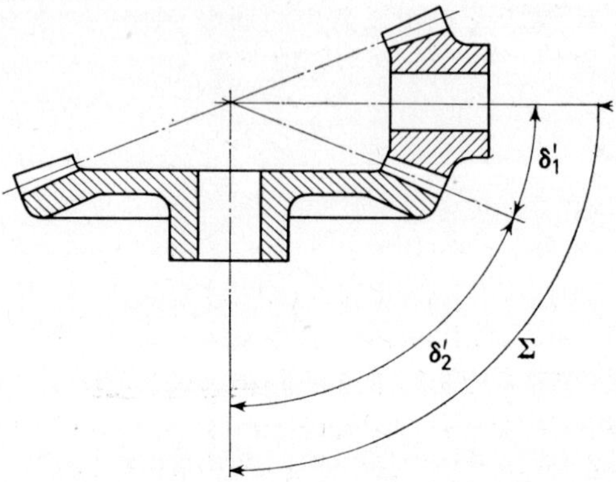

Fig. 6.14. Right angle bevel gears

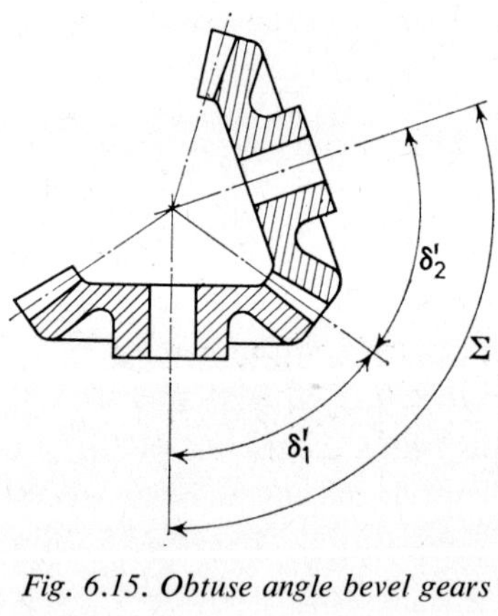

Fig. 6.15. Obtuse angle bevel gears

convenient form which can be produced mechanically with a straight sided cutter, and because of the conical form of bevel gears in general, it is unnecessary to adhere to standard pitches. By the nature of the conical form, the teeth of bevel gears taper, so that the pitch of a gear varies with face width.

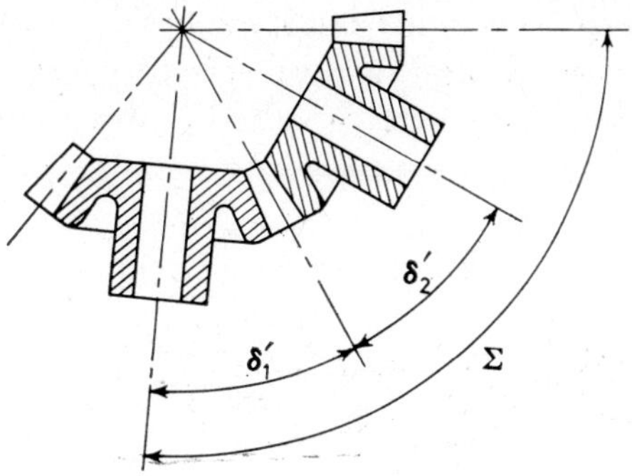

Fig. 6.16. Acute angle bevel gears

Fig. 6.21 indicates some of the more important dimensions and nomenclature associated with bevel gears; the symbols used are in accordance with the I.S.O. recommendations [10]. Many of the terms used with bevel gears are common to spur and helical gears, but because of the conical form, reference must be made to *pitch cones* rather than pitch cylinders. For this reason additional terms relating

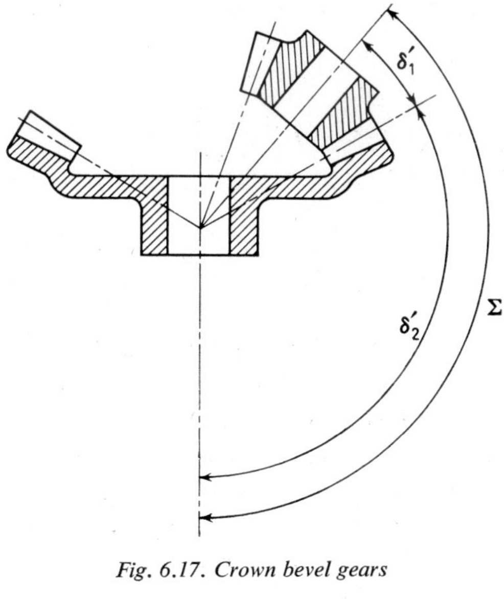

Fig. 6.17. Crown bevel gears

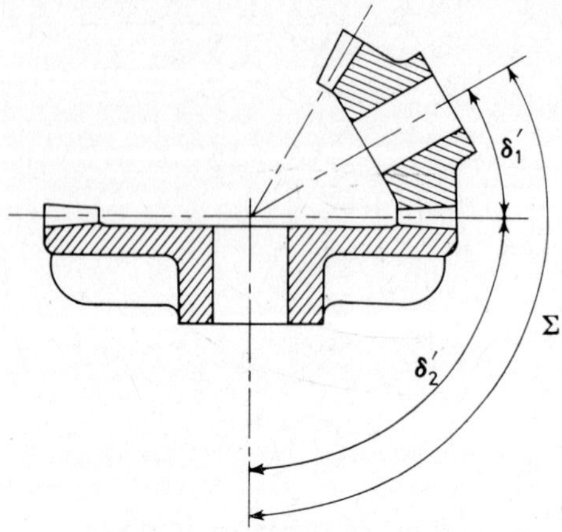

Fig. 6.18. Internal bevel gear

to cone angles are necessary. It will be convenient at this stage to define some of the terms associated with bevel gears.

Pitch cones are those cones which have their axes coincident with the axes of rotation of the gears and roll together in peripheral contact without slipping.

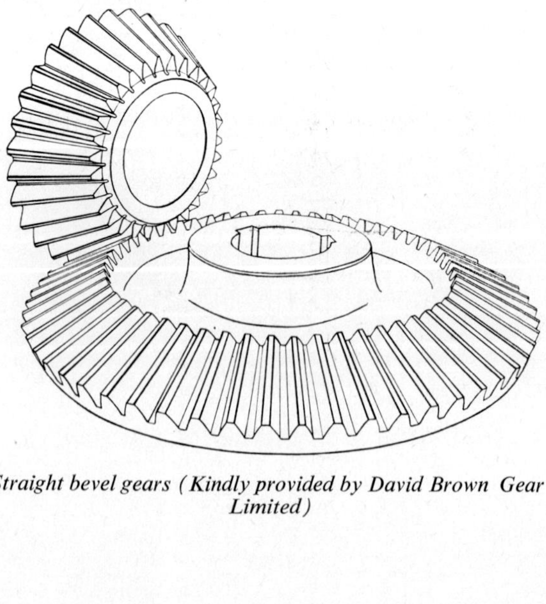

Fig. 6.19. Straight bevel gears (Kindly provided by David Brown Gear Industries Limited)

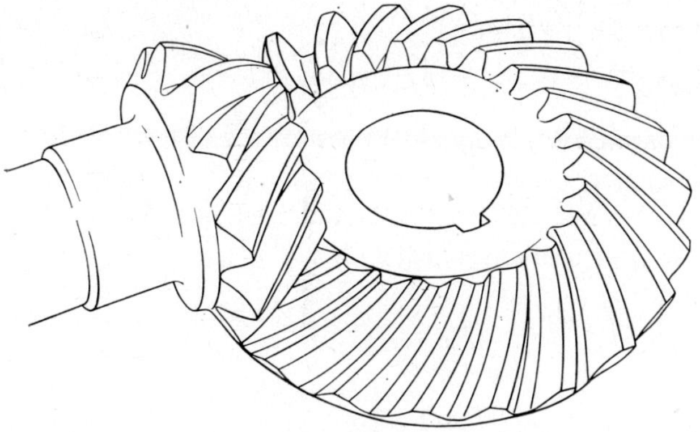

Fig. 6.20. Spiral bevel gears (Kindly provided by David Brown Gear Industries Limited)

Pitch circles refer to the circles of intersection of the pitch cones and the outer face of the teeth.

Pitch diameter (d_1' or d_2') is the diameter of the pitch circle (Fig. 6.21)

Circular pitch (p), when used, refers to the distance measured along the pitch circle between corresponding profiles of adjacent teeth, and equation 6.1 applies.

Normal pitch (p_n), when used, refers to the distance between similar faces of successive teeth measured normal to the teeth at the pitch circle. It is dependent upon the spiral angle at the pitch circle (β')

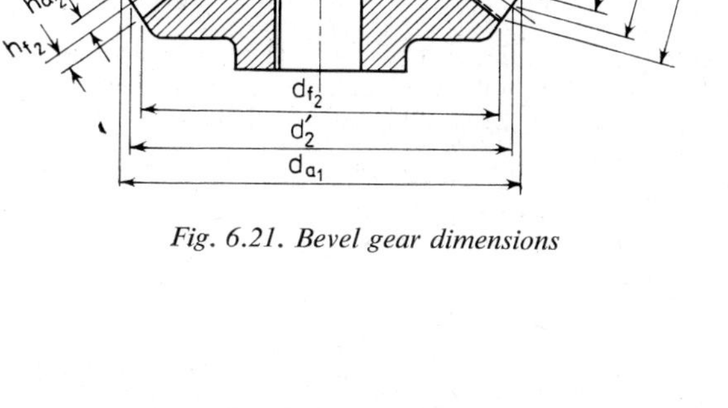

Fig. 6.21. Bevel gear dimensions

and is related to the circular pitch by the equation

$$p_n = p \cos \beta' \tag{6.81}$$

Diametral pitch (P) is equal to the number of teeth divided by the pitch diameter, i.e. equation 6.4 applies.
Normal diametral pitch (P_n) is the diametral pitch multiplied by the secant of the spiral angle at the pitch circle, i.e.

$$P_n = P \sec \beta' = \frac{\pi}{p_n} \tag{6.82}$$

o that

$$P_n = \frac{z_1}{d_1' \cos \beta'} = \frac{z_2}{d_2' \cos \beta} \tag{6.83}$$

Pitch angle (δ_1' or δ_2') is the angle between the shaft axis and the pitch cone generator (Fig. 6.21).
Shaft angle (Σ) is the angle between the intersecting shaft axes, and is equal to the sum of the pitch angles of the wheel and pinion (Fig. 6.21), i.e.

$$\Sigma = \delta_1' + \delta_2' \tag{6.84}$$

The pitch angles for the pinion and gear are related to the shaft angle and the gear ratio[4], refer to Fig. 6.21, by the equations

$$\tan \delta_1' = \frac{\sin \Sigma}{u + \cos \Sigma} \tag{6.85}$$

and

$$\delta_2' = \Sigma - \delta_1' \tag{6.86}$$

where u = gear ratio $= \dfrac{d_2'}{d_1'} = \dfrac{z_2}{z_1}$

Cone distance (R) is the length of the pitch cone generator from the pitch circle to the apex, and from the geometry of Fig. 6.21 it is seen that this is given by

$$R = \frac{d_1'}{2 \sin \delta_1'} = \frac{d_2'}{2 \sin \delta_2'} \tag{6.87}$$

Spiral angle (β) at any point on the pitch cone is the angle between the pitch cone generator passing through that point and the line formed by the intersection of the tooth surface and the pitch cone. It is usual to refer to the spiral angle for a spiral bevel gear as being that at mid-face, as indicated in Fig. 6.22, since it may vary across the face.

Addendum (h_{a1} or h_{a2}) is the height from the pitch circle to the tips of the teeth measured along the back cone generator (Fig. 6.21).

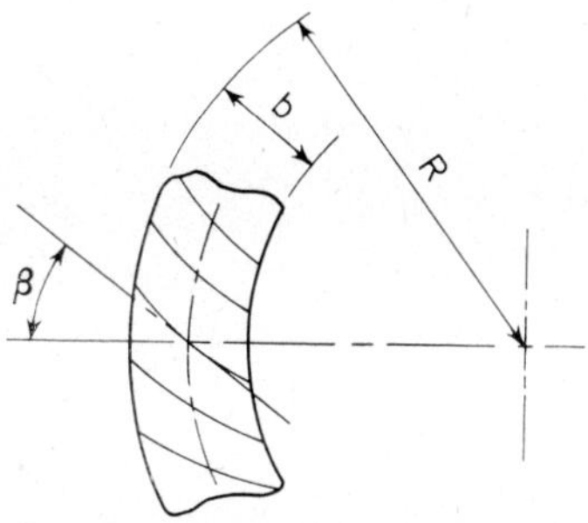

Fig. 6.22. Spiral bevel gear indicating spiral angle at mid-face

Dedendum (h_{f1} or h_{f2}) is the depth from the pitch circle to the roots of the teeth, measured along the back cone generator (Fig. 6.21).

Working depth is the sum of the addenda of the two gears, i.e. $h_{a1}+h_{a2}$.

Tip diameter (d_{a1} or d_{a2}) is the diameter of the circle of intersection of the blank cone and the back cone (Fig. 6.21).

Root diameter (d_{f1} or d_{f2}) is the diameter at the bottom of the tooth spaces at the large end of the tooth (Fig. 6.21)

Addendum angle (θ_{a1} or θ_{a2}) is the angle between the tip of the tooth and the pitch cone generator (Fig. 6.21).

Dedendum angle (θ_{f1} or θ_{f2}) is the angle subtended at the apex by the dedendum, Fig. 6.21.

Face width (b) is the length of the teeth measured along the pitch cone generator, Fig. 6.21.

6.9. GEAR FORCES AND STRENGTH AND WEAR REQUIREMENTS FOR STRAIGHT BEVEL GEARS

The forces acting upon the teeth of straight bevel gears when power is transmitted may be determined in a similar manner to that demonstrated for spur gears. However, in the case of bevel

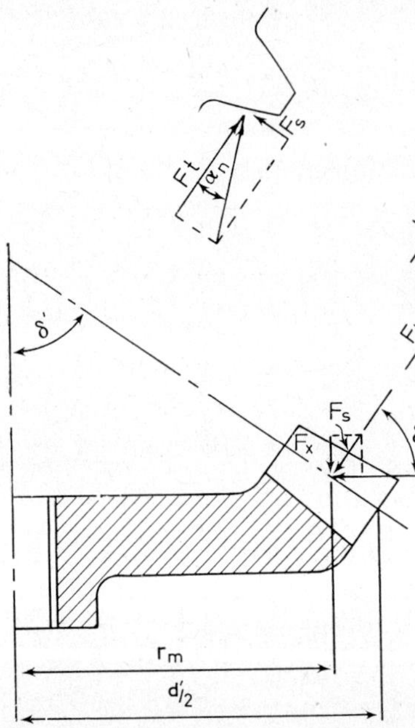

Fig. 6.23. Forces acting on straight bevel gear

gears, the force tending to separate the gears acts normal to the gear bevel angle (rather than at right angles to the shaft), and as a consequence an axial thrust component is induced in the gear and the pinion.

Fig. 6.23 indicates the forces acting on the tooth of a straight bevel gear. It is usually assumed that the tangential component of tooth load F_t acts at the mid-point of the tooth, although in reality it would act somewhere between this point and the large end; the error involved in making this assumption is small. Thus, the tangential force F_t is readily found from the equation

$$F_t = \frac{M_t}{r_m} \tag{6.88}$$

where M_t = torque transmitted (obtained from equation 3.1 or 3.2)

r_m = pitch radius of the tooth at the mid-point.

The value for the pitch radius at the mid-point of the tooth may be found[22] from the relationships

$$r_{m1} = \tfrac{1}{2}(d_1' - b \sin \delta_1') \tag{6.89}$$

$$r_{m2} = \tfrac{1}{2}(d_2' - b \sin \delta_2') \tag{6.90}$$

where r_{m1} = pitch radius at the mid-point of the pinion teeth.

r_{m2} = pitch radius at the mid-point of the gear teeth.

The separating force, say F_s, may be obtained from a consideration of Fig. 6.23, from which it may be deduced that

$$F_s = F_t \tan \alpha_n \tag{6.91}$$

where α_n = normal tooth pressure angle.

Further, the axial thrust acting on the tooth is given by the equation

$$F_x = F_s \sin \delta' = F_t \tan \alpha_n \sin \delta' \tag{6.92}$$

This thrust load F_x will also be responsible for a bending moment acting on the shaft of magnitude

$$M = F_x r_m \tag{6.93}$$

The radial load acting on the shaft, say F_r, is given by the vector sum of F_t and the radial component of F_s which, from Fig. 6.23 is seen to be

$$F_r = [F_t^2 + (F_s \cos \delta')^2]^{1/2}$$

so that

$$F_r = F_t[1 + \tan^2 \alpha_n \cos^2 \delta']^{1/2} \tag{6.94}$$

Thus, from the forces calculated above, the usual methods of statics may be applied to determine the loading on the supporting bearings, and the shaft may be analysed by the methods presented in Chapter 3.

In assessing the *strength of straight bevel gears*, one approach is to use a modification of the Lewis equation. The analysis is complicated by the fact that both the tooth thickness and the force acting

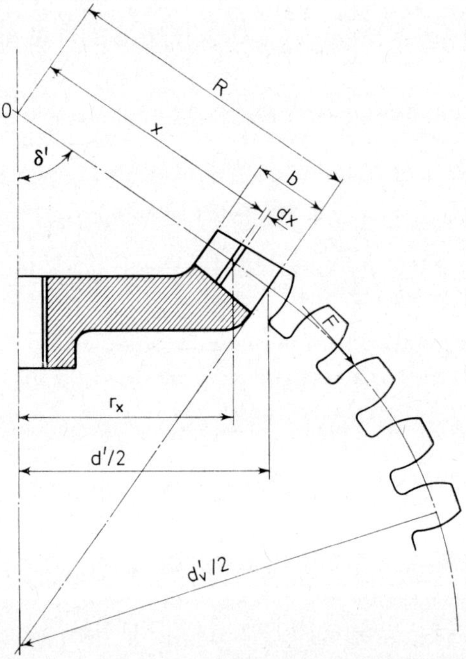

Fig. 6.24. Straight bevel gear and equivalent spur gear

on the tooth varies along the face width. However, if certain assumptions are made, a solution may be obtained which seems to give fair agreement with experiment.

Consider the half section of a straight bevel gear as shown in Fig. 6.24. Before proceeding to develop the modified Lewis equation for straight bevel gears, this is an opportune time to mention that when the shape of the teeth in a bevel gear are projected on to the back cone, as shown in Fig. 6.24, they have the same shape as in a spur gear whose radius is equal to the back cone radius. Thus, the virtual pitch diameters, i.e. the diameters of an equivalent spur gear, are given by the equations (from the geometry of Fig. 6.24)

$$d'_{v1} = d'_1 \sec \delta'_1 \tag{6.95}$$

$$d'_{v2} = d'_2 \sec \delta'_2 \tag{6.96}$$

Similarly, the virtual number of teeth in the equivalent spur gear may be obtained from the relationships

$$z_{v1} = z_1 \sec \delta'_1 \tag{6.97}$$

$$z_{v2} = z_2 \sec \delta'_2 \tag{6.98}$$

Now consider an elemental width of tooth in the bevel gear at some distance x measured from the cone centre 0, the width of the element being dx, and let the proportion of the tangential force acting on this element be dF. Applying the Lewis equation derived for spur gears (equation 6.16) to this element, we have

$$\mathrm{d}F = \frac{S_d(\pi y)\,\mathrm{d}x}{P_x} \tag{6.99}$$

where P_x = diametral pitch at the section under consideration. The torque acting on the element is given by the product of the elemental force dF and the radius r_x at which this force acts, i.e.

$$\mathrm{d}M_t = \mathrm{d}F\, r_x = \frac{S_d(\pi y) r_x \,\mathrm{d}x}{P_x}$$

and integrating across the face width, i.e. between the limits of $x = R - b$ and $x = R$, will give the total torque transmitted, i.e.

$$M_t = \int_{R-b}^{R} r_x\, \mathrm{d}F = \int_{R-b}^{R} \frac{S_d(\pi y) r_x\, \mathrm{d}x}{P_x} \tag{6.100}$$

By similar triangles, $r_x = (d'/2)(x/R)$ and $P_x = PR/x$, and substituting into equation 6.100 gives

$$M_t = \frac{S_d(\pi y)d'}{2PR^2}\int_{R-b}^{R} x^2\, dx$$

from which

$$M_t = \frac{S_d(\pi y)d'b}{2P}\left(1-\frac{b}{R}+\frac{b^2}{3R^2}\right)$$

If the face width b is small compared to the cone distance R (in practice, b is limited to being less than about one-third of R), then

$$M_t = \frac{S_d(\pi y)d'b}{2P}\left(\frac{R-b}{R}\right) \quad (6.101)$$

where the term in parenthesis is termed the *bevel gear factor*.

Dividing both sides of equation 6.101 by $d'/2$, and rearranging gives the result

$$F_b = \frac{F}{b} = \frac{S_d(\pi y)}{P}\left(\frac{R-b}{R}\right) \quad (6.102)$$

hwere F_b = allowable tangential tooth force per unit face width based on bending strength,
F = fictitious tangential force = $2\,M_t/d'$
b = face width,
M_t = torque transmitted (obtained from equations 3.1 or 3.2)
d' = pitch diameter of the gear,
S_d = design stress (equation 6.17),
P = diametral pitch at the pitch diameter = z/d'
R = cone distance (equation 6.87),
y = Lewis form factor based on virtual number of teeth (see equation 6.15)

The British Standards Institution in their Specification for Machine Cut Bevel Gears[23], recommends the use of the following equation

$$F_b = \frac{F_t}{b} = \frac{S_d Y X_b}{P}\left(\frac{R-b}{1{\cdot}1R}\right) \quad (6.103)$$

where S_b = bending strength factor for the material,
Y = strength factor,
X_b = speed factor for strength,
P = diametral pitch,
R = cone distance,
b = face width.

The reader will no doubt notice some similarity between equations 6.102 and 6.103. In equation 6.103 the bevel gear factor (in parenthesis) takes account of the tapered form of the tooth and the varying radius at which power is transmitted. It also includes a factor of 1.1 to allow for the mal-distribution. Charts for the determination of the strength factor (Y) and the speed factor for strength (X_b) are reproduced in Figs B.6. and B.7 respectively of Appendix B, by kind permission of the British Standards Institution. Values for the bending strength factor (S_b) for some commonly used gear materials are listed in Table B.1, of Appendix B.

The *wear characteristics of straight bevel gears* may be assessed using the Buckingham wear equation[6] and treating the bevel gear as a virtual spur gear taken at the mean radius of the bevel gear. Thus, the wear equation for straight bevel gears becomes

$$F_c = \frac{F_t}{b} = d_{vm}KQk \qquad (6.104)$$

where F_c = allowable tangential tooth force per unit face width based on wear,
d_{vm} = virtual mean diameter,
K = material and form factor,
Q = ratio factor,
k = face width factor depending upon method of mounting,
= 0·75 if pinion is mounted outboard of the bearing (i.e. overhung),
= 1.0 if the pinion is mounted inboard of the bearings (i.e. straddle mounted).

The material and form factor K is the same as for spur gears and may be obtained from equation 6.28. The ratio factor, Q, however, is based on the virtual number of teeth and is obtained from the equation

$$Q = \frac{2z_{v2}}{z_{v1}+z_{v2}} \tag{6.105}$$

where the virtual numbers of teeth are determined from equations 6.97 and 6.98. The virtual mean diameter is calculated using equations 6.95 and 6.96, by substituting the mean diameter for the pitch diameter. Thus,

$$d_{vm} = d_m \sec \delta' \tag{6.106}$$

where $d_m = 2\,r_m$ and r_m is calculated from equations 6.89 and 6.90.

Based on satisfying wear requirements for bevel gears, the British Standards Institution recommends[23] the following equation for determining the allowable tangential load per unit length of face width,

$$F_c = \frac{F_t}{b} = \frac{S_c Z X_c}{1{\cdot}33P}\left(\frac{R-b}{1{\cdot}1R}\right) \tag{6.107}$$

where S_c = surface strength factor for the material,
Z = zone factor,
X_c = speed factor for wear,
P = diametral pitch

and the term in parenthesis is the bevel gear factor mentioned previously. Charts for the determination of the zone factor (Z) and the speed factor for wear (X_c) are reproduced in Fig. B.8 and B.7 respectively of Appendix B, by kind permission of the British Standards Institution. Values for the surface strength factor for some commonly used gear materials are listed in Table B.1 of Appendix B.

6.10. GEAR FORCES AND STRENGTH AND WEAR REQUIREMENTS FOR SPIRAL BEVEL GEARS

The relationship between spiral bevel gears and straight bevel gears is similar to the relationship between helical gears and spur gears. It will be remembered that helical gears give rise to a thrust load not present with straight spur gears. Similarly, with spiral bevel

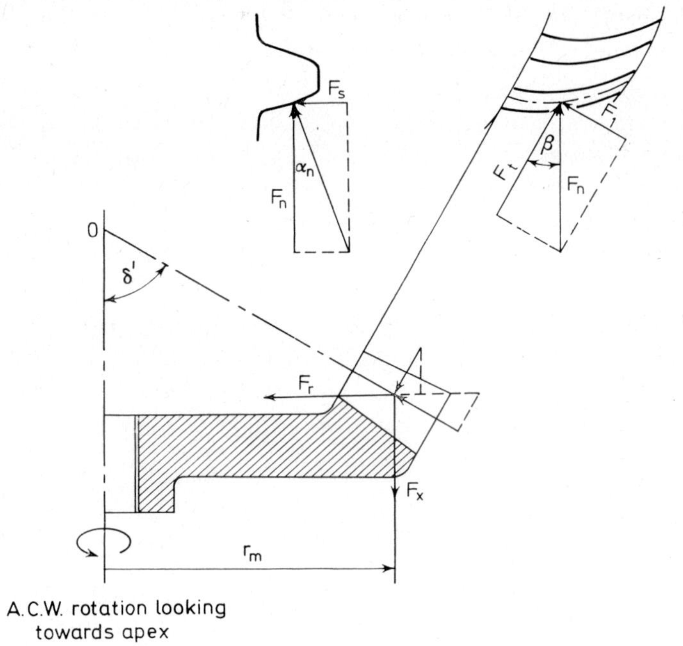

Fig. 6.25. Forces acting on spiral bever gear

gears, in addition to the tangential force F_t and the separating force F_s, a thrust load F_1 (say) acts on the teeth; these forces are shown in Fig. 6.25. The direction of the thrust component is dependent upon those factors discussed in connection with helical gears, namely the hand of the spiral, the direction of rotation, and whether the gear under consideration is the driver or the driven.

Probably one of the most important factors associated with spiral bevel gears is that of ensuring that the correct loading on the teeth is determined, for this has considerable significance with regard to the bearing loads and shaft stresses. The hand of spiral bevel gears may be determined by looking at the face of the gear; if the teeth recede in a clockwise direction the spiral is right hand, and if they recede in an anticlockwise direction the spiral is left hand. This is consistent with the notation used for defining the hand of helical gears. In a pair of mating spiral gears, one gear will be a right hand and the other a left hand. It is common practice to refer to the hand of a pair, and when this is done, the hand stated for the combination refers to the hand of the pinion.

Fig. 6.25 indicates the forces acting on the tooth of a spiral bevel gear, the spiral being left-hand, and it is assumed that rotation is in an anticlockwise direction when viewed towards the apex 0; it is further assumed that the gear is the driver.

The assumption is made that the tangential component of tooth load, F_t acts at the mid-point of the tooth, and its value may be calculated using equations 6.88, 6.89 and 6.90. However, the separating force F_s and the thrust load F_1 act neither parallel nor perpendicular to the shaft axis, and it is necessary to resolve both these forces into two mutually perpendicular components. From the geometry of Fig. 6.25,

$$F_1 = F_t \tan \beta$$
$$F_s = F_t \sec \beta \tan \alpha_n$$

Thus, the total axial component acting on the gear is

$$\begin{aligned} F_x &= F_s \sin \delta' - F_1 \cos \delta' \\ &= F_t(\sec \beta \tan \alpha_n \sin \delta' - \tan \beta \cos \delta') \qquad (6.108) \end{aligned}$$

Similarly, the total radial component acting on the gear is

$$\begin{aligned} F_r &= F_s \cos \delta' + F_1 \sin \delta' \\ &= F_t(\sec \beta \tan \alpha_n \cos \delta' + \tan \beta \sin \delta') \qquad (6.109) \end{aligned}$$

If the forces calculated are positive, they act in the directions indicated in Fig. 6.25, and if negative they act in opposite directions.

By changing any one of the factors, i.e. the hand of the spiral, the direction of rotation, or whether the gear is driving or being driven, the sign associated with the second term in parenthesis in equations 6.108 and 6.109 is also changed.

One final point concerning the forces acting on a spiral bevel gear must be mentioned. First of all note that the directions of the forces in Fig. 6.25 refer to the directions of forces acting on the gear and transmitted to the bearings. The direction of the axial force F_x is important from another point of view, since it will determine whether the gears are drawn tightly into mesh or pushed apart. If F_x acts in the direction shown in Fig. 6.25, this corresponds to the condition tending to separate the gears, and is the desirable condition to obtain in design. Naturally, the thrust must be reacted by suitable bearings. For the condition where a gear must operate in both directions of rotation, it becomes necessary to carry out a force analysis for both conditions, taking care to provide the necessary bearing reactions.

The *strength of spiral bevel gears* may be assessed using a modified form of the Lewis equation. Using the fact that the relationship between spiral bevel and straight bevel gears is similar to that between helical and spur gears enables the Lewis equation for straight bevel gears to be suitably modified. Thus, referring to equation 6.102, the modified Lewis equation for spiral bevel gears may be written

$$\mathrm{F}_b = \frac{F}{b} = \frac{S_d(\pi y)}{P \sec \beta}\left(\frac{R-b}{R}\right) \qquad (6.110)$$

where β = spiral angle at the mid-point of the gear face, and other quantities are as previously defined immediately below equation 6.102.

The British Standards Institution recommends[23] introducing a spiral angle factor, designated Ω_b, in assessing the bending strength of spiral bevel gears. Thus, equation 6.103 then becomes

$$F_b = \frac{F_t}{b} = \frac{S_b Y X_b}{P\Omega_b}\left(\frac{R-b}{1{\cdot}1R}\right) \qquad (6.111)$$

where, with the exception of Ω_b, the symbols are as previously defined and may be obtained as already discussed. The spiral angle factor Ω_b is dependent upon the overlap ratio ε_β, which will presently be discussed. Once the overlap ratio ε_β has been determined, the spiral angle factor for strength may be obtained from Fig. B.9 of Appendix B.

The *wear characteristics of spiral bevel gears* may be assessed using a modification of the Buckingham equation for wear, once again making use of the similarity between spiral bevel to straight bevel gears and helical to spur gears. Thus, equation 6.104 is modified for spiral bevel gears to read

$$F_c = \frac{F_t}{b} = \frac{d_{vm}KQk}{\cos^2 \beta} \tag{6.112}$$

where β = spiral angle at the mid-point of the gear face, and other quantities are as previously defined immediately following equation 6.104.

Based on satisfying wear requirements for spiral bevel gears, the British Standards Institution recommends[23] using the following equation for determining the allowable tangential load per unit length of face width,

$$F_c = \frac{F_t}{b} = \frac{S_c Z X_c}{P\Omega_c}\left(\frac{R-b}{1{\cdot}1R}\right) \tag{6.113}$$

where Ω_c is termed the overlap ratio factor, and is dependent upon the overlap ratio ε_β (to be presently discussed). The other quantities are as previously defined and may be obtained from Appendix B as already discussed. The overlap ratio factor Ω_c, may be obtained from Fig. B.10 of Appendix B.

6.11. TOOTH MODIFICATION OF BEVEL GEARS

In the United Kingdom, the British Standards Institution recommend[23] the use of a 20 degree normal pressure angle, but the form of the tooth spiral is left to the manufacturer, since this will depend upon the type of cutter and method of cutting being employed. In the United States the 20 degree normal pressure angle is also standard, but the American Gear Manufacturers Association recommended practice is to use the Gleason Works system. This suggests the use of pressure angles different from 20 degrees, according to the number of teeth in the pinion and the gear and the gear ratio. In order to avoid interference or undercutting, the sum of the *virtual* number of teeth in a bevel gear pair should not generally be less than 34 for a 20 degree pressure angle or 64 for a pressure angle of 14·5 degrees. For further details regarding the minimum number corresponding to different pressure angles and gear ratios, refer to Dean[21].

The British Standard recommendation[23] for the 20 degree full depth tooth is to make the addenda according to the following values:

$$h_{a1} = \frac{1+k_1}{P_n} \tag{6.114}$$

and

$$h_{a2} = \frac{1+k_2}{P_n} \tag{6.115}$$

$$k_2 = -k_1 \tag{6.116}$$

where h_{a1} = addendum of the pinion,
h_{a2} = addendum of the gear,
k_1 = addendum coefficient for the pinion,
k_2 = addendum coefficient for the gear.

The value for the pinion addendum coefficient is obtained from either

$$k_1 = 0{\cdot}4\left(1-\frac{z_1 \sec \delta_1'}{z_2 \sec \delta_2'}\right) \tag{6.117}$$

or

$$k_1 = 1-\frac{2z_1 \sec \delta_1'}{Q} \tag{6.118}$$

whichever is the greatest. The factor Q is a function of the spiral angle, and is termed the interference factor. It is a measure of the minimum total number of teeth in a gear pair required for satisfactory contact and freedom from undercutting. Its value may be obtained from Fig. B.11 of Appendix B. For straight bevel gears,

$$z_1 \sec \delta_1' + z_2 \sec \delta_2' \geq 34{\cdot}2 \tag{6.119}$$

Further details may be obtained by consulting the British Standard Specification[23].

6.12. FURTHER DESIGN CONSIDERATIONS FOR BEVEL GEARS

Although the equations relating to strength and wear of bevel gears indicate that, for a given pitch diameter and cone angle, the load carrying capacity increases with increase in face width, practical considerations limit the maximum permissible value. In the case of spur and helical gears, it will be remembered that the maximum face width was expressed as a function of the circular pitch, and that a face width factor was suggested to provide an indication of the maximum face width which could reasonably be used. It is more convenient in the case of bevel gears to relate the maximum face width to the cone distance R, and it is generally agreed in the literature that the maximum face width should never exceed one-third of the cone distance, a more acceptable value being in the region of one-quarter. Expressed as a limit equation

$$b \leq \frac{R}{3} \tag{6.120}$$

which is applicable to both straight and spiral bevel gears. Any increase in load carrying capacity which is apparently obtained by making the face width greater than this, is more than offset by the difficulty in manufacture and in maintaining uniform contact along the face of the tooth.

Another factor of importance in connection with spiral bevel gear is the overlap ratio ε_β, sometimes referred to as the face contact ratio (c.f. equation 6.71 in connection with helical gears). This is defined in the British Standard Specification[23] as the ratio of the greatest angle subtended at the apex of the developed pitch cone by any two points at the intersection of the tooth flank with the pitch cone (θ_γ), to the angle subtended at the apex of the developed pitch cone by two points equidistant from the apex of similar flanks of adjacent teeth (θ_α). This is illustrated in Fig. 6.26. Thus,

$$\varepsilon_\beta = \frac{\theta_\gamma}{\theta_\alpha} \tag{6.121}$$

In the case of straight bevel gears, the overlap ratio is always zero, but for spiral gears any value greater than zero may be obtained.

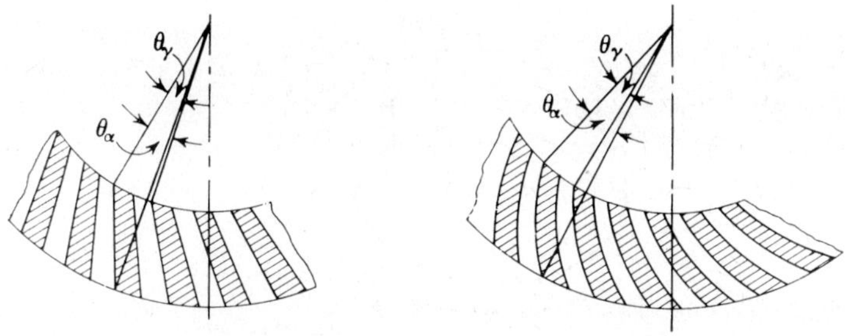

Fig. 6.26. Overlap ratio for spiral bevel gears

However, it is preferable to make the spiral angle such that an overlap ratio of at least unity is obtained where manufacturing methods permit. The minimum value of spiral angle to ensure an overlap ratio of not less than unity may be estimated[23] from the equation

$$\tan\beta' = \frac{1}{P}\left(\frac{1}{b} - \frac{1}{2R-b}\right) \tag{6.122}$$

Spiral bevel gears having a considerable spiral angle (say about 35 degrees) provide a large overlap of the teeth which results in the load being carried by a greater number of teeth with consequently smoother and more quiet operation. However, the thrust load carried by the bearings will be greatly increased, which might give rise to the operating conditions of the bearings being critical.

The reduction ratio for bevel gears should not generally exceed about 5 to 1, at which the typical efficiency is in the region of 98 per cent. Higher peripheral speeds are possible with spiral bevel gears than can be obtained with straight bevel gears.

6.13. WORM GEARING

Worm gears are used in the transmission of power between non-intersecting shafts, generally at right angles, providing a compact drive where large velocity ratios are required. Although they serve

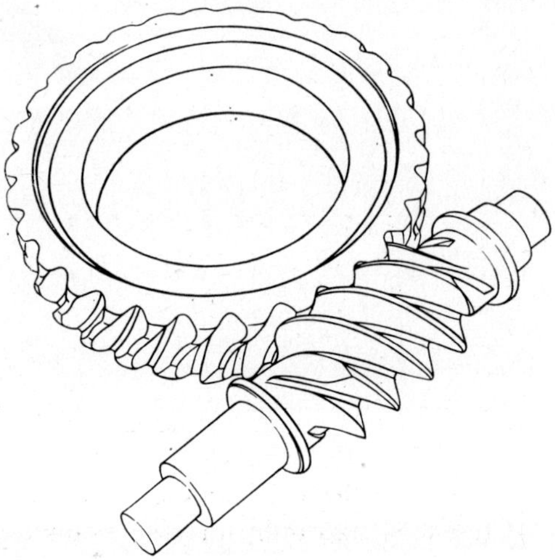

Fig. 6.27. Worm and wormwheel (Kindly reproduced by David Brown Gear Industries Limited)

essentially the same purpose as crossed-helical gears, line contact is provided (compared with hypothetical point contact in the case of crossed-helical gears) which makes them suitable for power transmission.

A worm gear drive consists essentially of two elements, namely a cylindrical worm having helical teeth or threads similar to a helical gear, and a wheel having a concave or hollow face; such a pair is indicated in Fig. 6.27. Compared with other types of gears worm gearing is less efficient, and due to the fact that the threads slide into contact with the gear teeth, although quiet operation ensues, under high loading conditions the heat generated in friction can be a problem.

Fig. 6.28 indicates some dimensions associated with a worm and worm gear. For a shaft angle of 90 degrees, the axial pitch of the worm is equal to the circular pitch of the wheel. Although a mating

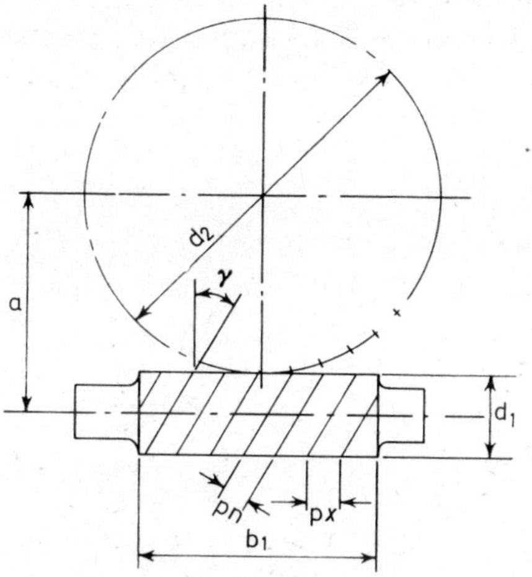

Fig. 6.28. Principal dimensions of worm and wormwheel

worm and worm gear have the same hand of helix, the actual helix angles are generally quite different, the helix angle on the worm being usually quite large compared with that of the gear wheel.

It has been the standard practice in the past to designate worm gears by their axial pitch p_x, and no doubt this practice will continue to be used to some extent in the future. The diametral pitch is not used at all. Merritt[4] has suggested the use of the module m when expressing the pitch of the worm, and this has been incorporated in the recommendations of the British Standards Institution[24]. The relationship between the module m and the axial pitch p_x is simply given by the equation

$$m = \frac{p_x}{\pi} \tag{6.123}$$

The term *reference diameter* is used in connection with worm and worm gears, and is preferred to the term pitch diameter; this refers to the diameter of the cylinder to which the detail dimensions of the wheel teeth or worm threads are calculated[4]. Since the efficiency of a worm gear drive increases as the worm diameter is reduced, the worm diameter should be as small as possible, consistent with adequate strength at the root section of the worm. The worm reference diameter, d_1, is related to the module by means of a so-called *diameter factor*[24] or *diameter quotient*[4], designated q, such that

$$q = \frac{d_1}{m}$$

$$\text{or} \quad d_1 = qm \tag{6.124}$$

Thus, the reference diameter of a worm is similar to the pitch diameter of a spur gear having q teeth and module m. The reference diameter of the worm wheel d_2, may be expressed in terms of the number of teeth on the wheel z_2 and the module m, in the same way as for a spur gear, i.e.

$$d_2 = z_2 m \tag{6.125}$$

For a worm gear drive operating at a standard centre distance, we have

$$a = \tfrac{1}{2}m(q+z_2) \tag{6.126}$$

The *lead* of a worm thread is the axial distance through which it advances in one revolution, and this will depend upon the number of threads in the worm. For a single thread the lead will obviously be equal to the axial pitch, for two threads twice the axial pitch, and so on, so that for the more general case where we have z_1 threads, the lead p_z is given by

$$p_z = z_1 p_x = \pi m z_1 \tag{6.127}$$

Fig. 6.29 indicates the development of the worm thread for one revolution, from which it is seen that the lead angle γ is given by

$$\tan \gamma = \frac{p_z}{pd_1} = \frac{z_1}{q} \tag{6.128}$$

For practical reasons, it is desirable to restrict the range of values recommended for the diameter factor q, and the standard values recommended in the British Standard Specification for Worm Gears[24] are 6, 6·5, 7, 7·5, 8, 8·5, 9, 9·5, 10, 11, 12, 13, 14, 17 and 20.

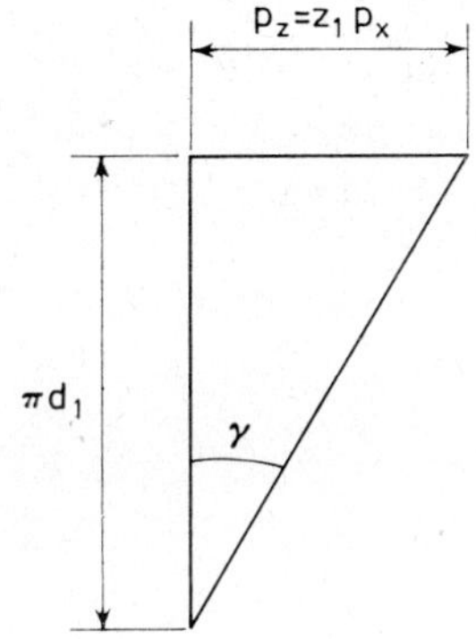

Fig. 6.29. Development of worm thread

The large worm diameter factors from 12 upwards are seldom required unless the worm is to be bored, in which case the thickness of the metal below the thread should be approximately equal to the thread depth.

6.14. GEAR FORCES AND STRENGTH AND WEAR REQUIREMENTS FOR WORM GEARING

The gear forces acting on the teeth of a worm driving a worm wheel are illustrated in Fig. 6.30, which shows the normal driving force resolved into three mutually perpendicular components, namely a tangential force F_t, a separating force F_s and a thrust force F_x; the forces acting on the worm gear teeth are in the opposite directions to those shown acting on the worm.

The tangential or driving force F_t is obtained from the usual

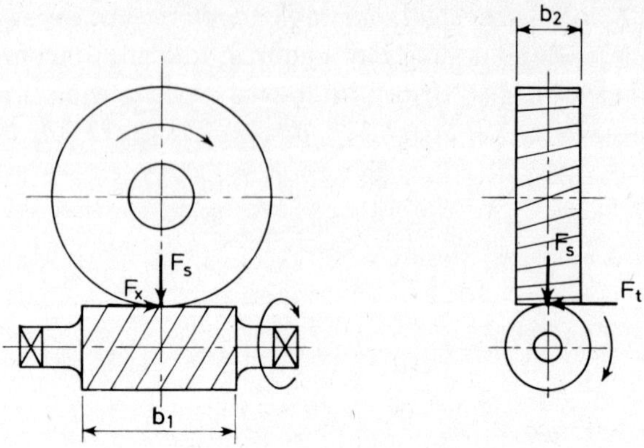

Fig. 6.30. Forces acting on worm thread

definition of torque, i.e.

$$F_t = \frac{2M_t}{d_1} \tag{6.129}$$

where M_t = torque transmitted obtained from the usual power equation (equation 3.1 or 3.2).

The axial force on the worm F_x (which is actually the driving force on the worm wheel) is dependent upon the lead angle γ, and the efficiency of the drive. Thus, the axial force is given by

$$F_x = \frac{F_t}{\tan \gamma} \times \text{Efficiency} \tag{6.130}$$

The separating (or radial) force F_s is dependent upon the worm thread pressure angle in the axial plane, say α_x, and is a function of the axial force, so that

$$F_s = F_x \tan \alpha_x \tag{6.131}$$

The directions of the forces are dependent upon the direction of rotation, the hand of the gears, and whether the worm or worm wheel is the driver. The directions indicated in Fig. 6.30 are correct for the forces acting on the worm for the conditions specified, and for the more usual case where the worm is the driver; the forces acting on the worm wheel will, of course, be in the opposite directions to those shown. Changing any one of the factors mentioned, i.e. direction of rotation, hand of helix or driver to driven, will change the directions of the tangential and axial forces. Once the forces acting have been calculated using equations 6.129–6.131, the forces transmitted to the supporting shafts and the necessary bearing

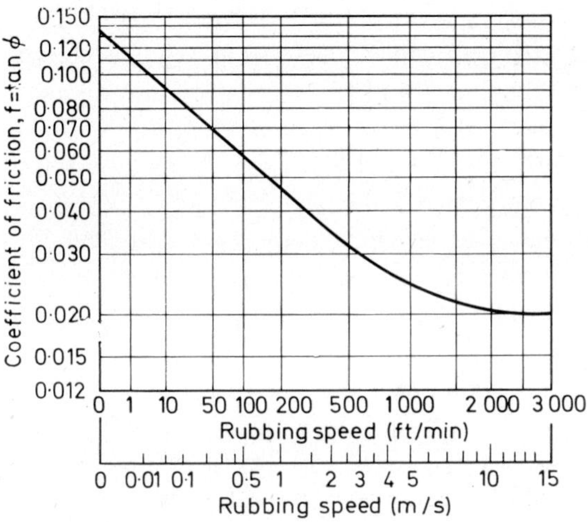

Fig. 6.31. Coefficient of friction (f) of worm gears (Reproduced from B.S. 721[2] by kind permission of the British Standards Institution)

reactions can be determined. Suitable bearings may be designed or selected as discussed in Chapter 4, and the shaft may be designed as demonstrated in Chapter 3.

The efficiency of worm gears is dependent upon the lead angle of the worm thread and, of course, the coefficient of friction. This latter quantity can only be determined experimentally, and based on the use of phosphor bronze wheels and case hardened ground and polished steel worms, lubricated by mineral oil, Fig. 6.31 represents reasonable values for the coefficient of friction f corresponding to rubbing velocity. This chart is based upon the recommendations of the British Standards Institution[24] and is reproduced by permission. Thus, excluding bearing and oil churning losses, the efficiency of worm gears may be estimated from the equations

$$\text{Efficiency (worm driving)} = \frac{\tan \gamma}{\tan (\gamma + \phi)} \qquad (6.132)$$

$$\text{Efficiency (wormwheel driving)} = \frac{\tan (\gamma - \phi)}{\tan \gamma} \qquad (6.133)$$

where $\tan \phi$ = coefficient of friction (from Fig. 6.31).

The load carrying capacity of worm gears is usually specified according to the amount of heat generated and dissipated and the surface durability. For worm gears operating at relatively slow speeds, *the strength of the wormwheel teeth* may be assessed using a modification suggested by Buckingham to the Lewis equation. Since in all cases the strength of the worm threads will be greater than that of the teeth in the wormwheel, it is only necessary to consider the wormwheel. The modified Lewis equation may be written in the form

$$F_b = \frac{F_t}{b} = S_d y p_x \cos \gamma \qquad (6.134)$$

where F_b = allowable tangential tooth load per unit face width based on bending strength,
S_d = design strength (see equation 6.17),
p_x = axial pitch ($= \pi m$),
γ = lead angle,
b = effective face width of the wormwheel,
y = form factor.

It will be evident that equation 6.134 is an oversimplification, and that the problem is somewhat complicated by the fact that the teeth of the wormwheel are not of uniform section across the face, but are thinner at the centre than they are at the edges. Further, in equation 6.134, the form factor is usually referred only to the normal pressure angle and the reference diameter, the number of teeth not being included. Values for y corresponding to different normal pressure angles are indicated in Table 6.3.

Table 6.3. FORM FACTORS FOR WORM GEARS

Normal pressure angle α_n	*Form factor* y
14·5	0·100
20	0·125
25	0·150
30	0·175

The British Standards Institution in their Specification for Worm Gears[24] recommends assessing the *load capacity based on strength* by use of the following equation

$$M_{t2} = 1{\cdot}8 X_b S_b m b_{r2} d_2 \cos\gamma \tag{6.135}$$

where M_{t2} = limiting torque which may be transmitted by the wormwheel,
X_b = speed factor for strength,
S_b = bending stress factor,
m = module,
b_{r2} = length of root of the wormwheel teeth,
d_2 = reference diameter of the wormwheel,
γ = lead angle.

The length of root of the wormwheel teeth is determined from the equation

$$b_{r2} = (d_{a1}+2a)\sin^{-1}\left(\frac{b_{e2}}{d_{a1}+2a}\right) \tag{6.136}$$

where b_{e2} = effective face width of the wormwheel,
d_{a1} = worm tip diameter,
a = centre distance.

The effective face width of the wormwheel is given by

$$b_{e2} = 2m\sqrt{(q+1)} \tag{6.137}$$

and the tip diameter of the worm by the equation

$$d_{a1} = d_1+2a = m(q+2) \tag{6.138}$$

In equation 6.135 the torque which may be transmitted by the wormwheel is limited to the lower of the two values obtained by considering X_b and S_b for the worm and wormwheel. Values for X_b and S_b may be obtained from Fig. B.12 and Table B.2 respectively of Appendix B, which are reproduced by kind permission of the British Standards Institution.

Based on *satisfying wear requirements*, Buckingham[6] suggests using the approximate equation

$$F_w = \frac{F_t}{b} = d_2 K \tag{6.139}$$

where F_w = allowable tangential tooth load per unit face width based on wear,
d_2 = reference diameter of the wormwheel,
K = material and form factor.

Suggested values for K corresponding to various material combinations can be obtained from the literature[14, 25].

The British Standards Institution recommends assessing the *load capacity based on wear*[24] by use of the following equation

$$M_{t2} = X_c S_c Z d_2^{1.8} m \tag{6.140}$$

where M_{t2} = limiting torque which may be transmitted by the wormwheel,
X_c = speed factor for wear, S_c = surface stress factor,
Z = zone factor,
d_2 = reference diameter of the wormwheel,
m = module.

In equation 6.140 the torque which may be transmitted to the wormwheel is limited to the lower of the two values obtained by considering X_c and S_c for the worm and the wormwheel. Values for X_c, S_c and Z may be obtained from Fig. B.13 and Tables B.2 and B.3 respectively, which are reproduced in Appendix B by kind permission of the British Standards Institution.

The British Standard Specification for Worm Gears[24] is very comprehensive, and provides a detailed document giving clear guidance on the proper selection of materials and the determination of geometry for worm gears. Much of the data is recorded in convenient tabular form, enabling the design to be carried out with the minimum of tedious calculations. This document should most certainly be consulted by designers in the United Kingdom. For convenience, Fig. B.14 of Appendix B (reproduced by kind permission of the British Standards Institution) may be used as a guide in determining a suitable value for the diameter factor q.

The load capacity determined using equations 6.135 and 6.140 are based upon a life of 26 000 hours. If it is required to design a drive based upon any other life, the wormwheel torque must be multiplied by a life correction factor. Further, where variable loading conditions are encountered, the equivalent running time may be assessed; the procedure is laid down in the relevant British Standard[24].

6.15. MODIFICATION OF WORM GEARS

In order to prevent undercutting of worms, particularly where large lead angles are used, it is desirable to carry out modification. This is achieved by withdrawing the basic rack so that the root diameter and thread thickness are increased and the dedendum decreased, but the tip diameter and reference diameter remain unaltered. The formulae and method of calculating the necessary dimensions are presented in the relevant British Standard[24].

6.16. FURTHER DESIGN CONSIDERATIONS FOR WORM GEARS

To provide adequate rigidity, to avoid undercutting, and to achieve maximum economy in material, it is desirable to establish the face width of the wormwheel within certain limits. Ideally, the overall face width b_2, may be made equal to the effective face width, but should not exceed it. However, to satisfy rigidity requirements, it is suggested[18] that the face width should be equal to at least one-tenth of the wormwheel reference diameter. Expressed as a limit equation

$$\frac{d_2}{10} \leq b_2 \leq b_{e2} \tag{6.141}$$

where b_{e2} may be obtained from equation 6.137.

The face width of the worm, b_1 (more appropriately referred to as the length of the worm), i.e. the length over the threads in an axial direction (refer to Fig. 6.30), must be sufficient to cover the axial length of tooth contact. This may be estimated[24] using the equation

$$b_1 \geq 14m \cos \gamma \tag{6.142}$$

The number of threads in the worm will depend upon the centre distance and the ratio, and it is suggested that for industrial gears[24] the number z_1 may be estimated as the nearest whole number to

$$\frac{7+12\sqrt{a}}{u} \tag{6.143}$$

The number of teeth in the wheel is then taken as the next whole number below uz_1, except for single thread worms, in which case $z_2 = u$.

Speaking generally, worm gears should not be used for ratios less than about 5 to 1 nor greater than about 50 to 1. Where ratios in excess of 50 to 1 are required, the use of multi-stage drives will usually give greater compactness and efficiency, and less expensive units will usually be achieved.

6.17. MATERIALS AND MANUFACTURE OF GEARS

Of obvious importance in the design of gears is a consideration of the materials and the methods of manufacture to be used in producing the gears.

The selection of a suitable gear material will depend upon the type of drive and the loading conditions. Ideally, it should combine the desirable properties of high strength and wear resistance; good machinability; low resistance to sliding contact; high damping capacity; the material should also be readily available at low cost. In practice, a compromise is usually required, but a wide variety of gear materials are available, and the designer should try to select those which give the optimum design.

Gear materials may be classified as ferrous, non-ferrous or non-metallic. The ferrous materials include a wide range of cast irons; cast steels; forged carbon steels; forged alloy steels; and case-hardening steels. For the majority of gears, selection of suitable ferrous materials provides the best solution, since such materials are less expensive than non-ferrous and non-metallic materials are only suitable for specific applications. Recommended non-ferrous materials include gunmetal, phosphor bronze, and aluminium alloys; such materials may also be used for special requirements. For example, if gears are to be used in applications requiring resistance to corrosion, then gunmetal or bronze may be used. Phosphor bronze is commonly used for wormwheels in conjunction with a hardened steel worm, due mainly to the fact that this combination offers little resistance to sliding under load, with high resistance to wear. Because of the high cost of phosphor bronze, for medium and large size wheels it becomes more economical to use a phosphor bronze rim shrunk or bolted on to a cast iron or steel centre. Suitable non-metallic materials may be used under certain circumstances and are particularly suitable for high speed drives where silent operation and low cost are important factors. The original so-called silent gears were made from rawhide, but gears made from synthetic bonded fabric and other plastic materials are generally used today[26].

Table B.1 of Appendix B lists a variety of commonly recommended metallic gear materials[15, 18, 23] which are applicable to spur, helical and bevel gears, and Table B.2 of Appendix B lists recommended[24] combination materials for worm and worm gears.

Gear teeth may be obtained by cutting the teeth in the blank

Fig. 6.32. Typical hobs (Reproduced by kind permission of David Brown Gear Industries Limited)

Fig. 6.33. Gear shaving cutters (Reproduced by kind permission of David Brown Gear Industries Limited)

Fig. 6.34. Milling a special pinion (Reproduced by kind permission of R. W. Transmissions Limited)

Fig. 6.35. Planing a double-helical wheel (Reproduced by kind permission of David Brown Gear Industries Limited)

Fig. 6.36. Cutting Cluster Gears on a Fellow Machine (Reproduced by kind permission of David Brown Gear Industries Limited)

Fig. 6.37. Milling a worm (Reproduced by kind permission of David Brown Gear Industries Limited)

Fig. 6.38. Grinding a spur gear (Reproduced by kind permission of David Brown Gear Industries Limited)

Fig. 6.39. Worm being ground with flat grinding wheel (Reproduced by kind permission of David Brown Gear Industries Limited)

using methods of milling, hobbing, shaping, shear cutting or broaching. The teeth may be used as cut if the surface finish is satisfactory, or they may be finished by grinding, shaving or lapping. A detailed discussion of the various gear cutting and finishing processes is beyond the scope of this chapter, and the reader is referred to the authoritative handbook of Dudley[27]. However, it must once again be emphasised that in the interest of standardisation and economy, an important factor to be considered in the design of gears is that relating to the use of standard cutters. Fig. 6.32 illustrates some typical hobs and Fig. 6.33 two gear shaving cutters. Various manufacturing procedures used in the production of gears are illustrated in Fig. 6.5 and Figs 6.34–6.39.

6.18. MOUNTING AND LUBRICATION OF GEARS

In order that gears perform satisfactorily in service, giving the desired life and quietness of operation, not only must the gears themselves be properly designed and accurately manufactured, but careful consideration must be given to the correct mounting and adjustment of the gears, and the provision of adequate lubrication.

It has already been mentioned in Chapters 3 and 4 that shaft dimensions and bearing sizes are not only determined through considerations of strength and life. Shafts and bearings must be so designed and positioned that they provide sufficient stiffness and rigidity to prevent excessive deflection and slope at the gear teeth. Whenever possible a gear should be straddle mounted, i.e. mounted between bearings rather than overhung. In any event, the bearings should be positioned as close to the gear as possible, making due allowance for lubrication and assembly, in order to provide a more rigid arrangement and reduce possible vibration problems. Where thrust loads are involved, it is common practice to pre-load the thrust bearings in order that the axial deflection under operating conditions is kept to a minimum.

Maldistribution of load across the gear teeth can be kept to a minimum by the provision of a suitably stiff shaft and carefully positioned bearings supported in rigid mountings, but the gear teeth themselves also require to be adequately supported by the proportioning of the gear blank. Hartman[28] suggests that, ideally, gears should be made integral with the shaft, but for a variety of reasons this is not always possible, and it is necessary to connect them using keys and press-fits, splines and serrations. In any case, the thickness of metal in the rim beneath the root of the teeth should be at least equal to the tooth depth, and preferably somewhat greater than this, and it should extend the full length of the teeth. The gear hub should be of adequate proportions to transmit the required torque, its thickness being sufficient to ensure proper fit on the shaft with provision for keyways, splines or serrations, and of length at least equal to the face width or longer if required to give adequate key length. The rim and hub are connected by a web or arms, usually integral, and in many instances the design is such that the web, rim and hub are formed from a solid cylinder, as indicated in Fig. 6.38. In any event, the web thickness or proportioning of the arms should provide reasonable rigidity and withstand the induced bending and tensile stresses due to the transmitted load under operating conditions. Fig. 6.40 illustrates a variety of finished gears.

The general principles mentioned above are applicable to gears in general, but the requirements for mounting worm and bevel gears are usually more exacting, and even greater care needs to be exercised than with other types of gears. Because of the desirability to keep the size of worm small, this can lead to relatively large deflections if an effort is not made to incorporate a stiff shaft and well supported bearings.

From the previous discussion concerning hydrodynamic lubrication (Chapter 4), it is evident that thick film or perfect lubrication between engaging gear teeth is a desirable condition, since the working surfaces will then be completely isolated from each other by a continuous film of lubricant. The actual mechanism of lubrication between gear teeth is complex, requiring consideration of elasto-hydrodynamic theory.

The extent to which hydrodynamic lubrication is attained in

practice is a matter of degree and depends upon a number of factors. It has been suggested[29] that the best that can normally be hoped for as far as gear teeth are concerned, is marginal hydrodynamic lubrication in which the film thickness is only slightly greater than the height of the asperities, and occasional contact persists. It is of obvious importance, then, that the gear teeth be accurately formed, and the degree of hydyodynamic lubrication increases greatly with

Fig. 6.40 Variety of finished gears (Reproduced by kind permission of R. W. Transmissions Limited)

running-in. A detailed discussion of the actual mechanism of lubrication between gear teeth is beyond the scope of this chapter, but an excellent account has been presented by Shell International[29], to which the reader is referred.

The manner in which gears are to be provided with lubricant, and the type of lubricant used, depends very much upon the particular conditions. The lubricant frequently performs a two-fold purpose of reducing friction and keeping the gears cool by carrying away heat. The application of lubricants to gears may be performed manually by hand application; by some means of mechanical applicator such as drip feed; by bath lubrication, i.e. by allowing the gear or gears to dip into a bath of oil; or by pressure circulating systems in which oil is applied to the gears by sprayers.

Hand application is only suitable for low speed gears, i.e. less than about 2·5 m/s (500 ft/min), and where it is convenient and economical for gears to be frequently lubricated in this manner. Generally speaking, this method of lubrication is not recommended and the number of gears requiring hand lubrication is progressively reducing. Where slow running gears can be adequately lubricated intermittently at a lower cost than by other methods, it is preferable to fit some kind of mechanical applicator.

One of the most frequently employed methods of lubricating gears is by enclosing them in an oil tight gear case and arranging for the teeth of one of the gears to dip into an oil bath; this method is suitable for peripheral speeds up to about 15 m/s (3 000 ft/min).

Where peripheral speeds are too high for bath lubrication, due to the heat generated by the gears and the consequential increase in temperature or power loss being prohibitive, lubrication by means of a pressure circulating oil system may be used. The oil is sprayed onto the gear teeth by means of a low pressure forced spray. The rate of spraying is usually dictated by the amount required to maintain a permissible temperature rise and to keep power losses to a minimum. It is suggested[29] that the rate of spraying be based upon an assumed temperature rise of not more than about 7 or 8 deg C (12–15 deg F). The oil may be cooled naturally or a cooler may be incorporated.

Finally, some mention should be made concerning the housing of the gears. The design of a suitable gear case is of obvious importance, and attention should be given to ensuring that it performs its function of providing adequate rigidity and support for the bearings and the associated shafts carrying the gears. Facilities should be provided for introducing the lubricant and cooling the oil if required, and suitable sealing arrangements to keep dirt and foreign particles out and prevent loss of lubricant. The actual design of the gear case will naturally be dependent upon the magnitude and

Fig. 6.41. Gear unit for rubber mixing plant (Reproduced by kind permission of R. W. Transmissions Limited)

Fig. 6.42. Speed increasing unit for refrigeration and air conditioning plant (Reproduced by kind permission of R. W. Transmissions Limited)

Fig. 6.43. Multiple gear drive (Reproduced by kind permission of R. W. Transmissions Limited)

Fig. 6.44. Speed increasing gear unit for a gas booster (Reproduced by kind permission of R. W. Transmissions Limited)

direction of the forces, and suitable ribbing or thickening up at critical sections may be necessary. If a single gear case is required then, depending upon the size, it may be more economical to use a welded construction rather than a casting. The reverse is likely to be the case for quantity production. Where weight is of no consequence, castings or fabrications in iron or steel will usually provide a suitabe choice of material, but for those special cases where weight is of some consequence, non-ferrous castings or fabrications may be advantagous.

Figs 6.41–6.43 illustrate some typical industrial gear units with the upper part of the gear casing removed and Fig. 6.44 an external view of a speed increasing unit for a gas booster.

6.19. CONCLUDING REMARKS

In this chapter an account has been given of the more common types of gears and their applications, with discussions relating to some of the more important factors which need to be considered.

For transmitting power between parallel shafts rotating in opposite directions, spur, single helical or double helical gears may be used, depending upon the speed, power transmitted and magnitude of the axial thrust. If it is necessary for the input and output shafts to rotate in the same direction, this may be achieved by a suitable arrangement incorporating an intermediate gear in the drive, or by the use of internal gears, as illustrated in Fig. 6.45.

For the transmission of power and motion between intersecting shafts, straight or spiral bevel gears are used, again depending upon the power requirements and speeds involved. If the shaft axes are to be off-set, and high speed, relatively silent operation, combined with high load carrying capacity are desired, the most suitable choice is the hypoid gear drive, as illustrated in Fig. 6.46; this type of gear is particularly suitable for automobile rear axle drives.

Crossed helical gears and worm gears essentially serve the same purpose, i.e. the connection of non-intersecting shafts. Crossed helical gears, however, are only suitable for situations requiring

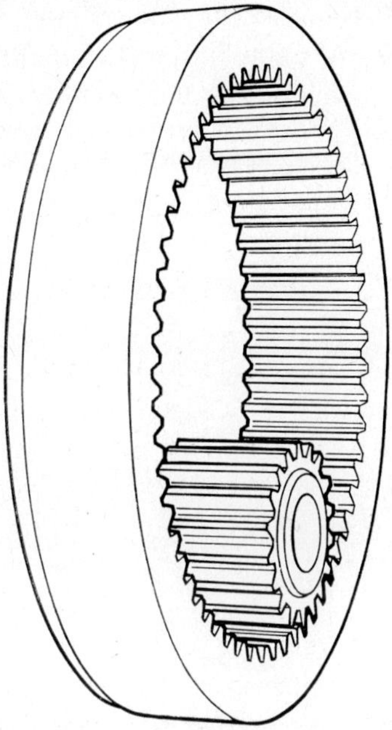

Fig. 6.45. *Internal gears (Kindly provided by David Brown Gear Industries Limited)*

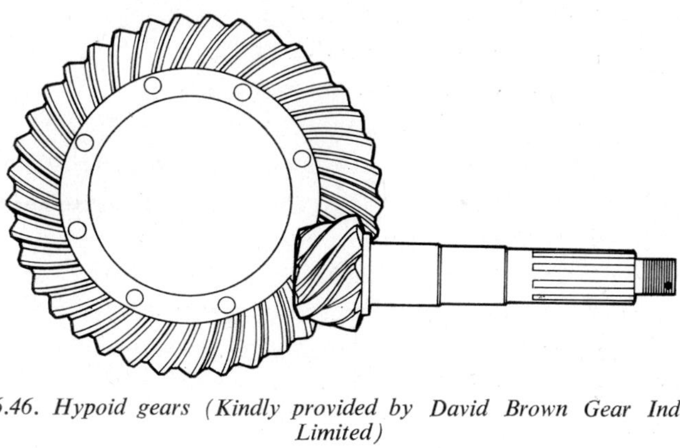

Fig. 6.46. *Hypoid gears (Kindly provided by David Brown Gear Industries Limited)*

light power transmission with relatively low reduction ratios, whereas worm gears are suitable for much higher ratings, and are used to provide a compact means of obtaining high gear ratios.

In this single chapter it has not been possible to study all the

Fig. 6.47. Back to back gear testing machine (Reproduced by kind permission of R. W. Transmissions Limited)

important aspects of gears and their associated problems, but a sufficient coverage has been attempted for the reader to be aware of the more important factors which need to be considered, and the list of references provide a comprehensive field of study for those requiring a more detailed account. At the present time, considerable research and experimental investigations are being undertaken in the study of gears, and Fig. 6.47 indicates a typical back to back test rig.

REFERENCES

1. ABBOTT, W., *Practical Geometry and Engineering Graphics*, Blackie and Son, Ltd., 7th edn. (1963)
2. DUGGAN, T. V., *Applied Engineering Design and Analysis*, Design Technology Series, Iliffe Books Ltd., London (1970)
3. JOHNSON, R. C., *Optimum Design of Mechanical Elements*, John Wiley and Sons. Inc., New York (1961)
4. MERRITT, H. E., *Gears*, Sir Isaac Pitman and Sons Ltd., London, 3rd edn. (1954)
5. TUPLIN, W. A., *Gear Design*, The Machinery Publishing Company Ltd., *London* (1962)
6. BUCKINGHAM, E., *Analytical Mechanics of Gears*, McGraw-Hill Book Company, New York (1949)
7. DUDLEY, D. W., *Practical Gear Design*, McGraw-Hill Book Company, New York (1954)
8. I. MECH. E., *International Conference on Gearing*, The Institution of Mechanical Engineers, London (1958)
9. I. MECH. E., *International Conference on Lubrication and Wear*, The Institution of Mechanical Engineers, London (1967)
10. I.S.O. RECOMMENDATION R. 701, *International Gear Notation—Symbols for Geometrical Data*, International Organisation for Standardization (1958) (Obtainable from the national standards organisations of countries which are members of I.S.O.)
11. LEWIS, W., *Investigation of the Strength of Gear Teeth*, Club of Philadelphia (1892)
12. BLACK, P. H. and ADAMS, O. E., *Machine Design*, McGraw-Hill Book Company, New York, 3rd edn. (1968)
13. BUCKINGHAM, E., *Dynamic Loads on Gear Teeth*, The Industrial Press, New York (1931)
14. SHIGLEY, J. E., *Machine Design*, McGraw-Hill Book Company, New York (1956)
15. B. S. 436: 1940, 'Machine Cut Gears—Helical and Straight Spur Gears', British Standards Institution, London (1940)
16. BUCKINGHAM, E., *Gear Tooth Loads*, Massachusetts Gear and Tool Company, Woburn, Mass. (1953)
17. PHELAN, R. M., *Fundamentals of Mechanical Design*, McGraw-Hill Book Company, New York, 2nd edn. (1962)
18. *Angus Gear Book*, George Angus and Co. Ltd., Gear Division, Hebburn-on-Tyne (1959) (Now R. W. Transmissions Ltd).
19. HIRSCHHORN, J., *Dynamics of Machinery*, Nelson, London (1967)
20. PARSONS, R. T., 'Numerical Data Tables', *Gear Handbook*, (Ed. Dudley, D. W.), McGraw-Hill Book Company, New York (1962)
21. DEAN, P. M., 'Gear Tooth Proportions', *Gear Handbook* (Ed. Dudley, D. W.), McGraw-Hill Book Company, New York (1962)
22. *Engineering Handbook*, Hy-Eng 1, Hyatt Bearing Division, General Motors Corporation, New Jersey (1951)
23. B.S. 545: 1949, 'Bevel Gears (Machine Cut)', British Standards Institution, London (1949)
24. B.S. 721: 1963, 'Worm Gearing', British Standards Institution, London (1963)

25. BUCKINGHAM, E. and RYFFEL, *Design of Worm and Spiral Gears*, The Industrial Press, New York,
26. *R. W. Transmissions:* 'Non-metallic Silent Gears', R.W. Transmissions Ltd., Hebburn, Co. Durham.
27. DUDLEY, D. W. (Ed.), *Gear Handbook*, McGraw-Hill Book Co., New York (1962)
28. HARTMAN, M. A., 'Gear Reactions and Mountings', *Gear Handbook* (Ed. Dudley, D. W.), McGraw-Hill Book Company, New York (1962)
29. *Shell:* 'The Lubrication of Industrial Gears', Shell International Petroleum Company Ltd., London (1964)

Chapter 7

Brakes and Clutches

Clutches and brakes are mechanical elements which are used to regulate or control the flow of mechanical power. Whereas a clutch connects one moving member to another moving member, a brake connects a moving member to a stationary part. They have in common the basic principle that, after engagement, the relative motion of the two elements is reduced, generally to zero. Thus, whilst a brake serves the purpose of arresting motion, a clutch transmits motion from a power source to a stationary component until the two are moving at the same speed.

The factors to be considered in the design of clutches and brakes are in many ways similar, but there are some differences, particularly with regard to structural details and the amount of heat which is generated. In this chapter we will look at some of the factors which need to be considered, but it is important to realise that a wide variety of units are commercially available as standard components, and apart from specialist designers, it is only very infrequently that the engineer will find it necessary to design clutches and brakes from first principles.

7.1. TYPES OF CLUTCHES AND BRAKES

Although simple positive clutches, such as illustrated in Fig. 7.1, may be used at low speeds where synchronous operation is required, the engagement is accompanied by shock, and frequent operation may lead to wear. Consequently, the more versatile and common types of clutches and brakes utilise frictional forces between the contacting surfaces for their successful operation. Whilst the positive clutches shown in Fig. 7.1 do not slip and do not generate heat, the

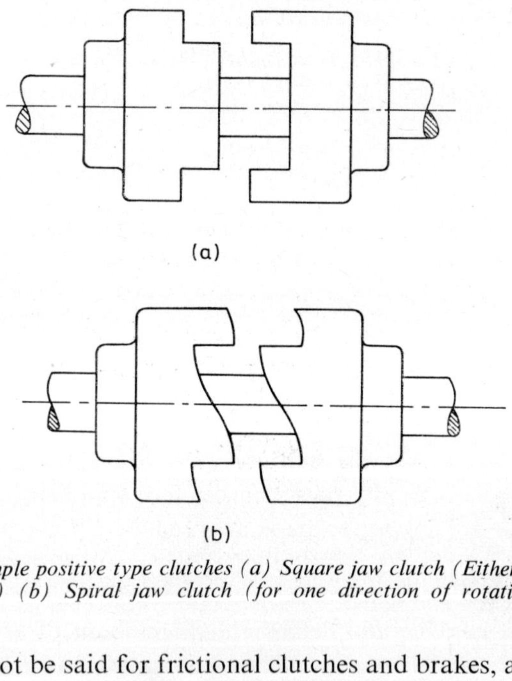

Fig. 7.1. Simple positive type clutches (a) Square jaw clutch (Either direction of rotation) (b) Spiral jaw clutch (for one direction of rotation only)

same cannot be said for frictional clutches and brakes, and considerable development and research has been carried out over the years to produce satisfactory frictional materials. The main factors to be considered in selecting a frictional material[1] are the basic friction value required, which determines the brake or clutch operating force; the fade resistance, i.e. the ability to maintain the friction value with increase in temperature; and the wear resistance of the lining.

Friction clutches and brakes consist essentially of three elements[2] namely, the mating frictional surfaces; the means of transmitting the torque to and from the surfaces; and the actuating mechanism. The arrangement of the frictional surfaces may be used as a simple method of classifying the type of brake or clutch, and in general they may be divided into three groups, namely:

1. Rim clutches and brakes,
2. Plate or Disk clutches and brakes,
3. Band clutches and brakes.

Where this method of classification is not sufficiently specific, the method of actuating the mechanism may also be specified, e.g. mechanical, magnetic, air operated, etc.

7.2 RIM CLUTCHES AND BRAKES

Clutches and brakes which are so designed that the contacting surfaces are on the inside or outside (sometimes both) of a rim, are referred to as rim clutches and brakes.

Fig. 7.2 illustrates a simple cone clutch in which engagement of the inner and outer cone is maintained by the force produced by

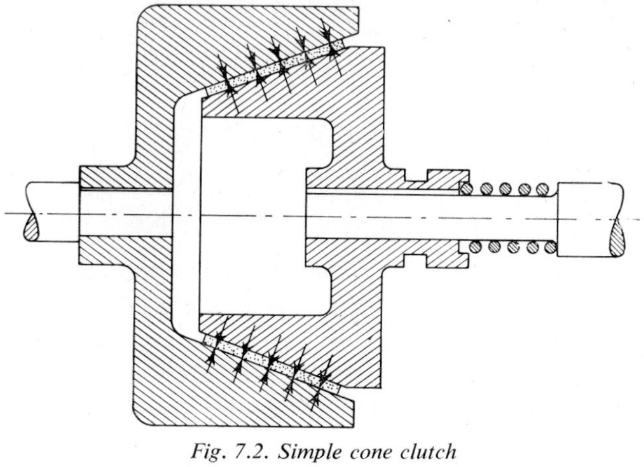

Fig. 7.2. Simple cone clutch

the compression spring. To disengage the clutch, a forked lever operates on a collar running in the groove, and is shifted so as to overcome the spring force. Such clutches are confined to small and medium power transmission in rugged machinery such as road rollers, concrete mixers, etc.

Rim clutches and brakes may be of either the external contracting or internal expanding type. An internal expanding rim clutch (or brake) specially designed for low speeds is illustrated in Fig. 7.3. Fig. 7.4 illustates an external contracting rim clutch (or brake)

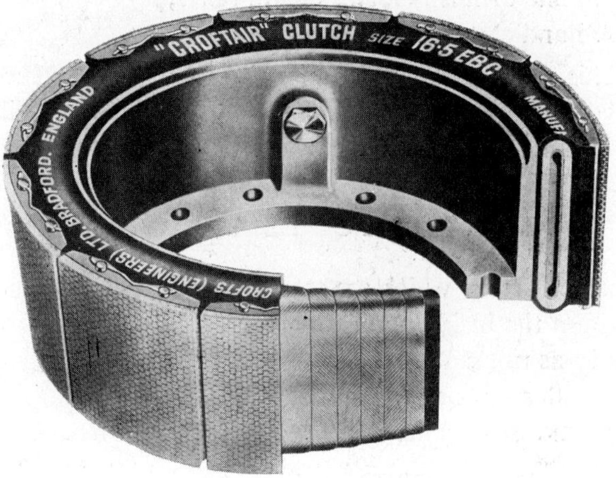

Fig. 7.3. 'Croftair' 'EBC' air operated flexible clutch (Manufactured by Crofts (Engineers) Ltd., Bradford)

specially designed for high speeds. Both types may be used for most normal clutch applications requiring frequent operation under light loads. These designs, which are manufactured by Crofts (Engineers) Ltd., consist of a ring-shaped cord reinforced rubber gland of patented design bonded to a steel housing. The rubber portion is prepared for the attachment of detachable friction slippers, the housing, gland and slippers forming the clutch element. To engage the clutch, compressed air is admitted to the rubber gland which expands and forces the specially designed friction slippers on to the friction rim; to disengage the clutch, air in the rubber gland is exhausted to the atmosphere. Air operated rim clutches of the external contracting type, designed for high speeds and frequent operation under heavy loads are also available. Complete details of these types of flexible clutches and brakes covering selection, capacity, etc., can be obtained by reference to the manufacturer's catalogue[3].

Fig. 7.4. 'Croftair' 'CBC' air operated flexible clutch (Manufactured by Crofts (Engineers) Ltd., Bradford)

A further type of rim clutch, used to obtain automatic and smooth engagement when the driving member is operated above a certain minimum speed, is the internal expanding centrifugal clutch illustrated in Fig. 7.5; this illustration shows the clutch shell with body and friction slippers assembled. In this instance, the actuating mechanism is due to centrifugal force overcoming the spring force which is present due to the provision of springs behind each slipper. Centrifugal clutches having blocks arranged to slide freely are also widely used.

Fig. 7.6 illustrates a rim friction clutch designed to meet present day requirements of speeds and arduous conditions. It is claimed[4] that, size for size, the power output of this clutch is greater than any other friction clutch of its type. The clutch body consists of a spider to which is fixed slides complete with inside and outside

Fig. 7.5. Centrifugal clutch (Manufactured by Crofts (Engineers) Ltd., Bradford)

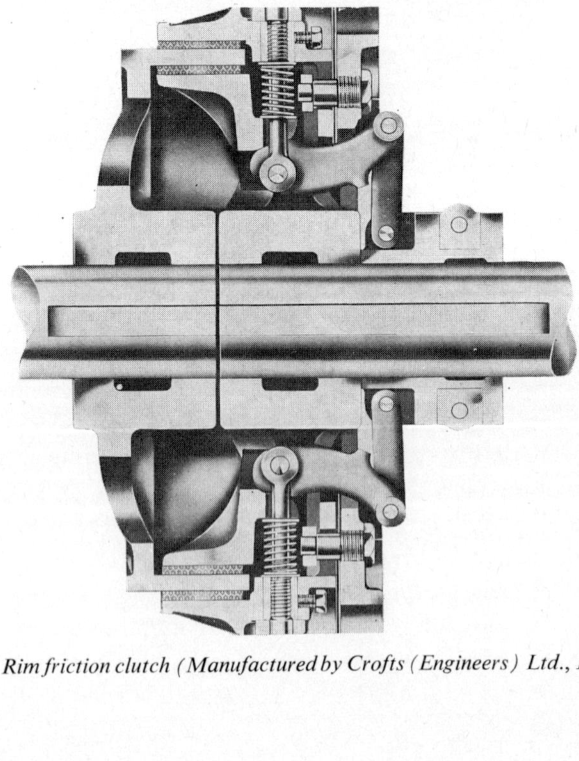

Fig. 7.6. Rim friction clutch (Manufactured by Crofts (Engineers) Ltd., Bradford)

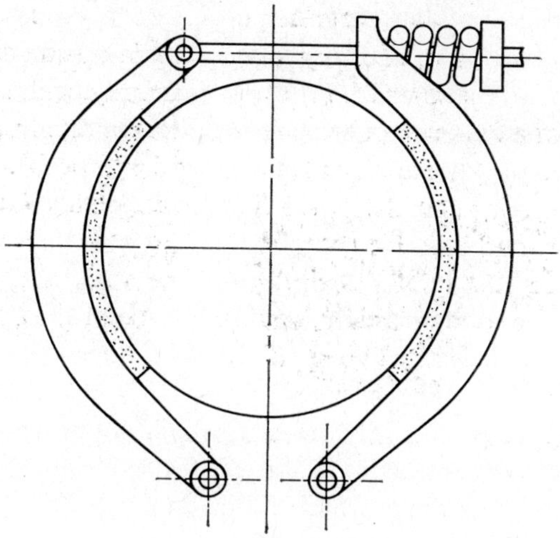

Fig. 7.7. External contracting two-shoe brake

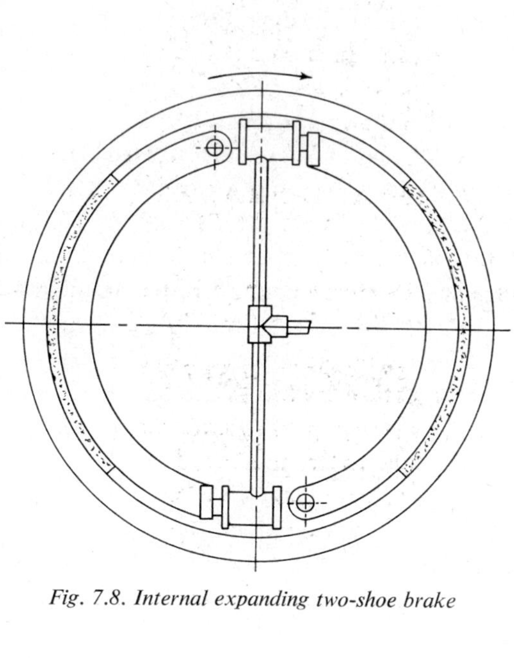

Fig. 7.8. Internal expanding two-shoe brake

slippers and their actuating mechanisms. The slippers are lined with bonded asbestos, and their action, when operated, is that of a number of vice jaws which grip simultaneously, with equal force, both sides of a cast iron or steel carrier rim. On disengagement the slippers move away from the rim, giving large positive clearance and disconnecting completely the rim and clutch body. For further details refer to Croft's Handbook[4].

Finally, in this brief description of the various types of rim clutches and brakes, mention should be made of the shoe type, of which a force analysis will be performed in the next section. Rigid and hinged shoe clutches and brakes normally consist of two shoes (this is not necessarily the case) which can be made to operate as an external contracting brake, i.e. the shoes are contracted on to the outside of a brake wheel, as illustrated in Fig. 7.7, or as an internal expanding brake or clutch, in which case the shoes are forced against the inner surface of a drum, as illustrated in Fig. 7.8. Individual shoes can be either leading or trailing, depending upon whether the frictional forces increase the effective force applied to the shoe, in which case it is, by definition, a leading shoe, or opposes the actuating force, in which case it is a trailing shoe.

7.3. FORCE ANALYSIS OF RIM CLUTCHES AND BRAKES

In carrying out a force analysis for clutches and brakes, it is necessary to assume or determine a pressure distribution for the frictional forces, and to establish a relationship between the maximum pressure and the pressure at any point. Once this has been done, the conditions for static equilibrium can be applied to calculate the actuating force and the support reactions, and the brake or clutch torque capacity can be determined.

The two types of shoe clutch or brake are the rigid shoe, i.e. where the shoe is an integral part of the post, or the hinged shoe, i.e. where the shoe is hinged or pivoted to the post. The hinged shoe brake or clutch is more complicated to analyse than the rigid shoe type, but has the advantage that by careful design it is possible to obtain uniform pressure distribution along the shoe length, with the consequence that wear is also more evenly distributed.

In this section a force analysis will be performed for external contracting and internal expanding rigid shoe rim clutches and brakes. A graphical construction for determining the position of maximum pressure and the pressure distribution has been proposed by Acres[5], and a force analysis for hinged shoe brakes has been performed by Hirschhorn[6]; both these methods are discussed in Ferodo's Handbook[1].

Fig. 7.9 indicates the notation for an external contracting rigid shoe brake. The shoe is rigidly attached to the actuating mechanism or lever arm, and this is pivoted at P. An actuating force is

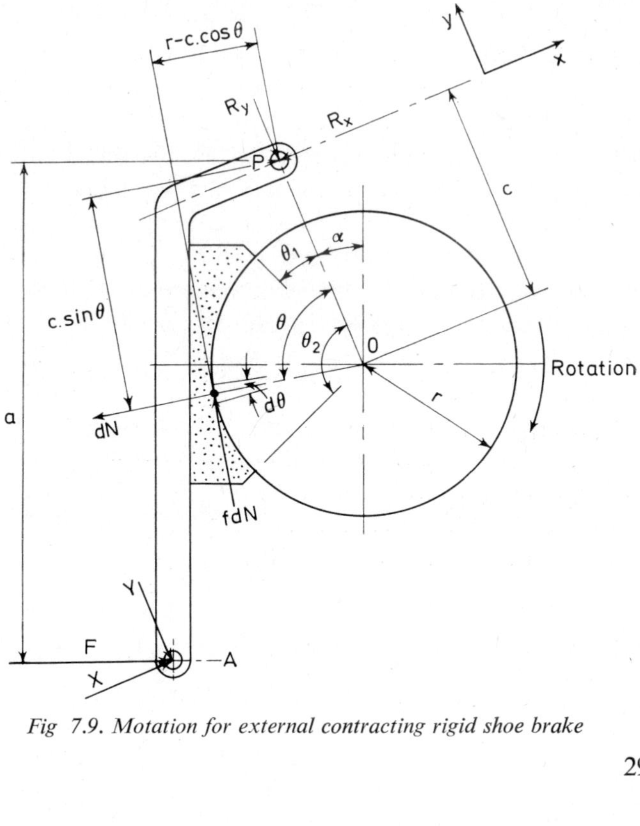

Fig 7.9. Motation for external contracting rigid shoe brake

applied at the other end of the arm A. The friction surface has an angle of contact $(\theta_2 - \theta_1)$, where the angle θ is measured from the datum *OP*, i.e. a line passing through the centre of rotation of the brake drum *O* and the pivot point *P*. In particular, consider an elemental arc of the frictional surface $d\theta$, at some angle θ measured from the line *OP*, and let the normal or radial pressure acting on this element be p. The normal and tangential forces acting on the element of the brake shoe are, respectively:

$$dN = prb \, d\theta \tag{7.1}$$

and

$$f \, dN = fprb \, d\theta \tag{7.2}$$

where f = coefficient of friction
b = width of friction surface in contact with the brake drum (i.e. perpendicular to the plane of the paper).

It now becomes necessary to consider the pressure distribution along the shoe, and Hirschhorn[7] suggests that the pressure between a brake drum and a given point on the lining is proportional to the radial component of the wear displacement of that point. Since frictional wear is usually assumed to be proportional to the product of pressure and velocity, and since the velocity will be the same for all points on the shoe, the suggestion that the wear of the lining will be proportional to pressure is obviously quite reasonable. This is the fundamental law of pressure distribution and assumes that wear is confined to the lining; that the coefficient of friction is independent of pressure and is not affected by temperature rise, and that the brake elements are perfectly rigid. Thus, expressed mathematically,

$$p = C\delta \tag{7.3}$$

where δ = radial wear,
C = constant of proportionality.

For a small angle of rotation of the lever, the radial wear displacement δ is proportional to $\sin\theta$, and the application of the fundamental law of pressure distribution therefore results in a sinusoidal distribution of pressure[8, 9] with respect to the datum *OP*, i.e.

$$p = C \sin\theta \tag{7.4}$$

The maximum pressure, say p_{max}, occurs when $\sin\theta$ is a maximum, i.e. when $\theta = \theta_m$ (say). Therefore,

$$p_{max} = C \sin\theta_m \tag{7.5}$$

Combining equations 7.4 and 7.5,

$$p = p_{max}\frac{\sin\theta}{\sin\theta_m} \tag{7.6}$$

Perhaps it should be mentioned that the above relationship for pressure distribution is valid for both long and short shoes, but if the shoe is short it may be reasonable to assume that the pressure is uniformly distributed.

Having established the relationship for pressure distribution along the shoe, the conditions for static equilibrium may be applied to determine the actuating force F, the brake torque M_t and the support reactions R_x and R_y. Thus,

$$\left.\begin{aligned} \Sigma M_P &= 0 \\ \Sigma F_x &= 0 \\ \Sigma F_y &= 0 \end{aligned}\right\} \tag{7.7}$$

Applying the first of equations 7.7, i.e. taking moments about the pivot P,

$$Fa = \int_{\theta_1}^{\theta_2} dN\, c \sin\theta + \int_{\theta_1}^{\theta_2} f\,dN\,(r - c\cos\theta) \tag{7.8}$$

The first term on the right hand side of equation 7.8 is the moment due to the normal pressure, say M_n, and this will always act in a direction which opposes the actuating force. However, the second term, which is the moment effect due to the frictional force, say M_f, may act either in the same direction or in the opposite direction to that of the actuating force, depending upon the direction of rotation of the brake drum and the geometry of the actuating mechanism. When the frictional forces are such that they act in the same direction as the actuating force, i.e. they assist in applying the brake, the brake is said to be *self-energising*. In the limit, if the moment due to the frictional forces became equal to the normal forces (a situation not very likely to arise for the external contracting rigid shoe brake), the degree of self-energisation would become so great that self-locking of the brake would occur when the shoe made contact with the brake drum, without any additional actuating force.

Returning now to equation 7.8 and substituting for dN from equation 7.1 and p from equation 7.6 yields

$$Fa = \frac{p_{\max} rbc}{\sin \theta_m} \int_{\theta_1}^{\theta_2} \sin^2 \theta \, d\theta + \frac{f p_{\max} rb}{\sin \theta_m} \left[r \int_{\theta_1}^{\theta_2} \sin \theta \, d\theta - c \int_{\theta_1}^{\theta_2} \sin \theta \cos \theta \, d\theta \right]$$

Integrating and transposing gives the actuating force as

$$F = \frac{p_{\max} rbc}{2a \sin \theta_m} \left[(\theta_2 - \theta_1) - \frac{1}{2} (\sin 2\theta_2 - \sin 2\theta_1) + \frac{2rf}{c} (\cos \theta_1 - \cos \theta_2) + \frac{f}{2} (\cos 2\theta_2 - \cos 2\theta_1) \right] \quad (7.9)$$

Thus, for a given geometry and frictional material (which means that $p_{\max}$ and f are known or can be assessed), the actuating force F can be determined. Of course, in a design situation it is the fixing of the geometry and the selection of suitable materials which require to be determined, and this will necessitate the making of decisions[10]; such decisions will invariably be based upon experience obtained on the success of previous designs.

For the brake to function satisfactorily, its braking capacity must be capable of absorbing the necessary torque, and the torque capacity can be easily obtained from the equation

$$M_t = \int_{\theta_1}^{\theta_2} f \, dN \, r = \frac{f b r^2 p_{\max}}{\sin \theta_m} (\cos \theta_1 - \cos \theta_2) \quad (7.10)$$

Finally, the reactions at the support can be obtained from equation 7.7 by resolving forces in two mutually perpendicular directions, say x and y, as indicated in Fig. 7.9. Thus,

$$R_x = \int_{\theta_1}^{\theta_2} f\,\mathrm{d}N \cos\theta - \int_{\theta_1}^{\theta_2} \mathrm{d}N \sin\theta + X \qquad (7.11)$$

and

$$R_y = \int_{\theta_1}^{\theta_2} f\,\mathrm{d}N \sin\theta + \int_{\theta_1}^{\theta_2} \mathrm{d}N \cos\theta - Y \qquad (7.12)$$

It is an easy matter to integrate equations 7.11 and 7.12, and this is left to the reader.

In order to increase the rating of the brake and at the same time reduce the bearing loads and consequential bending moments on the shaft carrying the brake drum, multiple shoes, more particularly two shoes are frequently employed. The linkage designs vary considerably, and the designer must take care to determine the magnitude and directions of the forces and moments acting.

A similar procedure to that presented above can be applied to the internal expanding rigid shoe rim clutch or brake, such as that

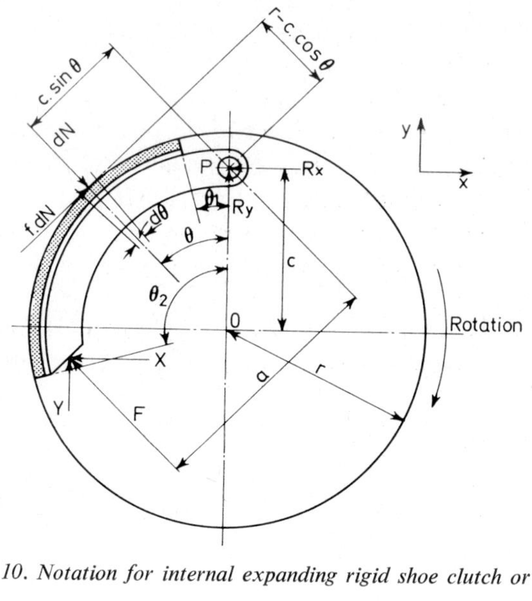

Fig. 7.10. Notation for internal expanding rigid shoe clutch or brake

indicated in Fig. 7.10. Taking moments about the pivot P,

$$Fa = \int_{\theta_1}^{\theta_2} \mathrm{d}N\, c \sin\theta - \int_{\theta_1}^{\theta_2} f\, \mathrm{d}N\, (r - c\cos\theta) \tag{7.13}$$

The first term on the right hand side of equation 7.13 represents the moment due to the normal forces and the second term the moment due to the frictional forces. Again, note that reversing the direction of rotation has the effect of reversing the direction of the frictional forces, and when these frictional forces act in the same direction as the actuating force a condition of self energisation of the clutch or brake exists. Self locking is much more readily achieved with internal expanding shoe clutches and brakes than it is with the external contracting shoe brake discussed previously.

Substituting for $\mathrm{d}N$ from equation 7.1 and p from equation 7.6 into equation 7.13, integrating and transposing gives the actuating force as

$$F = \frac{p_{\max} rbc}{2a \sin\theta_m}\Bigg[(\theta_2 - \theta_1) - \frac{1}{2}(\sin 2\theta_2 - \sin 2\theta_1)$$
$$+ \frac{2rf}{c}(\cos\theta_2 - \cos\theta_1) + \frac{f}{2}(\cos 2\theta_1 - \cos 2\theta_2)\Bigg] \tag{7.14}$$

which should be verified by the reader.

The torque transmission capability of the brake or clutch is identical with that presented for an external contracting shoe brake, and is given by equation 7.10. Similarly, the reactions at the hinge pin can be easily obtained from the condition that the summation of the horizontal and vertical forces must each be equal to zero.

7.4. DISK CLUTCHES AND BRAKES

In contrast to rim clutches and brakes previously discussed, in which the frictional surfaces are brought into contact by radial movement, disk clutches and brakes are designed so that the frictional surfaces are brought into contact by axial movement, and the rims are replaced by plates or disks.

Disk brakes and clutches are more compact than rim brakes and clutches, and generally remain free from fade at high temperature, this latter factor being attributable to the fact the friction material is flat rather than curved, which permits a wider range of organic bonds to be used in their manufacture[1]. Further, disk clutches and brakes can be designed to give more favourable pressure distribution and more effective heat dissipation surfaces.

Fig. 7.11 and 7.12 illustrate general purpose single and double disk clutches respectively. A similar multiple disk clutch unit ideally suited for machine tools and special machines, particularly

Fig. 7.11. Patent 'Ro' roller operated friction clutch with single friction disk and adaptor shell (Manufactured by Crofts (Engineers) Ltd., Bradford)

Fig. 7.12. Patent 'Ro' roller operated friction clutch with two friction disks and Adaptor shell (Manufactured by Crofts (Engineers) Ltd., Bradford)

where space is limited, is illustrated in Fig. 7.13. These clutches are operated by the smooth movement of precision spheroidal rollers on accurately machined inclined surfaces giving line contact with uniform distribution of the driving load, thus ensuring long life. They can be operated at low or high speeds in either direction of rotation, and are unaffected by centrifugal forces. It is claimed[11] that these patent roller operated disk friction clutches, manufactured by Crofts (Engineers) Ltd., England, are the most advanced clutch of their kind in the world. Fig. 7.14 indicates another multiple

Fig. 7.13. Patent 'RM' roller operated multiple disk clutch mechanism with Adaptor shell (Manufactured by Crofts (Engineers) Ltd., Bradford)

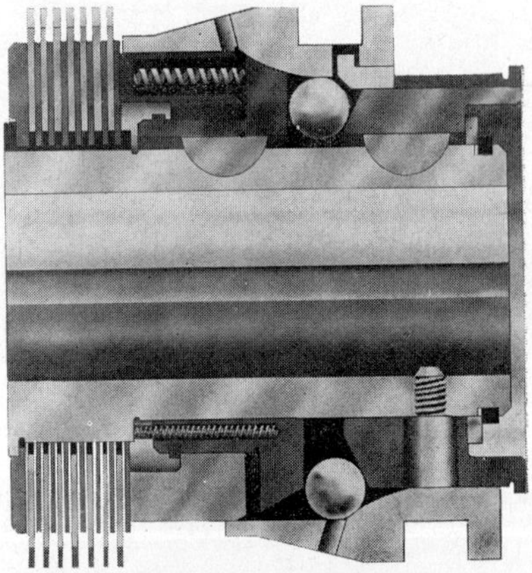

Fig. 7.14. Section of 'BOM-L' multiple disk clutch (Manufactured by Crofts Engineers) Ltd., Bradford)

disk clutch, the upper half section indicating the clutch in the engaged position and the lower half section in the disengaged position[12].

The basic principles involved in disk clutches can be indicated by referring to the typical single plate clutch illustrated in Fig. 7.15; this is similar to the type used in automobiles. It operates in the engaged position, in which it is held by forces exerted due to pressure springs, and to disengage the clutch it is necessary to apply an external actuating force; this external force is applied by depressing the clutch pedal. The release bearing (sometimes referred to as the throw-out bearing), which is coupled to the pressure plate by a series of toggles, then pulls the pressure plate against the pressure springs. There are many variations on this basic type.

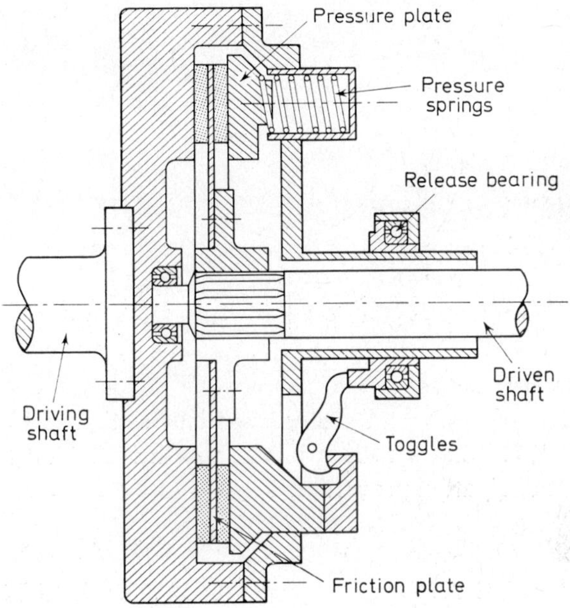

Fig. 7.15. Typical single plate clutch

Disk brakes may be either of the plate type, in which case they are similar to plate clutches, or of the spot type. Generally, they are of the spot type, i.e. they utilise pads rather than continuous friction plates, these pads generally covering about 45 degrees of the disk, leaving the remainder exposed for cooling purposes. Fig. 7.16 illustrates a typical arrangement of diametrically opposite pairs of pads placed one on either rubbing surface of the disc, thus reducing bearing loads to a minimum. These pads are brought into contact with the disk by means of a caliper action which provides the pressure necessary to slow down or stop the disk. Disk brake developments and their characteristics have been discussed by Kinchin[13] and Parker and Newcomb[14], to which the reader is referred for detailed discussions.

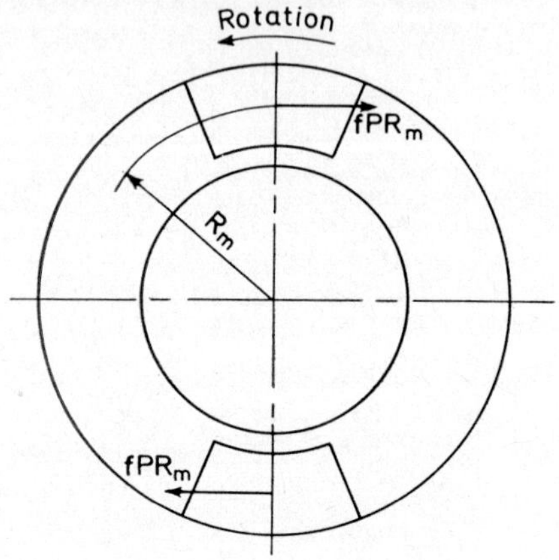

Fig. 7.16. Spot type disc brake

The plate type of disk brake is essentially similar to that of the disk clutch, and has the disadvantage that it is difficult to obtain good heat dissipation properties.

7.5. FORCE ANALYSIS FOR DISK CLUTCHES AND BRAKES

Referring to the spot type disk brake shown in Fig. 7.16, if each pad is pressed against the disk face with a force P, then the braking torque per pair of pads is given by

$$M_t = 2fPR_m \tag{7.15}$$

where f = coefficient of friction,
R_m = mean radius of the pads.

Koffman[15] has discussed the radius at which the brake force should be applied to give uniform wear of the pad, but it is suggested[1] that, provided the ratio of the external to internal radius does not exceed 2, for all practical purposes the radius at which the resultant friction force acts may be taken as the mean radius, R_m. It is further suggested that good practice is to ensure that the ratio of the external to internal radius is made equal to about 1.5.

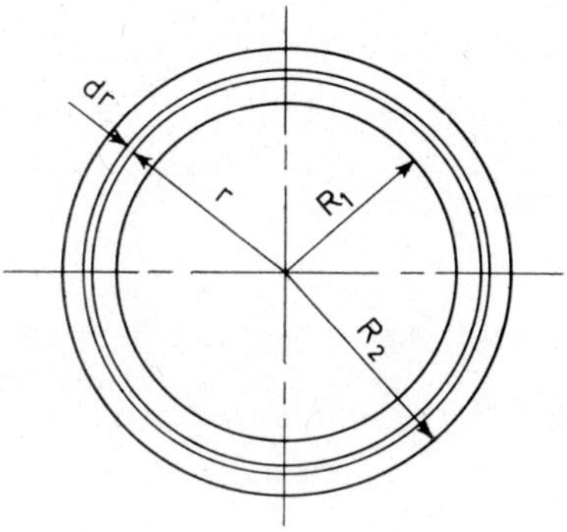

Fig. 7.17. Single plate clutch or brake facing

The area of the pads should be such that the average pressure intensity is kept to within reasonable limits, i.e.

$$p = \frac{P}{A} \tag{7.16}$$

where p = permissible average pressure intensity for the material,
A = contact area of the friction pads.

The torque capacity of a single plate brake or clutch can be analysed by referring to Fig. 7.17, and considering an elemental ring at some radius r and having radial thickness dr. If the pressure intensity acting on the elemental ring is p, then the total axial load P will be given by

$$P = 2\pi \int_{R_1}^{R_2} pr \, \mathrm{d}r \tag{7.17}$$

and the total torque capacity of the disk will be

$$M_t = 2\pi \int_{R_1}^{R_2} fpr^2 \, \mathrm{d}r \tag{7.18}$$

Both the pressure intensity p and the coefficient of friction f are likely to vary with respect to the surface velocity, and hence the radius, and it is usual to relate this variation with the rate of wear. Although the exact nature of such a relationship is not known, it is commonly assumed that wear is proportional to the product of pressure and velocity. Assuming uniform wear and a constant value for the coefficient of friction f, integration of equations 7.17 and 7.18 give the result

$$M_t = \frac{fP}{2}(R_1 + R_2) \tag{7.19}$$

If the pressure intensity is assumed uniformly distributed, integration of equations 7.17 and 7.18 lead to the result

$$M_t = \frac{2fP}{3}\left[\frac{R_2^3 - R_1^3}{R_2^2 - R_1^2}\right] \tag{7.20}$$

It is suggested that the ratio of R_2/R_1 should not exceed 2, and that good practice is to make this ratio approximately equal to 1.5.

7.6. BAND BRAKES

Band brakes operate by the action of a band wrapped around a rotating member, the braking action being achieved by tightening the band and consequently introducing friction which produces the tangential braking force, Fig. 7.18. Such brakes are generally of relatively simple construction, and are used extensively in machinery of all kinds, including large and rugged installations. The band itself may be made of steel, leather, canvas, or appropriate frictional materials. A common method of production is to use a flexible steel band with suitable friction linings bonded onto the inside surface. Since friction linings tend to reduce the flexibility of the steel band, it is good practice, when rigid moulded types of linings are to be used, to mount them in short segments with small gaps between them, thus helping to preserve the flexibility of the steel band[1].

In general, band brakes are capable of higher torque outputs for given operating forces, take up less space and are cheaper to produce than shoe brakes. The choice of a suitable band brake is usually a

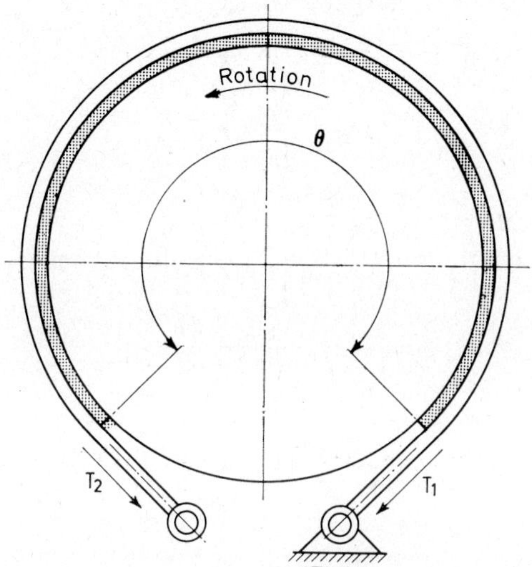

Fig. 7.18. Simple band brake—positive servo action

compromise between the degree of servo effect and the brake stability.

7.7. FORCE ANALYSIS FOR BAND BRAKES

For a simple band brake, such as that shown in Fig. 7.18, one end of the band is fixed and the operating force is applied at the other end. The similarity with the action of a flat belt, discussed in Chapter 5, will be evident, and neglecting the centrifugal force effect (which is obviously non-existent in the case of a band brake) we may write

$$\frac{T_1}{T_2} = \exp(f\theta) \tag{7.21}$$

where T_1 = tension on the tight side of the band,
T_2 = tension on the slack side of the band,
f = coefficient of friction,
θ = angle of contact.

The torque on the brake wheel is calculated from the fundamental equation

$$M_t = (T_1 - T_2)\frac{D}{2} \tag{7.22}$$

where D = diameter of the brake wheel.
Combining equations 7.21 and 7.22, we obtain the results

$$T_1 = \frac{2M_t}{D}\left[\frac{\exp(f\theta)}{\exp(f\theta)-1}\right] \tag{7.23}$$

and

$$T_2 = \frac{2M_t}{D}\left[\frac{1}{\exp(f\theta)-1}\right] \tag{7.24}$$

If the fixed end of the band is made the tight side, the force required at the operating end will be less, and this is of obvious advantage; such a condition is termed a positive servo action. For the case where the greater tension is at the operating end, a condition which would be obtained by reversing the direction of rotation in Fig. 7.18, a negative servo action is said to be obtained.

It is generally desirable to operate a band brake with positive servo action, but for the situation where reversible rotation is a requirement, the so-called reversible or equal acting band brake may be utilised. Such a band brake is indicated in Fig. 7.19, which shows the two ends of the band connected to the lever of the control mechanism, so that they are equi-distant on either side of the pivot. This arrangement permits the same torque for either direction of rotation, and equations 7.23 and 7.24 may be used to determine the tensions. The actuating force P, acting as indicated, may be determined by taking moments about the pivot, i.e.

$$Pl = (T_1 + T_2)\,a \tag{7.25}$$

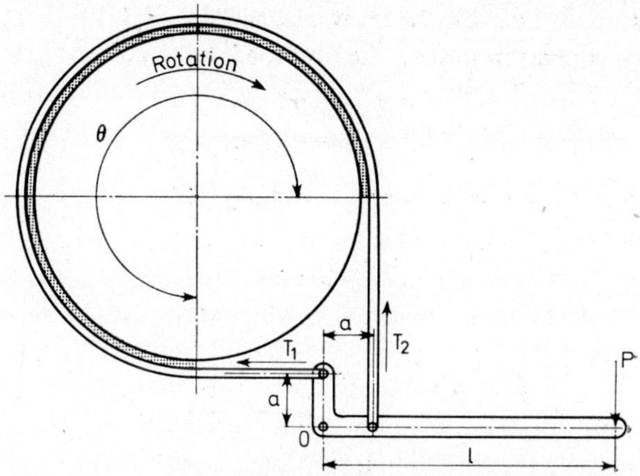

Fig. 7.19. Reversible or equal acting band brake

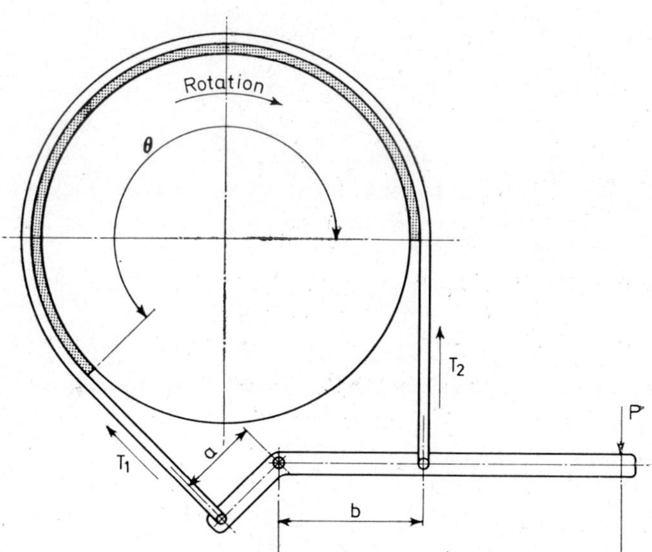

Fig. 7.20. Differential band brake

Substituting for T_1 and T_2 from equations 7.23 and 7.24 respectively, and transposing, gives the actuating force as

$$P = \frac{2aM_t}{Dl}\left[\frac{\exp(f\theta)+1}{\exp(f\theta)-1}\right] \tag{7.26}$$

A variation of the reversible or equal acting brake is the so-called differential band brake. In this arrangement, Fig. 7.20, each end of the band is connected to the lever mechanism, but the moment arm of the lever subjected to the greater tension is smaller than that of the moment arm of the lever subjected to the smaller tension. This type of brake has an extremely high servo action, making it unsuitable for any but static applications. The analysis for determining the actuating force may be performed in a similar manner to that demonstrated above for the equal acting band brake.

7.8. ENERGY DISSIPATION IN CLUTCHES AND BRAKES

In arresting motion in braking, or in bringing a machine up to speed in operating a clutch, the kinetic energy expended is converted into heat energy at the drum or disk and lining interface. This heat energy generated will be absorbed by the mass of the disk or drum, and if not dissipated efficiently, it can cause excessive temperature rise at the interface, which in turn may be responsible for a reduction in the coefficient of friction, high thermal stresses, breakdown of the lining bond, and rapid lining wear. Consequently, it is of obvious importance to give some consideration to this aspect, otherwise failure of the brake or clutch may occur due to the excessive temperature rise. The problem is similar in many respects to that discussed in connection with the thermal equilibrium of journal bearings (Chapter 4).

The friction energy required in a single braking or clutching operation in decreasing or increasing the speed of the system, may be computed by considering the change in the various energies required. In general, this may consist of a change in potential energy, a change in kinetic energy of rotation, and a change in the kinetic energy of translation. Thus, if a body of mass m_1 is subjected to a change in elevation, say $(h_2 - h_1)$, then the change in potential energy is

$$E_p = m_1 g(h_2 - h_1) \tag{7.27}$$

where g = gravitational constant.

Similarly, the change in kinetic energy of rotation of all the rotating parts in the system due to a change in the angular speed $(\omega_2 - \omega_1)$, is obtained from

$$E_r = \tfrac{1}{2} I(\omega_2 - \omega_1)^2 \tag{7.28}$$

where I = mass moment of inertia of all the rotating parts
$= m_2 k^2$,
m_2 = mass of the rotating parts,
k = radius of gyration of the rotating parts,
ω = angular velocity, rad/s,
$= 2\pi N/60$,
N = angular velocity or speed of rotation, rev/min.

So that

$$E_r = \frac{I\pi^2(N_2 - N_1)^2}{1\ 800}$$

$$= \frac{m_2 k^2 \pi^2 (N_2 - N_1)^2}{1\ 800} \tag{7.29}$$

For translatory motion, where the velocity of the system (or vehicle) is changed by an amount $(v_2 - v_1)$, the change in kinetic energy of translation for a mass m_3 is given by

$$E_t = \tfrac{1}{2} m_3 (v_2 - v_1)^2 \tag{7.30}$$

If SI units are used, the mass will be in kilogrammes (kg), the elevation in metres (m), the velocity in metres per second (m/s), and equations 7.27–7.30 will have the units of Newton metres (Nm) or Joules (J). If conventional English units are adopted, the energies calculated by equations 7.27–7.30 will be in ft lbf, and these units may be converted into heat units (Btu) by dividing by Joules equivalent, J, where J = 778 ft lbf/Btu.

Thus, the total frictional energy required for a single operation, in general, is given by

$$H_1 = E_p + E_r + E_t \tag{7.31}$$

The heat thus generated will cause the temperature at the interface to rise, as mentioned earlier, and this temperature rise depends primarily on the mass of the disk or drum absorbing the heat, and also upon the heat dissipation capacity of the clutch or brake Determining the temperature rise is consequently quite complex, involving heat transfer calculations and necessitating a knowledge of the heat transfer coefficients.

If it is assumed that all the frictional energy generated is absorbed by the metal drum or disk, then the temperature rise may be calculated by equating the heat generated to the heat absorbed, i.e.

$$H_1 = m_d C(\theta_2 - \theta_1) \tag{7.32}$$

where H_1 = heat generated in friction, J (or Btu),
m_d = mass of disk or drum absorbing heat, kg (or lbf s^2/ft), (lb),
C = thermal capacity per unit mass or specific heat, J/kg deg C (Btu/lb deg F),
θ_1 = initial temperature of disk or drum, °C (or °F),
θ_2 = final temperature of disk or drum, °C (or °F).

The value for the thermal capacity (specific heat) per unit mass may vary considerably for different applications and conditions, being dependent upon the metals used and operating temperatures attained. For cast iron and steels, an average value in the region of 630 J/kg deg C (0·15 Btu/lb deg F) is frequently used, and such a value is useful as an approximation.

Although the above simple approach may be reasonably satisfactory as a first approximation in determining the temperature rise for a single braking or clutch operation, it may be in serious error for the condition of repeated braking or clutch operation, where the operating frequency may be so rapid that some heat is retained following each operation. Consequently, there will be a build-up of heat until an equilibrium condition is attained, such a condition being very much dependent upon the heat dissipative properties of the brake or clutch. Under these conditions, an approximate method proposed by Gagne[16] may be found useful, and in equation form this may be written

$$K(\theta_e - \theta_1)\left[\frac{nt}{3\,600} + 1{\cdot}5\left(1 - \frac{nt}{3\,600}\right)\right] = \frac{H_1 n}{A_s} \qquad (7.33)$$

where K = coefficient of heat transfer, W/m² deg C (Btu/ft² h deg F),

θ_e = average equilibrium temperature, °C (°F),

θ_1 = initial or ambient temperature of disk or drum, °C (°F),

n = number of braking or clutching operations per hour,

t = time of single braking or clutching operation, s,

H_1 = Heat generated in friction during single operation, J (Btu),

A_s = exterior surface area dissipating heat, m² (ft²).

The coefficient of heat transfer, K, includes the effects of both conduction and radiation, and since this is a function of temperature, it will also be a function of rubbing velocity, the metals used, environment, and so on. The value for K may vary from about 8·5–29 W/m² deg C (1·5–5·1 Btu/ft² h deg F) depending upon the particular conditions which apply. For general design purposes, an average value in the region of about 15 W/m² deg C (2·6 Btu/ft² h deg F) may be used in order to assess whether the temperature rise is likely to be of major importance. It must be emphasised that the above approach is only approximate, and the exact performance of brakes and disks can only be determined by actual test.

Somewhat more sophisticated methods of determining the temperature rise in brakes and clutches have been developed by various workers, and the contributions made by Newcomb[1, 17] are of special importance. For design purposes, probably the best reference is *Friction Materials for Engineers*[1], which presents an excellent and most comprehensive discussion on all aspects associated with the design of clutches and brakes.

7.9. CONCLUDING REMARKS

It will be evident from the discussions in this Chapter, that associated with the design of clutches and brakes are certain factors which it is not always possible to assess with great reliability. Of these factors the coefficient of friction, maximum contact pressures and operating temperatures come to mind. These factors are themselves dependent upon each other and the materials under consideration, and great care must be exercised in putting numerical values to these factors. It would probably be unwise to quote values in the brief discussion given in this chapter, and the reader is recommended to consult the references appended. However, sufficient coverage has been given to make the reader aware of the problems associated with the design and selection of clutches and brakes, and of the necessity for continued testing and development.

REFERENCES

1. FERODO: 'Firctional Materials for Engineers', Ferodo Limited, Chapel-en-le-Frith, Stockport, 2nd edn. (1968)
2. SHIGLEY, J. E., *Machine Design*, McGraw-Hill Book Company, Inc., New York (1956)
3. CROFTS: 'Croftair Flexible Clutches and Brakes, (Publication 364)', Crofts (Engineers) Ltd., Bradford, England.
4. CROFTS: 'Rim Friction Clutches (Publication 154-B)', Crofts (Engineers) Ltd., Bradford, England.
5. ACRES, F. A. S., 'Some Problems of the Design of Braking Systems', *Proc. Instn. Auto. Engrs.*, **41,** 19 (1946)
6. HIRSCHHORN, J., 'An Analysis of the Forces and Pressure Distribution in Brake Shoes', *Engineering*, **173,** 418 (1952)
7. HIRSCHHORN, J., *Dynamics of Machinery*, Nelson, London (1967)
8. SPOTTS, M. F., *Design of Machine Elements*, Prentice-Hall, Inc., Englewood Cliffs, N. J., 3rd edn. (1961)
9. SLAYMAKER, R. R., *Mechanical Design and Analysis*, John Wiley and Sons, Inc., New York (1959)
10. DUGGAN, T. V., *Applied Engineering Design and Analysis*, Design Technology Series, Iliffe Books Ltd., London (1970)
11. CROFTS: 'Patent Roller Operated Friction Clutches, (Publication 5918 B)', Crofts (Engineers) Ltd., Bradford, England.
12. CROFTS: 'Bom-L Multiple Disc Clutches, (Publication 855 B)', Crofts (Engineers) Ltd., Bradford, England.
13. KINCHIN, J. W. 'Disc Brake Development and Anti-skid Braking Devices', *Soc. Automot. Engrs.* (1961)

14. PARKER, R. C. and NEWCOMB, T. P., 'The Performance and Characteristics of the Disc Brake', *S.A.E. National Automobile Week*, Detroit, Michigan, April (1964)
15. KOFFMAN, J. L., 'Ventilated Disc Brakes', *Auto. Engng.*, **46,** 277 (1956)
16. GAGNE, A. F., 'Torque Capacity and Design of Cone and Disc Clutches', *Machine Design*, **24,** no. 12 (1953)
17. NEWCOMB, T. P., 'Heat Dissipation in Braking', *Auto. Design Engng.*, **2,** no. 8 (1963)

Appendix A

Keys, Splines Serrations

The tables included in this Appendix are reproduced by kind permission of the British Standards Institution, from whom complete copies of the Standards may be obtained. It is important to note that these standards are revised from time to time, and care should be taken to ensure that the latest standard is being used.

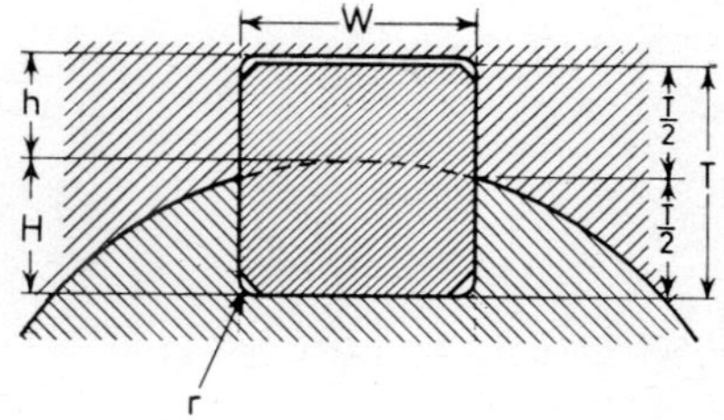

Table A. 1. SQUARE PARALLEL KEYS AND KEYWAYS

(Reproduced from B.S. 46 by kind permission of the British Standard Institution)

All dimensions are in inches

1	2	3	4	5	6	7	8	9	10	11	12	13	14	15	16
Diameter of Shaft		*Key size*	*Width W and Thickness T*		*Keyway in shaft*				*Keyway in hub*				*Nominal Keyway radius & key chamfer r**	*Bright keybar*	
Over	*Up to and including*				*Width W_s*		*Depth H*		*Width W_h*		*Depth h*			*Width W and thickness T*	
		W×T	*Max.*	*Min.*	*Min.*	*Max.*	*Min.*	*Max.*	*Min.*	*Max.*	*Min.*	*Max.*		*Max.*	*Min.*
1/4	1/2	1/8 × 1/8	0·127	0·125	0·124	0·125	0·072	0·078	0·125	0·126	0·060	0·066	0·010	0·127	0·125
1/2	3/4	3/16 × 3/16	0·190	0·188	0·187	0·188	0·107	0·113	0·188	0·189	0·088	0·094	0·010	0·190	0·188
3/4	1	1/4 × 1/4	0·252	0·250	0·249	0·250	0·142	0·148	0·250	0·251	0·115	0·121	0·010	0·252	0·250

(continued over page)

Table A. 1. *(continued)*

1	*2*	*3*	*4*	*5*	*6*	*7*	*8*	*9*	*10*	*11*	*12*	*13*	*14*	*15*	*16*
Diameter of Shaft		*Key size*	*Width W and Thickness*		*Keyway in shaft*				*Keyway in hub*				*Nominal Keyway radius & key chamfer*	*Bright keybar*	
					Width W_s		*Depth H*		*Width W_h*		*Depth h*			*Width W and thickness T*	
Over	*Up to and including*	*W×T*	*Max.*	*Min.*	*Min.*	*Max.*	*Min.*	*Max.*	*Min.*	*Max.*	*Min.*	*Max.*	*r**	*Max.*	*Min.*
1	$1\frac{1}{4}$	$\frac{5}{16}\times\frac{5}{16}$	0·314	0·312	0·311	0·312	0·177	0·183	0·312	0·313	0·142	0·148	0·010	0·314	0·312
$1\frac{1}{4}$	$1\frac{1}{2}$	$\frac{3}{8}\times\frac{3}{8}$	0·377	0·375	0·374	0·375	0·213	0·219	0·375	0·376	0·169	0·175	0·010	0·377	0·375
$1\frac{1}{2}$	$1\frac{3}{4}$	$\frac{7}{16}\times\frac{7}{16}$	0·440	0·438	0·437	0·438	0·248	0·254	0·438	0·439	0·197	0·203	0·020	0·440	0·438
$1\frac{3}{4}$	2	$\frac{1}{2}\times\frac{1}{2}$	0·502	0·500	0·499	0·500	0·283	0·289	0·500	0·501	0·224	0·230	0·020	0·502	0·500
2	$2\frac{1}{2}$	$\frac{5}{8}\times\frac{5}{8}$	0·627	0·625	0·624	0·625	0·354	0·360	0·625	0·626	0·278	0·284	0·020	0·627	0·625
$2\frac{1}{2}$	3	$\frac{3}{4}\times\frac{3}{4}$	0·752	0·750	0·749	0·750	0·424	0·430	0·750	0·751	0·333	0·339	0·020	0·752	0·750
3	$3\frac{1}{2}$	$\frac{7}{8}\times\frac{7}{8}$	0·877	0·875	0·874	0·875	0·495	0·501	0·875	0·876	0·387	0·393	0·062	0·877	0·875
$3\frac{1}{2}$	4	1×1	1·003	1·000	0·999	1·000	0·566	0·572	1·000	1·001	0·442	0·448	0·062	1·003	1·000
4	5	$1\frac{1}{4}\times1\frac{1}{4}$	1·253	1·250	1·248	1·250	0·707	0·713	1·250	1·252	0·551	0·557	0·062	1·253	1·250
5	6	$1\frac{1}{2}\times1\frac{1}{2}$	1·504	1·500	1·498	1·500	0·848	0·854	1·500	1·502	0·661	0·667	0·062	1·504	1·500

* NOTE. The key chamfer shall be the minimum to clear the keyway radius. Nominal values are given.

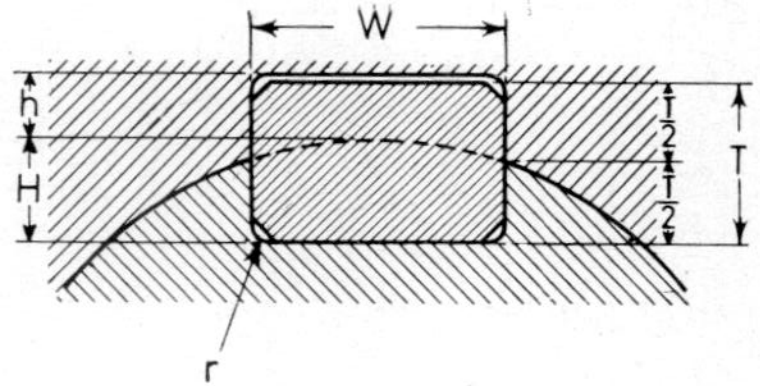

Table A. 2. RECTANGULAR PARALLEL KEYS AND KEYWAYS

(Reproduced from B.S. 46 by kind permission of the British Standards Institution)

All dimensions in mm

1	2	3	4	5	6	7	8	9	10	11	12	13	14	15	16	17	18	19	20
Diameter of shaft			*Key*				*Keyway in shaft*				*Keyway in hub*				*Nominal Keyway radius & key chamfer*	*Keybar*			
Over	*Up to and including*	*Key size* $W \times T$	*Width* W		*Thickness* T		*Width* W_s		*Depth* H		*Width* W_h		*Depth* h		r^*	*Width* W		*Thickness* T	
			Max.	*Min.*	*Max.*	*Min.*	*Min.*	*Max.*	*Min.*	*Max.*	*Min.*	*Max.*	*Min.*	*Max.*		*Max.*	*Min.*	*Max.*	*Min.*
1	$1\frac{1}{4}$	$\frac{5}{16} \times \frac{1}{4}$	0·314	0·312	0·253	0·250	0·311	0·312	0·146	0·152	0·312	0·313	0·112	0·118	0·010	0·314	0·312	0·253	0·250
$1\frac{1}{4}$	$1\frac{1}{2}$	$\frac{3}{8} \times \frac{1}{4}$	0·377	0·375	0·253	0·250	0·374	0·375	0·150	0·156	0·375	0·376	0·108	0·114	0·010	0·377	0·375	0·253	0·250
$1\frac{1}{2}$	$1\frac{3}{4}$	$\frac{7}{16} \times \frac{5}{16}$	0·440	0·438	0·315	0·312	0·437	0·438	0·186	0·192	0·438	0·439	0·135	0·141	0·020	0·440	0·438	0·315	0·312

(continued over page)

Table A. 2. *(continued)*

1	2	3	4	5	6	7	8	9	10	11	12	13	14	15	16	17	18	19	20
Diameter of shaft			*Key*				*Keyway in shaft*				*Keyway in hub*				*Nominal Keyway radius & key chamfer r**	*Keybar*			
Over	*Up to and including*	*Key size W×T*	*Width W*		*Thickness T*		*Width W_s*		*Depth H*		*Width W_h*		*Depth h*			*Width W*		*Thickness T*	
			Max.	*Min.*	*Max.*	*Min.*	*Min.*	*Max.*	*Min.*	*Max.*	*Min.*	*Max.*	*Min.*	*Max.*		*Max.*	*Min.*	*Max.*	*Min.*
1 3/4	2	1/2 × 5/16	0·502	0·500	0·315	0·312	0·499	0·500	0·190	0·196	0·500	0·501	0·131	0·137	0·020	0·502	0·500	0·315	0·312
2	2 1/2	5/8 × 7/16	0·627	0·625	0·441	0·438	0·624	0·625	0·260	0·266	0·625	0·626	0·185	0·191	0·020	0·627	0·625	0·441	0·438
2 1/2	3	3/4 × 1/2	0·752	0·750	0·503	0·500	0·749	0·750	0·299	0·305	0·750	0·751	0·209	0·215	0·020	0·752	0·750	0·503	0·500
3	3 1/2	7/8 × 5/8	0·877	0·875	0·629	0·625	0·874	0·875	0·370	0·376	0·875	0·876	0·264	0·270	0·062	0·877	0·875	0·629	0·625
3 1/2	4	1 × 3/4	1·003	1·000	0·754	0·750	0·999	1·000	0·441	0·447	1·000	1·001	0·318	0·324	0·062	1·003	1·000	0·754	0·750
4	5	1 1/4 × 7/8	1·253	1·250	0·879	0·875	1·248	1·250	0·518	0·524	1·250	1·252	0·366	0·372	0·062	1·253	1·250	0·879	0·875
5	6	1 1/2 × 1	1·504	1·500	1·006	1·000	1·498	1·500	0·599	0·605	1·500	1·502	0·412	0·418	0·062	1·504	1·500	1·006	1·000
6	7	1 3/4 × 1 1/4	1·754	1·750	1·256	1·250	1·748	1·750	0·740	0·746	1·750	1·752	0·526	0·532	0·125				
7	8	2 × 1 3/8	2·005	2·000	1·381	1·375	1·998	2·000	0·818	0·824	2·000	2·002	0·573	0·579	0·125	Bright keybar is not normally available in sections larger than the above.			
8	9	2 1/4 × 1 1/2	2·255	2·250	1·506	1·500	2·248	2·250	0·897	0·905	2·250	2·252	0·619	0·627	0·125				
9	10	2 1/2 × 1 5/8	2·505	2·500	1·631	1·625	2·498	2·500	0·975	0·983	2·500	2·502	0·666	0·674	0·187				
10	11	2 3/4 × 1 7/8	2·755	2·750	1·881	1·875	2·748	2·750	1·114	1·122	2·750	2·752	0·777	0·785	0·187				

* NOTE, The key chamfer shall be the minimum to clear the keyway radius. Nominal values are given,

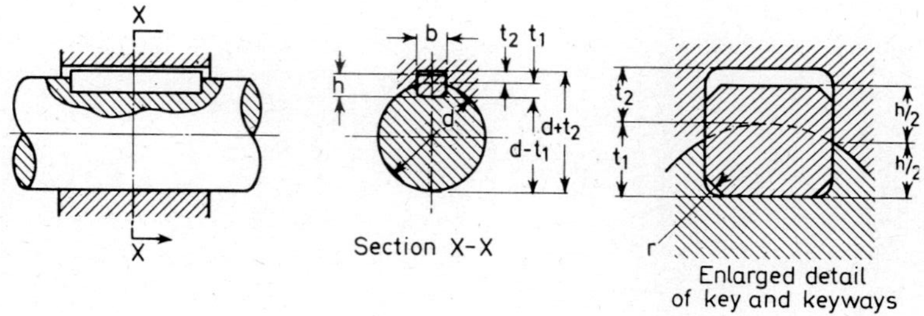

Table A. 3. SQUARE PARALLEL KEYS AND KEYWAYS

(Reproduced from B.S. 4235 by kind permission of the British Standards Institution)

All dimensions in mm

1	*2*	*3*	*4*	*5*	*6*	*7*	*8*	*9*	*10*	*11*	*12*	*13*	*14*	*15*
Shaft		*Key (see note)*	*Keyway*											
Nominal diameter (see note) d		*Section b×h width × thickness*	*Width b*						*Depth*				*Radius r*	
				Tolerance for class of fit					*Shaft t_1*		*Hub t_2*			
				Free		*Normal*		*Close*						
Over	*To*		*Nom.*	*Shaft (H9)*	*Hub (D10)*	*Shaft (N9)*	*Hub (JS9)*	*Shaft and hub (P9)*	*Nom.*	*Tol.*	*Nom.*	*Tol.*	*Max.*	*Min.*
6	8	2×2	2	+0·025	+0·060	−0·004	+0·0125	−0·006	1·2		1		0·16	0·08
8	10	3×3	3	0	+0·020	−0·029	−0·0125	−0·031	1·8	+0·1	1·4	+0·1	0·16	0·08

(continued over page)

Table A. 3. *(continued)*

1	*2*	*3*	*4*	*5*	*6*	*7*	*8*	*9*	*10*	*11*	*12*	*13*	*14*	*15*
Shaft		*Key (see note)*	*Keyway*											
Nominal diameter (see note) d		*Section b×h width × thickness*	*Width b*						*Depth*				*Radius r*	
				Tolerance for class of fit					*Shaft t_1*		*Hub t_2*			
				Free		*Normal*		*Close*						
Over	*To*		*Nom.*	*Shaft (H9)*	*Hub (D10)*	*Shaft (N9)*	*Hub (JS9)*	*Shaft and hub (P9)*	*Nom.*	*Tol.*	*Nom.*	*Tol.*	*Max.*	*Min.*
10	12	4×4	4						2·5	0	1·8	0	0·16	0·08
12	17	5×5	5	+0·030 0	+0·078 +0·030	0 −0·030	+0·015 −0·015	−0·012 −0·042	3		2·3		0·25	0·16
17	22	6×6	6						3·5		2·8		0·25	0·16

NOTE. The relations between shaft diameter and key section given above are for general applications. The use of smaller key sections is permitted if suitable for the torque transmitted. In cases such as stepped shafts when larger diameters are required, for example to resist bending, and when fans, gears and impellers are fitted with a smaller key than normal, an unequal disposition of key in shaft with relation to the hub results. Therefore, dimensions $d-t_1$ and $d+t_2$ should be recalculated to maintain the $h/2$ relationship.

The use of larger key sections is not permitted.

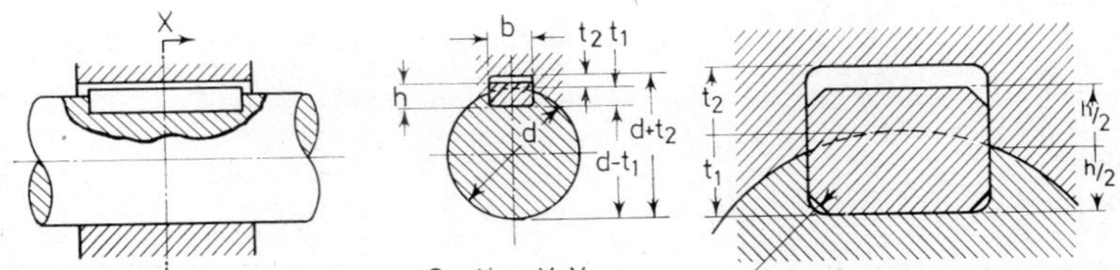

Section X-X

Enlarged detail of key and keyways

Table A. 4. RECTANGULAR PARALLEL KEYS AND KEYWAYS

(Reproduced from B.S. 4235 by kind permission of the British Standards Institution)

All dimensions in mm

1	2	3	4	5	6	7	8	9	10	11	12	13	14	15
Shaft		*Key (see note)*	*Keyway*											
Nominal diameter (see note) d		*Section b×h width × thickness*	*Width b*						*Depth*				*Radius r*	
			Tolerance for class of fit						*Shaft t_1*		*Hub t_2*			
				Free		*Normal*		*Close*						
Over	*To*		*Nom.*	*Shaft (H9)*	*Hub (D10)*	*Shaft (N9)*	*Hub (JS9)*	*Shaft and hub (P9)*	*Nom.*	*Tol.*	*Nom.*	*Tol.*	*Max.*	*Min.*
22	30	8×7	8	+0·036	+0·098	0	+0·018	−0·015	4		3·3		0·25	0·16
30	38	10×8	10	0	+0·040	−0·036	−0·018	−0·051	5		3·3		0·40	0·25
38	44	12×8	12						5		3·3		0·40	0·25
44	50	14×9	14	+0·043	+0·120	0	+0·0215	−0·018	5·5		3·8		0·40	0·25
50	58	16×10	16	0	+0·050	−0·043	−0·0215	−0·061	6	+0·2	4·3	+0·2	0·40	0·25
58	65	18×11	18						7	0	4·4	0	0·40	0·25

(continued over page)

Table A. 4. *(continued)*

1	2	3	4	5	6	7	8	9	10	11	12	13	14	15
Shaft		*Key (see note)*	*Keyway*											
Nominal diameter (see note) d		*Section b×h width × thickness*	*Width b*						*Depth*				*Radius r*	
				Toüerance for class of fit					*Shaft t_1*		*Hub t_2*			
				Free		*Normal*		*Close*						
Over	*To*		*Nom.*	*Shaft (H9)*	*Hub (D10)*	*Shaft (N9)*	*Hub (JS9)*	*Shaft and hub (P9)*	*Nom.*	*Tol.*	*Nom.*	*Tol.*	*Max.*	*Min.*
65	75	20×12	20						7·5		4·9		0·60	0·40
75	85	22×14	22	+0·052	+0·149	0	+0·026	−0·022	9		5·4		0·60	0·40
85	95	25×14	25	0	+0·065	−0·052	−0·026	−0·074	9		5·4		0·60	0·40
95	110	28×16	28						10		6·4		0·60	0·40
110	130	32×18	32						11		7·4		0·60	0·40
130	150	36×20	36	+0·062	+0·180	0	+0·031	−0·026	12		8·4		1·00	0·70
150	170	40×22	40	0	+0·080	−0·062	−0·031	−0·088	13		9·4		1·00	0·70
170	200	45×25	45						15		10·4		1·00	0·70
200	230	50×28	50						17	+0·3	11·4	+0·3	1·00	0·70
230	260	56×32	56						20	0	12·4	0	1·60	1·20
260	290	63×32	63	+0·074	+0·220	0	+0·037	−0·032	20		12·4		1·60	1·20
290	330	70×36	70	0	+0·100	−0·074	−0·037	−0·106	22		14·4		1·60	1·20
330	380	80×40	80						25		15·4		2·50	2·00
380	440	90×45	90	+0·087	+0·260	0	+0·0435	−0·037	28		17·4		2·50	2·00
440	500	100×50	100	0	+0·120	−0·087	−0·0435	−0·124	31		19·5		2·50	2·00

NOTE. The relations between shaft diameter and key section given above are for general applications. The use of smaller key sections is permitted if suitable for the torque transmitted. In cases such as stepped shafts when larger diameters are required, for example to resist bending, and when fans, gears and impellers are fitted with a smaller key than normal, an unequal disposition of key in shaft with relation to the hub results. Therefore, dimensions $d-t_1$ and $d+t_2$ should be recalculated to maintain the $h/2$ relationship.

The use of larger key sections is not permitted.

22*

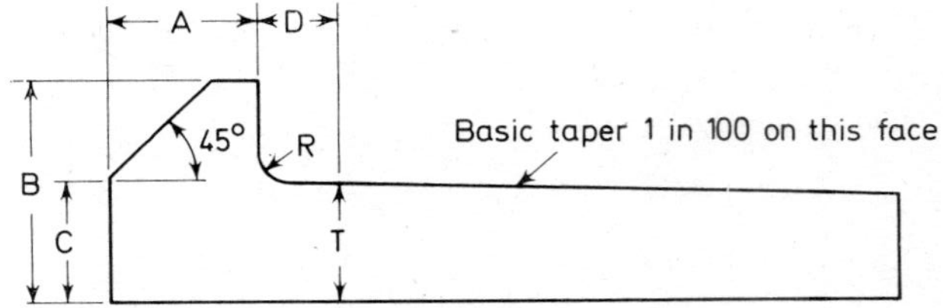

Gib-head key

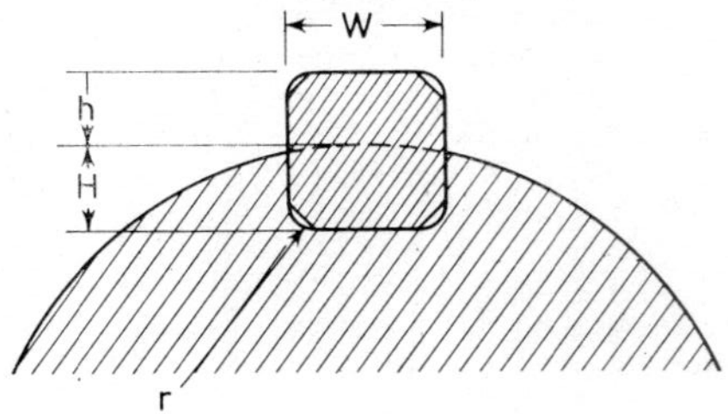

Section at deep end of keyway in hub

Table A.5. SQUARE TAPER KEYS AND KEYWAYS, GIB-HEAD OR PLAIN *(Reproduced from B.S. 46 by kind Permission of the British Standards Institution)*

All dimensions are in inches

1	*2*	*3*	*4*	*5*	*6*	*7*
Diameter of shaft		*Key size*	*Key*			
Over	*Up to and includ-ing*		*Width W*		*Thickness T*	
		W×T	*Max.*	*Min.*	*Max.*	*Min.*
1/4	1/2	1/8 × 1/8	0·127	0·125	0·129	0·124
1/2	3/4	3/16 × 3/16	0·190	0·188	0·192	0·188
3/4	1	1/4 × 1/4	0·252	0·250	0·254	0·249
1	1 1/4	5/16 × 5/16	0·314	0·312	0·316	0·311
1 1/4	1 1/2	3/8 × 3/8	0·377	0·375	0·379	0·374
1 1/2	1 3/4	7/16 × 7/16	0·440	0·438	0·442	0·437
1 3/4	2	1/2 × 1/2	0·502	0·500	0·504	0·499
2	2 1/2	5/8 × 5/8	0·627	0·625	0·630	0·624
2 1/2	3	3/4 × 3/4	0·752	0·750	0·755	0·749
3	3 1/2	7/8 × 7/8	0·877	0·875	0·880	0·874
3 1/2	4	1 × 1	1·003	1·000	1·007	0·999
4	5	1 1/4 × 1 1/4	1·253	1·250	1·257	1·249
5	6	1 1/2 × 1 1/2	1·504	1·500	1·509	1·499

NOTE.* The key chamfer shall be the minimum to clear the keyway radius. Nominal values are given. For plain keys Columns 17 to 21 do not apply.

(Columns 8 to 21 over page)

Table A.5 *(continued)*

8	9	10	11	12	13	14	15	16	17	18	19	20	21
Keyway in shaft and hub								*Nominal keyway radius and key chamfer r**	*Gib-head*				
Keyway in shaft		*Keyway in hub*		*Depth in shaft H*		*Depth in hub at deep end of keyway h*							
Width W_s		*Width W_h*											
Min.	*Max.*	*Min.*	*Max.*	*Min.*	*Max.*	*Min.*	*Max.*		*A*	*B*	*C*	*D*	*Radius R*
0·124	0·125	0·125	0·126	0·072	0·078	0·039	0·045	0·010	3/16	1/4	5/32	0·1	1/32
0·187	0·188	0·188	0·189	0·107	0·113	0·067	0·073	0·010	1/4	3/8	7/32	0·2	1/32
0·249	0·250	0·250	0·251	0·142	0·148	0·094	0·100	0·010	5/16	7/16	9/32	0·2	1/16
0·311	0·312	0·312	0·313	0·177	0·183	0·121	0·127	0·010	3/8	9/16	11/32	0·3	1/16
0·374	0·375	0·375	0·376	0·213	0·219	0·148	0·154	0·010	7/16	5/8	13/32	0·3	1/16
0·437	0·438	0·438	0·439	0·248	0·254	0·175	0·181	0·020	1/2	3/4	15/32	0·4	1/16
0·499	0·500	0·500	0·501	0·283	0·289	0·202	0·208	0·020	9/16	7/8	17/32	0·4	1/16
0·624	0·625	0·625	0·626	0·354	0·360	0·256	0·262	0·020	11/16	1	21/32	0·5	1/8
0·749	0·750	0·750	0·751	0·424	0·430	0·310	0·316	0·020	13/16	1 1/4	25/32	0·5	1/8
0·874	0·875	0·875	0·876	0·495	0·501	0·364	0·370	0·062	15/16	1 3/8	29/32	0·6	1/8
0·999	1·000	1·000	1·001	0·566	0·572	0·418	0·424	0·062	1 1/16	1 5/8	1 1/32	0·6	1/8
1·248	1·250	1·250	1·252	0·707	0·713	0·526	0·532	0·062	1 5/16	2	1 9/32	0·7	1/4
1·498	1·500	1·500	1·502	0·848	0·854	0·635	0·641	0·062	1 9/16	2 1/2	1 17/32	0·7	1/4

NOTE.* The key chamfer shall be the minimum to clear the keyway radius. Nominal values are given. For plain keys Columns 17 to 21 do not apply.

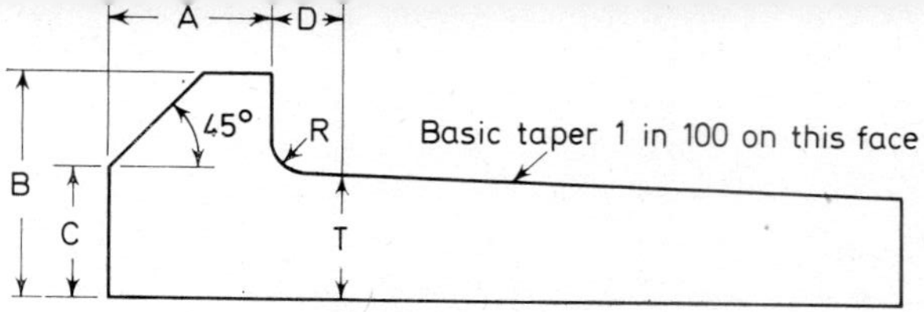

Gib-head key

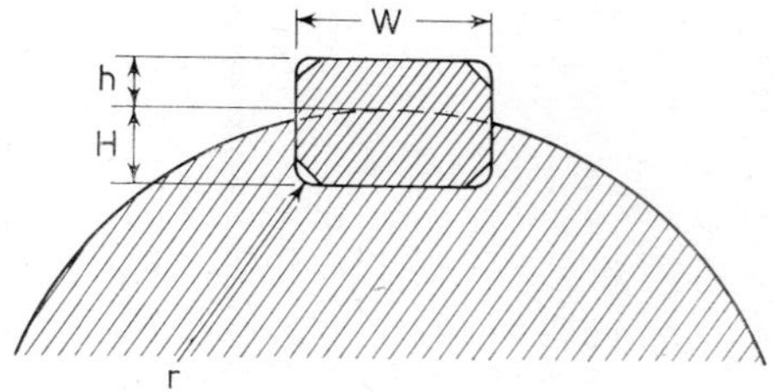

Section at deep end of keyway in hub

Table A.6. RECTANGULAR TAPER KEYS AND KEYWAYS, GIB-HEAD OR PLAIN *(Reproduced from B.S. 46 by kind permission of the British Standards Institution)*

All dimensions are in inches

1	2	3	4	5	6	7
Diameter of shaft		*Key size*	*Key*			
Over	*Up to and including*	$W \times T$	*Width W Max.*	*Width W Min.*	*Thickness T Max.*	*Thickness T Min.*
1	$1\frac{1}{4}$	$\frac{5}{16} \times \frac{1}{4}$	0·314	0·312	0·254	0·249
$1\frac{1}{4}$	$1\frac{1}{2}$	$\frac{3}{8} \times \frac{1}{4}$	0·377	0·375	0·254	0·249
$1\frac{1}{2}$	$1\frac{3}{4}$	$\frac{7}{16} \times \frac{5}{16}$	0·440	0·438	0·316	0·311
$1\frac{3}{4}$	2	$\frac{1}{2} \times \frac{5}{16}$	0·502	0·500	0·316	0·311
2	$2\frac{1}{2}$	$\frac{5}{8} \times \frac{7}{16}$	0·627	0·625	0·442	0·437
$2\frac{1}{2}$	3	$\frac{3}{4} \times \frac{1}{2}$	0·752	0·750	0·504	0·499
3	$3\frac{1}{2}$	$\frac{7}{8} \times \frac{5}{8}$	0·877	0·875	0·630	0·624
$3\frac{1}{2}$	4	$1 \times \frac{3}{4}$	1·003	1·000	0·755	0·749
4	5	$1\frac{1}{4} \times \frac{7}{8}$	1·253	1·250	0·880	0·874
5	6	$1\frac{1}{2} \times 1$	1·504	1·500	1·007	0·999
6	7	$1\frac{3}{4} \times 1\frac{1}{4}$	1·754	1·750	1·257	1·249
7	8	$2 \times 1\frac{3}{8}$	2·005	2·000	1·382	1·374
8	9	$2\frac{1}{4} \times 1\frac{1}{2}$	2·255	2·250	1·509	1·499
9	10	$2\frac{1}{2} \times 1\frac{5}{8}$	2·505	2·500	1·634	1·624
10	11	$2\frac{3}{4} \times 1\frac{7}{8}$	2·755	2·750	1·884	1·874
11	12	3×2	3·006	3·000	2·014	1·999

NOTE.* The key chamfer shall be the minimum to clear the keyway radius. Nominal values shall be given. For plain keys Columns 17 to 21 do not apply.

(Colums 8 to 21 over page)

Table A.6 *(continued)*

8	9	10	11	12	13	14	15	16	17	18	19	20	21
Keyway in shaft and hub								*Nominal keyway radius and key chamfer r**	*Gib-head*				
Keyway in shaft		*Keyway in hub*		*Depth in shaft H*		*Depth in hub at deep end of keyway h*							
Width W_s		*Width W_h*											
Min.	*Max.*	*Min.*	*Max.*	*Min.*	*Max.*	*Min.*	*Max.*		*A*	*B*	*C*	*D*	*Radius R*
0·311	0·312	0·312	0·313	0·146	0·152	0·090	0·096	0·010	3/8	7/16	1/4	0·3	1/16
0·374	0·375	0·375	0·376	0·150	0·156	0·086	0·092	0·010	7/16	7/16	9/32	0·3	1/16
0·437	0·438	0·438	0·439	0·186	0·192	0·112	0·118	0·020	1/2	9/16	5/16	0·4	1/16
0·499	0·500	0·500	0·501	0·190	0·196	0·108	0·114	0·020	9/16	5/8	3/8	0·4	1/16
0·624	0·625	0·625	0·626	0·260	0·266	0·162	0·168	0·020	11/16	3/4	7/16	0·5	1/8
0·749	0·750	0·750	0·751	0·299	0·305	0·185	0·191	0·020	13/16	7/8	17/32	0·5	1/8
0·874	0·875	0·875	0·876	0·370	0·376	0·239	0·245	0·062	15/16	1	21/32	0·6	1/8
0·999	1·000	1·000	1·001	0·441	0·447	0·293	0·299	0·062	1 1/16	1 1/4	23/32	0·6	1/8
1·248	1·250	1·250	1·252	0·518	0·524	0·340	0·346	0·062	1 5/16	1 1/2	27/32	0·7	1/4
1·498	1·500	1·500	1·502	0·599	0·605	0·384	0·390	0·062	1 9/16	1 5/8	1 1/32	0·7	1/4
1·748	1·750	1·750	1·752	0·740	0·746	0·493	0·499	0·125	1 13/16	2	1 7/32	0·8	1/4
1·998	2·000	2·000	2·002	0·818	0·824	0·539	0·545	0·125	2 1/16	2 1/4	1 13/32	0·8	1/4
2·248	2·250	2·250	2·252	0·897	0·905	0·581	0·589	0·125	2 5/16	2 1/2	1 9/16	0·9	3/8
2·498	2·500	2·500	2·502	0·975	0·983	0·628	0·636	0·187	2 9/16	2 3/4	1 11/16	0·9	3/8
2·748	2·750	2·750	2·752	1·114	1·122	0·738	0·746	0·187	2 13/16	3	1 15/16	1·0	3/8
2·998	3·000	3·000	3·002	1·195	1·203	0·782	0·790	0·187	3 1/16	3 1/4	2 1/16	1·0	3/8

NOTE. * The key chamfer shall be the minimum to clear the keyway radius. Nominal values shall be given. For plain keys Columns 17 to 21 do not apply.

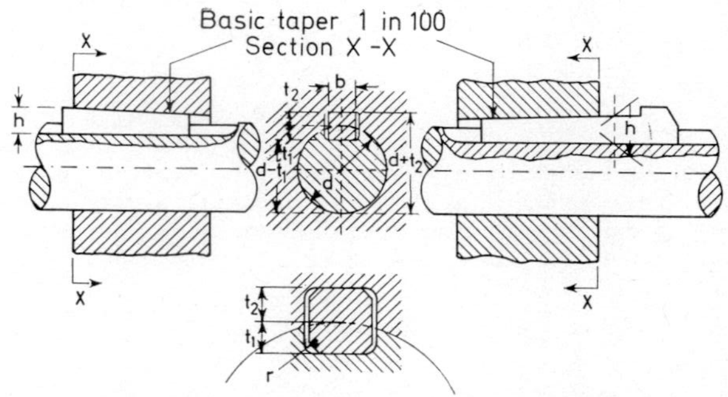

Enlarged view of key and keyway on section X-X

Table A.7. *SQUARE TAPER KEYS AND KEYWAY*

(Reproduced from B.S. 4235 by kind permission of the British Standards Institution)

All dimensions in mm

1	*2*	*3*	*4*	*5*	*6*	*7*	*8*	*9*	*10*	*11*
Shaft (see note 1)		*Key (see note 1)*	*Keyway*							
Nominal diameter d		*Section b×h width × thickness*	*Width b shaft and hub*		*Depth*				*Radius r*	
					Shaft t_1		*Hub t_2*			
Over	*To*		*Nom.*	*Tol. (D10)*	*Nom.*	*Tol.*	*Nom.*	*Tol.*	*Max.*	*Min.*
6	8	2×2	2	+0·060 +0·020	1·2		0·5		0·16	0·08

(continued over page)

Table A. 7. *(continued)*

1	2	3	4	5	6	7	8	9	10	11
Shaft (see note 1)		*Key (see note 1*	*Keyway*							
Nominal diameter d		*Section b×h width × thickness*	*Width b shaft and hub*		*Depth*				*Radius r*	
					Shaft t_1		*Hub t_2*			
Over	*To*		*Nom.*	*Tol. (D10)*	*Nom.*	*Tol.*	*Nom.*	*Tol.*	*Max.*	*Min.*
8	10	3×3	3		1·8	+0·1 0	0·9	+0·1 0	0·16	0·08
10	12	4×4	4	+0·078 +0·030	2·5		1·2		0·16	0·08
12	17	5×5	5		3		1·7		0·25	0·16
17	22	6×6	6		3·5	+0·2 0	2·2	+0·2 0	0·25	0·16

NOTE 1. The relation between shaft diameter and key section given above are for general applications. The use of smaller key sections is permitted if suitable for the torque transmitted. In cases such as stepped shafts when larger diameters are required, for example to resist bending, and when fans, gears and impellers are fitted with a smaller key than normal, an unequal disposisition of key in shaft with relation to the hub results, therefore, dimensions $d-t_1$ and $d+t_2$ should be re-calculated to maintain the $h/2$ relationship.

The use of larger key sections is not permitted.

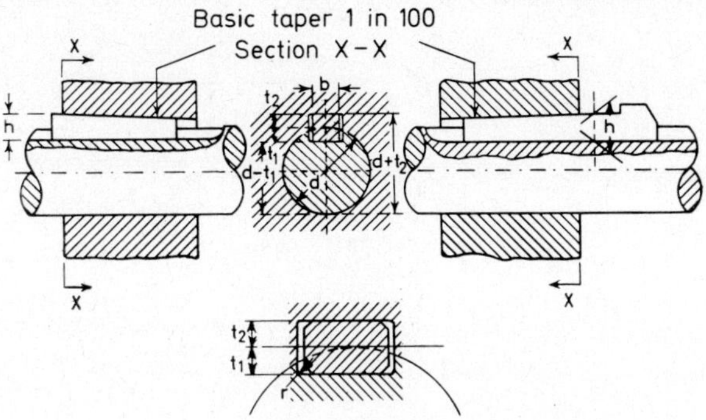

Enlarged veiw of key and keyway on section X-X

Table A.8. RECTANGULAR TAPER KEYS AND KEYWAYS

(Reproduced from B.S. 4235 by kind permission of the British Standards Institution)

All dimensions in mm

1	*2*	*3*	*4*	*5*	*6*	*7*	*8*	*9*	*10*	*11*
Shaft		*Key (see note 1)*	*Keyway*							
Nominal diameter (see note 1) d		*Section b×h width × thickness*	*Width b shaft and hub*		*Depth*				*Radius r*	
					Shaft t_1		*Hub t_2*			
Over	*To*		*Nom.*	*Tol. (D10)*	*Nom.*	*Tol.*	*Nom.*	*Tol.*	*Max.*	*Min.*
22	30	8×7	8	+0·098	4		2·4		0·25	0·16
30	38	10×8	10	+0·040	5		2·4		0·40	0·25
38	44	12×8	12		5		2·4		0·40	0·25
44	50	14×9	14	+0·120	5·5		2·9		0·40	0·25
50	58	16×10	16	+0·050	6	+0·2	3·4	+0·2	0·40	0·25
58	65	18×11	18		7	0	3·4	0	0·40	0·25
65	75	20×12	20		7·5		3·9		0·60	0·40
75	85	22×14	22	+0·149	9		4·4		0·60	0·40
85	95	25×14	25	+0·065	9		4·4		0·60	0·40
95	110	28×16	28		10		5·4		0·60	0·40

(continued over page)

Tabl e A. 8. *(continued)*

1	*2*	*3*	*4*	*5*	*6*	*7*	*8*	*9*	*10*	*11*
Shaft		*Key* (*see note 1*	*Keyway*							
Nominal diameter (*see note 1*) *d*		*Section* $b \times h$ *width* $\times$ *thickness*	*Width* *b* *shaft and hub*		*Depth*				*Radius* *r*	
					Shaft t_1		*Hub* t_2			
Over	*To*		*Nom.*	*Tol.* (*D10*)	*Nom.*	*Tol.*	*Nom.*	*Tol.*	*Max.*	*Min.*
110	130	32×18	32		11		6·4		0·60	0·40
130	150	36×20	36		12		7·1		1·00	0·70
150	170	40×22	40	+0·180	13		8·1		1·00	0·70
170	200	45×25	45	+0·080	15		9·1		1·00	0·70
200	230	50×28	50		17		10·1		1·00	0·70
230	260	56×32	56		20	+0·3	11·1	+0·3	1·60	1·20
260	290	63×32	63	+0·220	20	0	11·1	0	1·60	1·20
290	330	70×36	70	+0·120	22		13·1		1·60	1·20
330	380	80×40	80		25		14·1		2·50	2·00
380	440	90×45	90	+0·260	28		16·1		2·50	2·00
440	500	100×50	100	+0·120	31		18·1		2·50	2·00

NOTE 1. The relations between shaft diameter and key section given above are for general applications. The use of smaller key sections is permitted if suitable for the torque transmitted, in cases such as stepped shafts when larger diameters are required, for example to resist bending, and when fans, gears and impellers are fitted with a smaller key than normal, an unequal disposition of key in shaft with relation to the hub results. Therefore, dimensions $d-t_1$ and $d+t_2$ should be recalculated to maintain the $h/2$ relationship.

The use of larger key sections is not permitted.

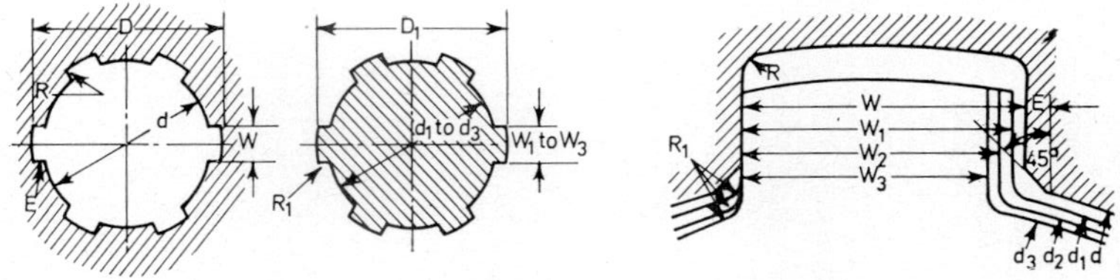

Table A.9. STRAIGHT SIDED SPLINES, BOTTOM FITTING, SHALLOW

(Reproduced from B.S. 2059 by kind permission of the British Standards Institution)

1	2	3	4	5	6	7	8	9	10	11	12	13	14	15	16	17	18	19	20	21
	Hole					Shaft								Resultant clearance fit in thousanths of an inch						
							Fit 1		Fit 2		Fit 3			1		2		3		
Nom. size	D	d	W	R	E Chamfer at 45°	D_1	d_1	W_1	d_2	W_2	d_3	W_3	R_1	on dia. $d-d_1$	on spline $W-W_1$	on dia. $d-d_2$	on spline $W-W_2$	on dia. $d-d_3$	on spline $W-W_3$	Nom size
1/2	0·520	0·4500	0·1250	0·010	0·007	0·4990	0·4500	0·1240	0·4485	0·1232	0·4470	0·1224	0·006	0	1·0	1·5	1·8	3·0	2·6	1/2
	0·530	0·4510	0·1265	0·005	0·010	0·4970	0·4485	0·1225	0·4465	0·1215	0·4445	0·1205	0·003	2·5	4·0	4·5	5·0	6·5	6·0	
5/8	0·645	0·5630	0·1560	0·010	0·007	0·6240	0·5630	0·1550	0·5615	0·1542	0·5600	0·1534	0·006	0	1·0	1·5	1·8	3·0	2·6	5/8
	0·655	0·5640	0·1575	0·005	0·010	0·6210	0·5615	0·1535	0·5595	0·1525	0·5575	0·1515	0·003	2·5	4·0	4·5	5·0	6·5	6·0	
3/4	0·770	0·6750	0·8170	0·010	0·008	0·7490	0·6750	0·1860	0·6735	0·1852	0·6720	0·1844	0·007	0	1·0	1·5	1·8	3·0	2·6	3/4
	0·780	0·6760	0·1885	0·005	0·011	0·7460	0·6735	0·1845	0·6715	0·1835	0·6695	0·1825	0·004	2·5	4·0	4·5	5·0	6·5	6·0	
7/8	0·905	0·7880	0·2180	0·015	0·009	0·8740	0·7880	0·2170	0·7865	0·2162	0·7850	0·2145	0·008	0	1·0	1·5	1·8	3·0	2·6	7/8
	0·915	0·7890	0·2195	0·010	0·012	0·8710	0·7865	0·2155	0·7845	0·2145	0·7825	0·2135	0·005	2·5	4·0	4·5	5·0	6·5	6·0	
1	1·030	0·9000	0·2490	0·015	0·010	0·9990	0·9000	0·2480	0·8985	0·2472	0·8970	0·2464	0·009	0	1·0	1·5	1·8	3·0	2·6	1
	1·040	0·9010	0·2510	0·010	0·013	0·9950	0·8985	0·2460	0·8965	0·2450	0·8945	0·2440	0·005	2·5	5·0	4·5	6·0	6·5	7·0	
1 1/8	1·155	1·0130	0·2800	0·015	0·012	1·1240	1·0130	0·2790	1·0115	0·2782	1·0100	0·2774	0·011	0	1·0	1·5	1·8	3·0	2·6	1 1/8
	1·165	1·0140	0·2820	0·010	0·015	1·1190	1·0115	0·2770	1·0090	0·2757	1·0065	0·2744	0·005	2·5	5·0	5·0	6·3	7·5	7·6	
1 1/4	1·280	1·1250	0·3120	0·015	0·013	1·2490	1·1250	0·3110	1·1235	0·3102	1·1220	0·3094	0·012	0	1·0	1·5	1·8	3·0	2·6	1 1/4
	1·290	1·1260	0·3140	0·010	0·017	1·2450	1·1235	0·3090	1·1210	0·3077	1·1185	0·3064	0·005	2·5	5·0	5·0	6·3	7·5	7·6	
1 3/8	1·405	1·2380	0·3430	0·015	0·014	1·3740	1·2380	0·3420	1·2365	0·3412	1·2350	0·3404	0·013	0	1·0	1·5	1·8	3·0	2·6	1 3/8
	1·415	1·2390	0·3450	0·010	0·018	1·3700	1·2365	0·3400	1·2340	0·3387	1·2315	0·3374	0·005	2·5	5·0	5·0	6·3	7·5	7·6	
1 1/2	1·560	1·3500	0·3740	0·030	0·015	1·4990	1·3500	0·3730	1·3485	0·3722	1·3470	0·3714	0·014	0	1·0	1·5	1·8	3·0	2·6	1 1/2
	1·570	1·3510	0·3765	0·015	0·019	1·4940	1·3485	0·3705	1·3460	0·3692	1·3435	0·3679	0·005	2·5	6·0	5·0	7·3	7·5	8·6	

(Continued over page)

Table A.9 *(continued)*

1 5/8	1·685	1·4630	0·4050	0·030	0·017	1·6240	1·4630	0·4035	1·4615	0·4027	1·4600	0·4019	0·015	0	1·5	1·5	2·3	3·0	3·1	1 5/8
	1·695	1·4645	0·4075	0·015	0·021	1·6190	1·4615	0·4010	1·4585	0·3995	1·4555	0·3930	0·005	3·0	6·5	6·0	8·0	9·0	9·5	
1 3/4	1·810	1·5750	0·4370	0·030	0·018	1·7490	1·5750	0·4355	1·5735	0·4347	1·5720	0·4339	0·016	0	1·5	1·5	2·3	·3·0	3·1	1 3/4
	1·820	1·5765	0·4395	0·015	0·022	1·7440	1·5735	0·4330	1·5705	0·4315	1·5675	0·4300	0·005	3·0	6·5	6·0	8·0	9·0	9·5	
1 7/8	1·935	1·6880	0·4680	0·030	0·019	1·8740	1·6880	0·4665	1·6865	0·4657	1·6850	0·4649	0·017	0	1·5	1·5	2·3	3·0	3·1	1 7/8
	1·945	1·6895	0·4705	0·015	0·023	1·8690	1·6865	0·4640	1·6835	0·4625	1·6805	0·4610	0·005	3·0	6·5	6·0	8·0	9·0	9·5	
2	2·060	1·8000	0·4980	0·030	0·021	1·9990	1·8000	0·4965	1·7985	0·4957	1·7970	0·4949	0·019	0	1·5	1·5	2·3	3·0	3·1	2
	2·070	1·8015	0·5010	0·015	0·026	1·9940	1·7985	0·4935	1·7955	0·4920	1·7925	0·4905	0·005	3·0	7·5	6·0	9·0	9·0	10·5	
2 1/4	2·330	2·0250	0·5610	0·040	0·023	2·2490	2·0250	0·5595	2·0235	0·5587	2·0220	0·5579	0·021	0	1·5	1·5	2·3	3·0	3·1	2 1/4
	2·340	2·0265	0·5640	0·020	0·028	2·2440	2·0235	0·5565	2·0200	0·5547	2·0165	0·5529	0·005	3·0	7·5	6·5	9·3	10·0	11·1	
2 1/2	2·580	2·2500	0·6230	0·040	0·025	2·4990	2·2500	0·6215	2·2485	0·6207	2·2470	0·6199	0·023	0	1·5	1·5	2·3	3·0	3·1	2 1/2
	2·590	2·2515	0·6260	0·020	0·030	2·4940	2·2485	0·6185	2·2450	0·6167	2·2415	0·6149	0·005	3·0	7·5	6·5	9·3	10·0	11·1	
2 3/4	2·830	2·4750	0·6860	0·040	0·028	2·7490	2·4750	0·6845	2·4735	0·6837	2·4720	0·6829	0·026	0	1·5	1·5	2·3	3·0	3·1	2 3/4
	2·840	2·4765	0·6895	0·020	0·033	2·7440	2·4735	0·6810	2·4700	0·6792	2·4665	0·6774	0·005	3·0	8·5	6·5	10·3	10·0	12·1	
3	3·080	2·7000	0·7480	0·040	0·031	2·9990	2·7000	0·7465	2·6985	0·7457	2·6970	0·7449	0·029	0	1·5	1·5	2·3	3·0	3·1	3
	3·090	2·7015	0·7515	0·020	0·036	2·9930	2·6985	0·7430	2·6950	0·7412	0·6915	0·7394	0·010	3·0	8·5	6·5	10·3	10·0	12·1	
3 1/4	3·330	2·9250	0·8110	0·040	0·033	3·2490	2·9250	0·8090	2·9230	0·8080	2·9210	0·8070	0·030	0	2·0	2·0	3·0	4·0	4·0	3 1/4
	3·340	2·9265	0·8145	0·020	0·038	3·2430	2·9230	0·8055	2·9190	0·8035	2·9150	0·3015	0·010	3·5	9·0	7·5	11·0	11·5	13·0	
3 1/2	3·580	3·1500	0·8730	0·040	0·036	3·4990	3·1500	0·8710	3·1480	0·8700	3·1460	0·8690	0·033	0	2·0	2·0	3·0	4·0	4·0	3 1/2
	3·590	3·1515	0·8765	0·020	0·042	3·4930	3·1480	0·8675	3·1440	0·8655	3·1400	0·8635	0·010	3·5	9·0	7·5	11·0	11·5	13·0	
3 3/4	3·830	3·3750	0·9360	0·040	0·039	3·7490	3·3750	0·9340	3·3730	0·9330	3·3710	0·9320	0·036	0	2·0	2·0	3·0	4·0	4·0	3 3/4
	3·840	3·3765	0·9400	0·020	0·045	3·7420	3·3730	0·9300	3·3685	0·9277	3·3640	0·9254	0·010	3·5	10·0	8·0	12·3	12·5	14·6	
4	4·080	3·6000	0·9980	0·040	0·041	3·9990	3·6000	0·9960	3·5980	0·9950	3·5960	0·9940	0·038	0	2·0	2·0	3·0	4·0	4·0	4
	4·090	3·6015	1·0020	0·020	0·047	3·9920	3·5980	0·9920	3·5935	0·9897	3·5890	0·9874	0·010	3·5	10·0	8·0	12·3	12·5	14·6	
4 1/4	4·330	3·8250	1·0600	0·040	0·043	4·2490	3·8250	1·0580	3·8230	1·0570	3·8210	1·0560	0·040	0	2·0	2·0	3·0	4·0	4·0	4 1/4
	4·340	3·8270	1·0640	0·020	0·049	4·2420	3·8230	1·0540	3·8185	1·0517	3·8140	1·0494	0·010	4·0	10·0	8·5	12·3	13·0	14·6	
4 1/2	4·580	4·0500	1·1220	0·040	0·045	4·4990	4·0500	1·1200	4·0480	1·1190	4·0460	1·1180	0·042	0	2·0	2·0	3·0	4·0	4·0	4 1/2
	4·590	4·0520	1·1260	0·020	0·051	4·4910	4·0480	1·1160	4·0435	1·1137	4·0390	1·1114	0·010	4·0	10·0	8·5	12·3	13·0	14·6	
4 3/4	4·830	4·2750	1·1850	0·040	0·048	4·7490	4·2750	1·1825	4·2730	1·1815	4·2710	1·1805	0·044	0	2·5	2·0	3·5	4·0	4·5	4 3/4
	4·840	4·2770	1·1895	0·020	0·054	4·7410	4·2730	1·1780	4·2680	1·1755	4·2630	1·1730	0·010	4·0	11·5	9·0	14·0	14·0	16·5	
5	5·080	4·5000	1·2470	0·040	0·051	4·9990	4·5000	1·2445	4·4980	1·2435	4·4960	1·2425	0·047	0	2·5	2·0	3·5	4·0	4·5	5
	5·090	4·5020	1·2515	0·020	0·057	4·9910	4·4980	1·2400	4·4930	1·2375	4·4880	1·2350	0·010	4·0	11·5	9·0	14·0	14·0	16·5	
5 1/4	5·330	4·7250	1·3100	0·040	0·053	5·2490	4·7250	1·3075	4·7230	1·3065	4·7210	1·3055	0·049	0	2·5	2·0	3·5	4·0	4·5	5 1/4
	5·340	4·7270	1·3145	0·020	0·060	5·2410	4·7230	1·3030	4·7180	1·3005	4·7130	1·2980	0·010	4·0	11·5	9·0	14·0	14·0	16·5	
5 1/2	5·580	4·9500	1·3720	0·040	0·055	5·4990	4·9500	1·3695	4·9480	1·3685	4·9460	1·3675	0·051	0	2·5	2·0	3·5	4·0	4·5	5 1/2
	5·590	4·9520	1·3765	0·020	0·062	5·4900	4·9480	1·3650	4·9430	1·3625	4·9380	1·3600	0·010	4·0	11·5	9·0	14·0	14·0	16·5	
5 3/4	5·830	5·1750	1·4350	0·040	0·060	5·7490	5·1750	1·4325	5·1730	1·4315	5·1710	1·4305	0·056	0	2·5	2·0	3·5	4·0	4·5	5 3/4
	5·840	5·1770	1·4395	0·020	0·067	5·7400	5·1730	1·4280	5·1680	1·4255	5·1630	1·4230	0·015	4·0	11·5	9·0	14·0	14·0	16·5	
6	6·080	5·4000	1·4970	0·040	0·061	5·9990	5·4000	1·4945	5·3980	1·4935	5·3960	1·4925	0·057	0	2·5	2·0	3·5	4·0	4·5	6
	6·090	5·0420	1·5015	0·020	0·068	5·9900	5·3980	1·4900	5·3930	1·4875	5·3880	1·4850	0·015	4·0	11·5	9·0	14·0	14·0	16·5	

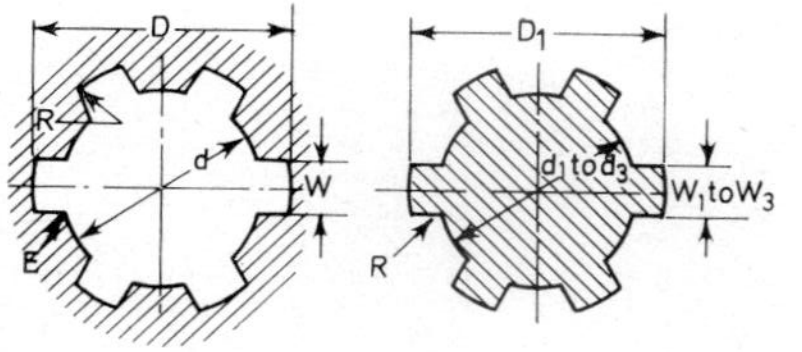

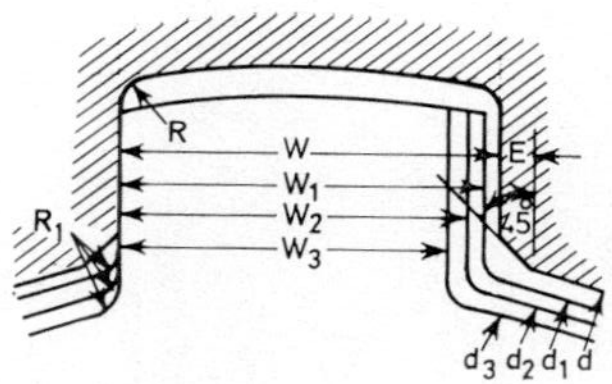

Table A.10. STRAIGHT SIDED SPLINES, BOTTOM FITTING, DEEP.

(Reproduced from B.S. 2059 by kind permission of the British Standards Institution)

1	2	3	4	5	6	7	8	9	10	11	12	13	14	15	16	17	18	19	20	21
	Hole					Shaft								Resultant clearance fit in thousandths of an inch						
							Fit 1		Fit 2		Fit 3			1		2		3		
Nom. size	D	d	W	R	E Chamfer at 45°	D_1	d_1	W_1	d_2	W_2	d_3	W_3	R_1	on dia. $d-d_1$	on spline $W-W_1$	on dia. $d-d_2$	on spline $W-W_2$	on dia. $d-d_3$	on spline $W-W_3$	Nom. size
1/2	0·520	0·4250	0·1250	0·010	0·010	0·4990	0·4250	0·1240	0·4235	0·1232	0·4220	0·1224	0·009	0	1·0	1·5	1·8	3·0	2·6	1/2
	0·530	0·4260	0·1265	0·005	0·013	0·4970	0·4235	0·1225	0·4215	0·1215	0·4195	0·1205	0·005	2·5	4·0	4·5	5·0	6·5	6·0	
5/8	0·645	0·5320	0·1560	0·010	0·010	0·6240	0·5320	0·1550	0·5305	0·1542	0·5290	0·1534	0·009	0	1·0	1·5	1·8	3·0	2·6	5/8
	0·655	0·5330	0·1575	0·005	0·013	0·6210	0·5305	0·1535	0·5285	0·1525	0·5265	0·1515	0·005	2·5	4·0	4·5	5·0	6·5	6·0	
3/4	0·770	0·6380	0·1870	0·010	0·012	0·7490	0·6380	0·1860	0·6365	0·1852	0·6350	0·1844	0·010	0	1·0	1·5	1·8	3·0	2·6	3/4
	0·780	0·6390	0·1885	0·005	0·015	0·7460	0·6365	0·1845	0·6345	0·1835	0·6325	0·1825	0·005	2·5	4·0	4·5	5·0	6·5	6·0	
7/8	0·905	0·7440	0·2180	0·015	0·014	0·8740	0·7440	0·2170	0·7425	0·2162	0·7410	0·2154	0·012	0	1·0	1·5	1·8	3·0	2·6	7/8
	0·915	0·7450	0·2195	0·010	0·018	0·8710	0·7425	0·2155	0·7405	0·2145	0·7385	0·2135	0·005	2·5	4·0	4·5	5·0	6·5	6·0	
1	1·030	0·8500	0·2490	0·015	0·016	0·9990	0·8500	0·2480	0·8485	0·2472	0·8470	0·2464	0·014	0	1·0	1·5	1·8	3·0	2·6	1
	1·040	0·8510	0·2510	0·010	0·020	0·9950	0·8485	0·2460	0·8465	0·2450	0·8445	0·2440	0·005	2·5	5·0	4·5	6·0	6·5	7·0	
1 1/8	1·155	0·9560	0·2800	0·015	0·018	1·1240	0·9560	0·2790	0·9545	0·2782	0·9530	0·2774	0·016	0	1·0	1·5	1·8	3·0	2·6	1 1/8
	1·165	0·9570	0·2820	0·010	0·022	1·1190	0·9545	0·2770	0·9520	0·2757	0·9495	0·2744	0·005	2·5	5·0	5·0	6·3	7·5	7·6	
1 1/4	1·280	1·0630	0·3120	0·015	0·020	1·2490	1·0630	0·3110	1·0615	0·3102	1·0600	0·3094	0·018	0	1·0	1·5	1·8	3·0	2·6	1 1/4
	1·290	1·0640	0·3140	0·010	0·025	1·2450	1·0615	0·3090	1·0590	0·3077	1·0565	0·3064	0·005	2·5	5·0	5·0	6·3	7·5	7·6	
1 3/8	1·405	1·1690	0·3430	0·015	0·020	1·3740	1·1690	0·3420	1·1675	0·3412	1·1660	0·3404	0·018	0	1·0	1·5	1·8	3·0	2·6	1 3/8
	1·415	1·1700	0·3450	0·010	0·025	1·3700	1·1675	0·3400	1·1650	0·3387	1·1625	0·3374	0·005	2·5	5·0	5·0	6·3	7·5	7·6	
1 1/2	1·560	1·2750	0·3740	0·030	0·023	1·4990	1·2750	0·3730	1·2735	0·3722	1·2720	0·3714	0·021	0	1·0	1·5	1·8	3·0	2·6	1 1/2
	1·570	1·2760	0·3765	0·015	0·028	1·4940	1·2735	0·3705	1·2710	0·3692	1·2685	0·3679	0·005	2·5	6·0	5·0	7·3	7·5	8·6	

(Continued over page)

Table A.10 *(continued)*

1 5/8	1·685 1·695	1·3810 1·3825	0·4050 0·4075	0·030 0·015	0·025 0·030	1·6240 1·6190	1·3810 1·3795	0·4035 0·4010	1·3795 1·3765	0·4027 0·3995	1·3780 1·3735	0·4019 0·3980	0·023 0·005	0 3·0	1·5 6·5	1·5 6·0	2·3 8·0	3·0 9·0	3·1 9·5	1 5/8
1 3/4	1·810 1·820	1·4880 1·4895	0·4370 0·4395	0·030 0·015	0·027 0·032	1·7490 1·7440	1·4880 1·4865	0·4355 0·4330	1·4865 1·4835	0·4347 0·4315	1·4850 1·4805	0·4339 0·4300	0·024 0·005	0 3·0	1·5 6·5	1·5 6·0	2·3 8·0	3·0 9·0	3·1 9·5	1 3/4
1 7/8	1·935 1·945	1·5940 1·5955	0·4680 0·4705	0·030 0·015	0·030 0·035	1·8740 1·8690	1·5940 1·5925	0·4665 0·4640	1·5925 1·5895	0·4657 0·4625	1·5910 1·5865	0·4649 0·4610	0·027 0·010	0 3·0	1·5 6·5	1·5 6·0	2·3 8·0	3·0 9·0	3·1 9·5	1 7/8
2	2·060 2·070	1·7000 1·7015	0·4980 0·5010	0·030 0·015	0·030 0·035	1·9990 1·9940	1·7000 1·6985	0·4965 0·4935	1·6985 1·6955	0·4957 0·4920	1·6970 1·6925	0·4949 0·4905	0·027 0·010	0 3·0	1·5 7·5	1·5 6·0	2·3 9·0	3·0 9·0	3·1 10·5	2
2 1/4	2·330 2·340	1·9130 1·9145	0·5610 0·5640	0·040 0·020	0·034 0·040	2·2490 2·2440	1·9130 1·9115	0·5595 0·5565	1·9115 1·9080	0·5587 0·5547	1·9100 1·9045	0·5579 0·5529	0·031 0·010	0 3·0	1·5 7·5	1·5 6·5	2·3 9·3	3·0 10·0	3·1 11·1	2 1/4
2 1/2	2·580 2·590	2·1250 2·1265	0·6230 0·6260	0·040 0·020	0·038 0·044	2·4990 2·4940	2·1250 2·1235	0·6215 0·6185	2·1235 2·1200	0·6207 0·6167	2·1220 2·1165	0·6199 0·6149	0·034 0·010	0 3·0	1·5 7·5	1·5 6·5	2·3 9·3	3·0 10·0	3·1 11·1	2 1/2
2 3/4	2·830 2·840	2·3380 2·3395	0·6860 0·6895	0·040 0·020	0·042 0·048	2·7490 2·7440	2·3380 2·3365	0·5845 0·5810	2·3365 2·3330	0·6837 0·6792	2·3350 2·3295	0·6829 0·6774	0·038 0·010	0 3·0	1·5 8·5	1·5 6·5	2·3 14·3	3·0 10·0	3·1 12·1	2 3/4
3	3·080 3·090	2·5500 2·5515	0·7480 0·7515	0·040 0·020	0·046 0·052	2·9990 2·9930	2·5500 2·5485	0·7465 0·7430	2·5485 2·5450	0·7457 0·7412	2·5470 2·5415	0·7449 0·7394	0·042 0·010	0 3·0	1·5 8·5	1·5 6·5	2·3 14·3	3·0 10·0	3·1 12·1	3
3 1/4	3·330 3·340	2·7630 2·7645	0·8110 0·8145	0·040 0·020	0·050 0·056	3·2490 3·2430	2·7630 2·7610	0·8090 0·8055	2·7610 2·7570	0·8080 0·8035	2·7590 2·7530	0·8070 0·8015	0·046 0·010	0 3·5	2·0 9·0	2·0 7·5	3·0 11·0	4·0 11·5	4·0 13·0	3 1/4
3 1/2	3·580 3·590	2·9750 2·9765	0·8730 0·8765	0·040 0·020	0·053 0·060	3·4990 3·4930	2·9750 2·9730	0·8710 0·8675	2·9730 2·9690	0·8700 0·8655	2·9710 2·9650	0·8690 0·9635	0·048 0·010	0 3·5	2·0 9·0	2·0 7·5	3·0 11·0	4·0 11·5	4·0 13·0	3 1/2
3 3/4	3·830 3·840	3·1880 3·1895	0·9360 0·9400	0·040 0·020	0·056 0·063	3·7490 3·7420	3·1880 3·1860	0·9340 0·9300	3·1860 3·1815	0·9330 0·9277	3·1840 3·1770	0·9320 0·9254	0·051 0·010	0 3·5	2·0 10·0	2·0 8·0	3·0 12·3	4·0 12·5	4·0 14·6	3 3/4
4	4·080 4·090	3·4000 3·4015	0·9980 1·0020	0·040 0·020	0·061 0·068	3·9990 3·9920	3·4000 3·3980	0·9960 0·9920	3·3980 3·3935	0·9950 0·9897	3·3960 3·3890	0·9940 0·9874	0·055 0·015	0 3·5	2·0 10·0	2·0 8·0	3·0 12·3	4·0 12·5	4·0 14·6	4
4 1/4	4·330 4·340	3·6130 3·6150	1·0600 1·0640	0·040 0·020	0·064 0·072	4·2490 4·2420	3·6130 3·6110	1·0580 1·0540	3·6110 3·6065	1·0570 1·0517	3·6090 3·6020	1·0560 1·0494	0·058 0·015	0 4·0	2·0 10·0	2·0 8·5	3·0 12·3	4·0 13·0	4·0 14·6	4 1/4
4 1/2	4·580 4·590	3·8250 3·8270	1·1220 1·1260	0·040 0·020	0·068 0·076	4·4990 4·4910	3·8250 3·8230	1·1200 1·1160	3·8230 3·8185	1·1190 1·1137	3·8210 3·8140	1·1180 1·1114	0·062 0·015	0 4·0	2·0 10·0	2·0 8·5	3·0 12·3	4·0 13·0	4·0 14·6	4 1/2
4 3/4	4·830 4·840	4·0380 4·0400	1·1850 1·1895	0·040 0·020	0·072 0·080	4·7490 4·7410	4·0380 4·0360	1·1825 1·1780	4·0360 4·0310	1·1815 1·1755	4·0340 4·0260	1·1805 1·1730	0·066 0·015	0 4·0	2·5 11·5	2·0 9·0	3·5 14·0	4·0 14·0	4·5 16·5	4 3/4
5	5·080 5·090	4·2500 4·2520	1·2470 1·2515	0·040 0·020	0·076 0·084	4·9990 4·9910	4·2500 4·2480	1·2445 1·2400	4·2480 4·2430	1·2435 1·2375	4·2460 4·2380	1·2425 1·2350	0·069 0·020	0 4·0	2·5 11·5	2·0 9·0	3·5 14·0	4·0 14·0	4·5 16·5	5
5 1/4	5·330 5·340	4·4630 4·4650	1·3100 1·3145	0·040 0·020	0·080 0·088	5·2490 5·2410	4·4630 4·4610	1·3075 1·3030	4·4610 4·4560	1·3065 1·3005	4·4590 4·4510	1·3055 1·2980	0·073 0·020	0 4·0	2·5 11·5	2·0 9·0	3·5 14·0	4·0 14·0	4·5 16·5	5 1/4
5 1/2	5·580 5·590	4·6750 4·6770	1·3720 1·3765	0·040 0·020	0·083 0·091	5·4990 5·4900	4·6750 4·6730	1·3695 1·3650	4·6730 4·6680	1·3685 1·3625	4·6710 4·6630	1·3675 1·3600	0·076 0·020	0 4·0	2·5 11·5	2·0 9·0	3·5 14·0	4·0 14·0	4·5 16·5	5 1/2
5 3/4	5·830 5·840	4·8880 4·8900	1·4350 1·4395	0·040 0·020	0·087 0·095	5·7490 5·7400	4·8880 4·8860	1·4325 1·4280	4·8860 4·8810	1·4315 1·4255	4·8840 4·8760	1·4305 1·4230	0·080 0·020	0 4·0	2·5 11·5	2·0 9·0	3·5 14·0	4·0 14·0	4·5 16·5	5 3/4
6	6·080 6·090	5·1000 5·1020	1·4970 1·5015	0·040 0·020	0·090 0·100	5·9990 5· 900	5·1000 5·0980	1·4945 1·4900	5·0980 5·0930	1·4935 1·4875	5·0960 5·0880	1·4925 1·4850	0·082 0·020	0 4·0	2·5 11·5	2·0 9·0	3·5 14·0	4·0 14·0	4·5 16·5	6

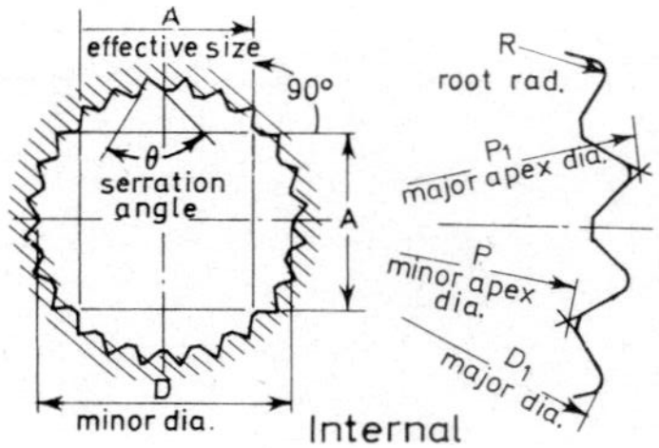

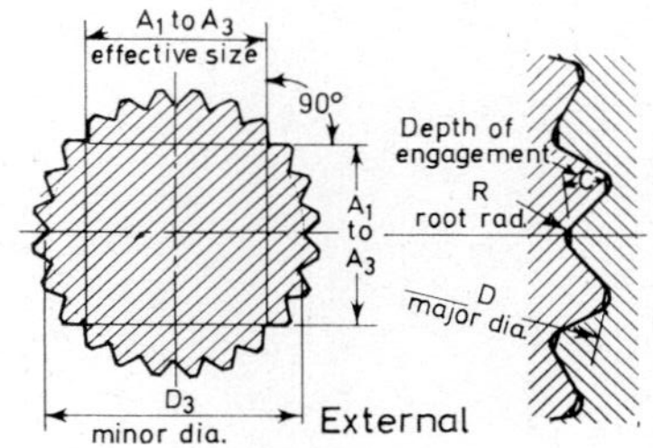

Table A.11. STRAIGHT SIDED SERRATIONS

(Reproduced from B.S. 2059 by kind permission of the British Standards Institution)

1	2	3	4	5	6	7	8	9	10	11	12	13	14	15	16	17	18	19	20
									Hole				Shaft			Resultant fits*			
																Fit 1	Fit 2	Fit 3	
Nom. size	Number of serrations															Inter-fer-ence	Trans ition	Clear-ance	No size
	S	θ	P_1	P	C (min.)	C min. $\times S$	R	D_1 (max.)	D	A	D_2	D_3 (min.)	A_1	A_2	A_3	$A-A_1$	$A-A_2$	$A-A_3$	
$^1/_4$	12	degs. 60	0·2780	0·1966	0·0172	0·2064	0·009 0·006	0·268	0·2114 0·2126	0·1390 0·1399	0·248 0·247	0·199	0·1405 0·1399	0·1393 0·1384	0·1381 0·1369	−1·5 0	−0·3 +1·5	+ 0·9 + 3·0	$^1/_4$
$^5/_{16}$	12	60	0·3476	0·2458	0·0214	0·2568	0·012 0·007	0·335	0·2649 0·2661	0·1738 0·1747	0·310 0·309	0·249	0·1753 0·1747	0·1741 0·1732	0·1729 0·1717	−1·5 0	−0·3 +1·5	+ 0·9 + 3·0	$^5/_{16}$
$^3/_8$	16	67½	0·3974	0·3123	0·0216	0·3456	0·009 0·006	0·389	0·3271 0·3287	0·2208 0·2217	0·373 0·372	0·314	0·2223 0·2217	0·2211 0·2202	0·2199 0·2187	−1·5 0	−0·3 +1·5	+ 0·9 + 3·0	$^3/_8$
$^7/_{16}$	16	67½	0·4637	0·3643	0·0252	0·4032	0·011 0·006	0·456	0·3820 0·3836	0·2576 0·2585	0·435 0·434	0·366	0·2591 0·2585	0·2579 0·2570	0·2567 0·2555	−1·5 0	−0·3 +1·5	+ 0·9 + 3·0	$^7/_{16}$
$^1/_2$	16	67½	0·5299	0·4163	0·0300	0·4800	0·012 0·007	0·520	0·4354 0·4370	0·2944 0·2953	0·498 0·497	0·419	0·2959 0·2953	0·2947 0·2938	0·2935 0·2923	−1·5 0	−0·3 +1·5	+ 0·9 + 3·0	$^1/_2$
$^5/_8$	20	72	0·6489	0·5394	0·0313	0·6260	0·011 0·006	0·642	0·5578 0·5594	0·3814 0·3826	0·623 0·622	0·540	0·3834 0·3826	0·3818 0·3806	0·3802 0·3786	−2·0 0	−0·4 +2·0	+ 1·2 + 4·0	$^5/_8$
$^3/_4$	20	72	0·7787	0·6473	0·0378	0·7560	0·013 0·008	0·769	0·6685 0·6705	0·4577 0·4589	0·748 0·746	0·650	0·4597 0·4589	0·4581 0·4569	0·4565 0·4549	−2·0 0	−0·4 +2·0	+ 1·2 + 4·0	$^3/_4$
$^7/_8$	20	72	0·9085	0·7552	0·0449	0·8980	0·015 0·010	0·896	0·7792 0·7812	0·5340 0·5352	0·873 0·871	0·760	0·5360 0·5352	0·5344 0·5332	0·5328 0·5312	−2·0 0	−0·4 +2·0	+ 1·2 + 4·0	$^7/_8$
1	24	75	1·0277	0·8847	0·0437	1·0488	0·013 0·008	1·020	0·9066 0·9086	0·6256 0·6271	0·998 0·996	0·886	0·6281 0·6271	0·6261 0·6246	0·6241 0·6221	−2·5 0	−0·5 +2·5	+ 1·5 + 5·0	1

(Continued over page)

Table A.11 *(continued)*

1 1/8	24	75	1·1561	0·9953	0·0492	1·1808	0·015 0·010	1·146	1·0201 1·0225	0·7038 0·7053	1·123 1·121	0·999	0·7063 0·7053	0·7043 0·7028	0·7023 0·7003	−2·5 0	−0·5 +2·5	+ 1·5 + 5·0	1 1/8
1 1/4	24	75	1·2846	1·1059	0·0550	1·3200	0·017 0·012	1·271	1·1335 1·1359	0·7820 0·7835	1·248 1·246	1·11	0·7845 0·7835	0·7825 0·7810	0·7805 0·7785	−2·5 0	−0·5 +2·5	+ 1·5 + 5·0	1 1/4
1 3/8	24	75	1·4130	1·2165	0·0616	1·4784	0·018 0·013	1·399	1·2455 1·2479	0·8602 0·8617	1·373 1·371	1·222	0·8627 0·8617	0·8607 0·8592	0·8587 0·8567	−2·5 0	−0·5 +2·5	+ 1·5 + 5·0	1 3/8
1 1/2	24	75	1·5415	1·3271	0·0674	1·6176	0·020 0·015	1·525	1·3589 1·3613	0·9384 0·9399	1·498 1·496	1·335	0·9409 0·9399	0·9389 0·9374	0·9369 0·9349	−2·5 0	−0·5 +2·5	+ 1·5 + 5·0	1 1/2
1 5/8	36	80	1·6503	1·5002	0·0470	1·6920	0·014 0·009	1·643	1·5242 1·5270	1·0508 1·0526	1·623 1·621	1·502	1·0638 1·0626	1·0614 1·0596	1·0590 1·0566	−3·0 0	−0·6 +3·0	+ 1·8 + 6·0	1 5/8
1 3/4	36	80	1·7773	1·6156	0·0511	1·8396	0·015 0·010	1·769	1·6410 1·6438	1·1424 1·1442	1·748 1·746	1·618	1·1454 1·1442	1·1430 1·1412	1·1406 1·1382	−3·0 0	−0·6 +3·0	+ 1·8 + 6·0	1 3/4
1 7/8	36	80	1·9044	1·7311	0·0551	1·9836	0·016 0·011	1·895	1·7580 1·7608	1·2241 1·2259	1·873 1·871	1·734	1·2271 1·2259	1·2247 1·2229	1·2223 1·2199	−3·0 0	−0·6 +3·0	+ 1·8 + 6·0	1 7/8
2	36	80	2·0313	1·8465	0·0592	2·1312	0·017 0·012	2·021	1·8748 1·8776	1·3057 1·3075	1·998 1·996	1·851	1·3087 1·3075	1·3063 1·3045	1·3039 1·3015	−3·0 0	−0·6 +3·0	+ 1·8 + 6·0	2
2 1/4	36	80	2·2852	2·0773	0·0672	2·4192	0·019 0·014	2·272	2·1084 2·1116	1·4689 1·4707	2·248 2·246	2·083	1·4719 1·4707	1·4695 1·4677	1·4671 1·4647	−3·0 0	−0·6 +3·0	+ 1·8 + 6·0	2 1/4
2 1/2	36	80	2·5391	2·3081	0·0750	2·7000	0·021 0·016	2·525	2·3428 2·3460	1·6321 1·6342	2·498 2·496	2·314	1·6356 1·6342	1·6328 1·6307	1·6300 1·6272	−3·5 0	−0·7 +3·5	+ 2·1 + 7·0	2 1/2
2 3/4	40	81	2·7885	2·5611	0·0735	2·9400	0·022 0·017	2·773	2·5972 2·6004	1·8110 1·8131	2·748 2·746	2·568	1·8145 1·8131	1·8117 1·8096	1·8089 1·8061	−3·5 0	−0·7 +3·5	+ 2·1 + 7·0	2 3/4
3	40	81	3·0420	2·7939	0·0798	3·1920	0·024 0·019	3·025	2·8328 2·8364	1·9755 1·9777	2·998 2·996	2·803	1·9791 1·9777	1·9763 1·9742	1·9735 1·9707	−3·5 0	−0·7 +3·5	+ 2·1 + 7·0	3
3 1/4	48	82 1/2	3·2893	3·0671	0·0686	3·2928	0·023 0·018	3·274	3·1053 3·1089	2·1688 2·1712	3·248 3·246	3·074	2·1728 2·1712	2·1696 2·1672	2·1664 2·1632	−4·0 0	−0·8 +4·0	+ 2·4 + 8·0	3 1/4
3 1/2	48	82 1/2	3·5424	3·3032	0·0741	3·5568	0·025 0·020	3·525	3·3442 3·3478	2·3357 2·3381	3·498 3·496	3·312	2·3397 2·3381	2·3365 2·3341	2·3333 2·3301	−4·0 0	−0·8 +4·0	+ 2·4 + 8·0	3 1/2
3 3/4	48	82 1/2	3·7954	3·5391	0·0804	3·8592	0·026 0·021	3·777	3·5815 3·5851	2·5025 2·5049	3·748 3·746	3·548	2·5065 2·5049	2·5033 2·5009	2·5001 2·4969	−4·0 0	−0·8 +4·0	+ 2·4 + 8·0	3 3/4
4	48	82 1/2	4·0484	3·7750	0·0859	4·1232	0·028 0·023	4·028	3·8202 3·8242	2·6693 2·6717	3·998 3·996	3·876	2·6733 2·6717	2·6701 2·6677	2·6669 2·6637	−4·0 0	−0·8 +4·0	+ 2·4 + 8·0	4
4 1/4	48	82 1/2	4·3015	4·0110	0·0911	4·3728	0·030 0·025	4·280	4·0598 4·0638	2·8362 2·8389	4·248 4·246	4·023	2·8407 2·8389	2·8371 2·8344	2·8335 2·8299	−4·5 0	−0·9 +4·5	+ 2·7 + 9·0	4 1/4
4 1/2	60	84	4·5486	4·3043	0·0702	4·2120	0·029 0·024	4·529	4·3517 4·3557	3·0436 3·0463	4·498 4·496	4·315	3·0481 3·0463	3·0445 3·0418	3·0409 3·0373	−4·5 0	−0·9 +4·5	+ 2·7 + 9·0	4 1/2
4 3/4	60	84	4·8013	4·5434	0·0742	4·4520	0·031 0·026	4·779	4·5936 4·5976	3·2127 3·2154	4·748 4·746	4·556	3·2172 3·2154	3·2136 3·2109	3·2100 3·2064	−4·5 0	−0·9 +4·5	+ 2·7 + 9·0	4 3/4
5	60	84	5·0540	4·7826	0·0782	4·6920	0·033 0·028	5·030	4·8356 4·8396	3·3818 3·3845	4·998 4·996	4·797	3·3863 3·3845	3·3827 3·3800	3·3791 3·3755	−4·5 0	−0·9 +4·5	+ 2·7 + 9·0	5
5 1/4	60	84	5·3067	5·0217	0·0829	4·9740	0·034 0·029	5·282	5·0762 5·0802	3·5509 3·5536	5·248 5·246	5·037	3·5554 3·5536	3·5518 3·5491	3·5482 3·5446	−4·5 0	−0·9 +4·5	+ 2·7 + 9·0	5 1/4
5 1/2	60	84	5·5594	5·2609	0·0861	5·1660	0·036 0·031	5·533	5·3188 5·3238	3·7200 3·7230	5·498 5·496	5·277	3·7250 3·7230	3·7210 3·7180	3·7170 3·7130	−5·0 0	−1·0 +5·0	+ 3·0 +10·0	5 1/2
5 3/4	60	84	5·8122	5·4999	0·0901	5·4060	0·038 0·033	5·784	5·5608 5·5658	3·8891 3·8921	5·748 5·746	5·518	3·8941 3·8921	3·8901 3·8871	3·8861 3·8821	−5·0 0	−1·0 +5·0	+ 3·0 +10·0	5 3/4
6	60	84	6·0649	5·7390	0·0949	5·6940	0·039 0·034	6·036	5·8014 5·8064	4·0582 4·0612	5·998 5·996	5·757	4·0632 4·0612	4·0592 4·0562	4·0552 4·0512	−5·0 0	−1·0 +5·0	+ 3·0 +10·0	6

• The figures shown for resultant fits are thousandths of an inch. The + refers to a clearance. The − refers to interference.

Appendix B

Material Properties and Design Data Sheets for Gears

The tables and design charts included in this Appendix are based upon the various British Standards Specifications for gears, and this information is reproduced by kind permission of the British Standards Institution, from whom complete copies of the standards may be obtained. A constant effort is made to keep the standards up to date, and revisions take place from time to time. Consequently, care should be taken to ensure that the latest standard is being used.

Table B.1. STRESS FACTORS FOR RECOMMENDED METALLIC MATERIALS FOR SPUR, HELICAL AND BEVEL GEARS, IN UNITS OF LBF/IN² AND (MN/M²) *T. V. Duggan*

CASTINGS

B.S. Specification Ref.	*Material*	*Condition*	*Minimum U.T.S. Tonf/sq.in.* (MN/m²)	*Minimum Brinell Hardness*	*Surface Sress Factor Sc*	*Bending Stress Factor Sb*	*Remarks*
B.S. 821	Cast Iron, Medium Grade	As Cast	16 (247)	210	1350 (9·3)	7 600 (52·4)	Medium quality
B.S. 592	Cast Steel, Grade B	N	32 (494)	—	—	—	General C.S. gears
B.S. 592	Cast Steel, Grade C	N	35 (540)	152	1 400 (9·65)	19 000 (131)	Suitable for surface hardening

FORGED CARBON STEEL

B.S. Specification Ref.	*Material*	*Condition*	*Minimum U.T.S. Tonf/sq.in.* (MN/m²)	*Minimum Brinell Hardness*	*Surface Stress Factor Sc*	*Bending Stress Factor Sb*	*Remarks*
En. 7 B.S. 970	Free cutting Carbon	—	—	—	—	—	For rapid machining Light duty only
En. 8 B.S. 970	0·4% Carbon	N	35 (540)	152	1 400 (9·65)	19 000 (131)	Good general gear steel Suitable for surface hardening
En. 8 B.S. 970	0·4% Carbon	H.T.	40 (617)	179	2 000 (13·8)	24 500 (169)	Good general gear steel Suitable for surface hardening
En. 9 B.S. 970	0·55% Carbon	N	45 (695)	201	2 300 (15·85)	24 000 (165)	Good gear steel, but more difficult to machine than En.8
En. 9 B.S. 970	0·55% Carbon	H.T.	50 (771)	223	2 650 (18·3)	30 500 (210)	Good gear steel, but more difficult to machine than En.8

(Continued on facing page)

Table B.1 *(continued)*

FORGED ALLOY STEELS

B.S. Specification Ref.	*Material*	*Condition*	*Minimum U.T.S. Tonf/sq.in.* (MN/m^2)	*Minimum Brinell Hardness*	*Surface Stress Factor Sc*	*Bending Stress Factor Sb*	*Remarks*
En. 19 B.S. 970	1% Chrome Molybdenum	H.T.	50 (771)	223	2 650 (18·3)	30 500 (210)	
En. 19 B.S. 970	1% Chrome Molybdenum	H.T.	70 (1 080)	311	4 000 (27·5)	41 000 (282)	Difficult to machine
En. 24 B.S. 970	1½% N.C. Molybdenum	H.T.	55 (850)	248	3 000 (20·7)	33 500 (231)	Good machineability
En. 26 B.S. 970	2½% N.C. Molybdenum	H.T.	60 (925)	269	3 300 (22·8)	36 500 (252)	Preferable to En. 24 for larger Sections
En. 28 B.S. 970	3½% N.C. Molybdenum	H.T.	60 (925)	269	3 300 (22·8)	36 500 (252)	

CASE-HARDENING STEEL

B.S. Specification Ref.	*Material*	*Condition*	*Minimum U.T.S. Tonf/sq.in.* (MN/m^2)	*Minimum Brinell Hardness*	*Surface Stress Factor Sc*	*Bending Stress Factor Sb*	*Remarks*
En. 32 B.S. 970	Carbon	C.H.	32 (Core) (495)	140 (Core)	9 200 (63·4)	40 000 (276)	Good machineability. Useful when localised areas are to be left soft
En. 33 B.S. 970	3% Nickel	C.H.	45 (Core) (695)	201 (Core)	10 500 (72·3)	47 000 (324)	Good machineability. Useful when localised areas are to be left soft
En. 36 B.S. 970	3% Nickel Chrome	C.H.	55 (Core) (850)	248 (Core)	11 000 (75·7)	50 000 (345)	Good machineability. Avoid when localised areas are to be left soft

N = Normalised H.T. = Heat Treated C.H. = Case-Hardened

(Continued over page)

Table B.1 *(continned)*

RECOMMENDED NON-FERROUS MATERIALS

B.S. Specification Ref.	*Material*	*Condition*	*Minimum U.T.S. Tonf/sq.in.* (MN/m^2)	*Minimum Brinell Hardness*	*Surface Stress Factor Sc*	*Bending Stress Factor Sb*	*Remarks*
B.S. 1400-G1	Gunmetal	—	—	—	—	—	
B.S. 1400-PB2-C	Phosphor Bronze	Sand Cast	12 (185)	69	710 (4·9)	7 200 (49·5)	
B.S. 1400-PB2-C	Phosphor Bronze	Chill Cast	15 (232)	82	880 (6·05)	9 100 (62·5)	
B.S. 1400-PB2-C	Phosphor Bronze	Centrifugally Cast	17 (262)	90	1 000 (6·9)	10 000 (69)	

Table B.2. STRESS FACTORS FOR WORM GEARS IN UNITS OF LBF/IN² AND (MN/m²)

(Based on B.S. 721 and reproduced by kind permission of the **British Standards Institution***)*

	Material	B.S. reference	Bending stress factor S_b	Surface stress factor S_c when running with: A	B	C	D	E
A	Phosphor bronze centrifugally cast	**1 400** PB2—C	10 000 (69)	—	1 200† (8·3)	1 200 (8·3)	1 300 (8·95)	2 200 (15·2)
	Phosphor bronze sand cast, chilled	—	9 100 (63)	—	900† (6·2)	900 (6·2)	1 000 (6·9)	1 800 (12·4)
	Phosphor bronze sand cast	—	7 200 (50)	—	660† (4·5)	660 (4·5)	770 (5·3)	1 500 (10·35)
B	Grey cast iron	**1 452** Grade 12	5 800 (40)	900† (6·2)	600† (4·5)	600‡ (4·5)	600‡ (4·5)	750‡ (5·16)
C	0·40 per cent carbon steel, normalized	**970** En 8	20 000 (138)	1 550 (10·7)	1 000 (6·9)	—	—	—
D	0·55 per cent carbon steel, normalized	**970** En 9	25 000 (173)	2 200 (15·2)	1 200‡ (8·3)	—	—	—
E	Carbon case-hardening steel	**970** En 32	40 000 (276)	7 000 (48·1)	4 400‡ (30·3)	—	—	2 200‡ (15·2)
	Nickel and nickel-molybdenum case-hardening steels	En 33, En 34, En 35 *	47 000 (324)	7 700 (53)	4 400‡ (30·3)	—	—	2 200‡ (15·2)
	Nickel-chromium and nickel-chromium-molybdenum case-hardening steels	En 36, En 39	50 000 (345)	8 800 (60·5)	4 400‡ (30·3)	—	—	2 200‡ (15·2)

* equivalent grades of steel are permissible
† maximum permissible rubbings speed = 500 ft/min (2·5 m/s)
‡ permissible for hand motions only

Table B.3. WORM GEAR ZONE FACTORS (Z)

(Reproduced from B.S. 721 by kind permission of the British Standards Institution)

z_1 \ q	6	6·5	7	7·5	8	8·5	9	9·5	10	11	12	13	14	17	20
1	1·045	1·048	1·052	1·065	1·084	1·107	1·128	1·137	1·143	1·160	1·202	1·260	1·318	1·402	1·508
2	0·991	1·028	1·055	1·099	1·144	1·183	1·214	1·223	1·231	1·250	1·280	1·320	1·360	1·447	1·575
3	0·822	0·890	0·989	1·109	1·209	1·260	1·305	1·333	1·350	1·365	1·393	1·422	1·442	1·532	1·674
4	0·826	0·883	0·981	1·098	1·204	1·301	1·380	1·428	1·460	1·490	1·515	1·545	1·570	1·666	1·798
5	0·947	0·991	1·050	1·122	1·216	1·315	1·417	1·490	1·550	1·610	1·632	1·652	1·675	1·765	1·886
6	1·132	1·145	1·172	1·220	1·287	1·350	1·438	1·521	1·588	1·675	1·694	1·714	1·733	1·818	1·928
7			1·316	1·340	1·370	1·405	1·452	1·540	1·614	1·704	1·725	1·740	1·760	1·846	1·950
8					1·437	1·462	1·500	1·557	1·623	1·715	1·738	1·753	1·778	1·868	1·960
9							1·573	1·604	1·648	1·720	1·743	1·767	1·790	1·880	1·970
10									1·680	1·728	1·748	1·773	1·798	1·888	1·980
11										1·732	1·753	1·777	1·802	1·892	1·987
12											1·760	1·780	1·806	1·895	1·992
13												1·784	1·806	1·898	1·998
14													1·811	1·900	2·000

The values are based on $be_2 = 2m\sqrt{q+1}$ symmetrical about the centre plane of the wheel.

For smaller face widths, the value of Z must be reduced proportionately.

When it is necessary to obtain greater load capacity, the wormwheel face width may be increased up to a maximum of $2{\cdot}3m\sqrt{q+1}$, and the zone factor increased proportionately.

The table applies to wormwheels having 30 teeth; variations in the number of teeth produce negligible changes in the values of Z.

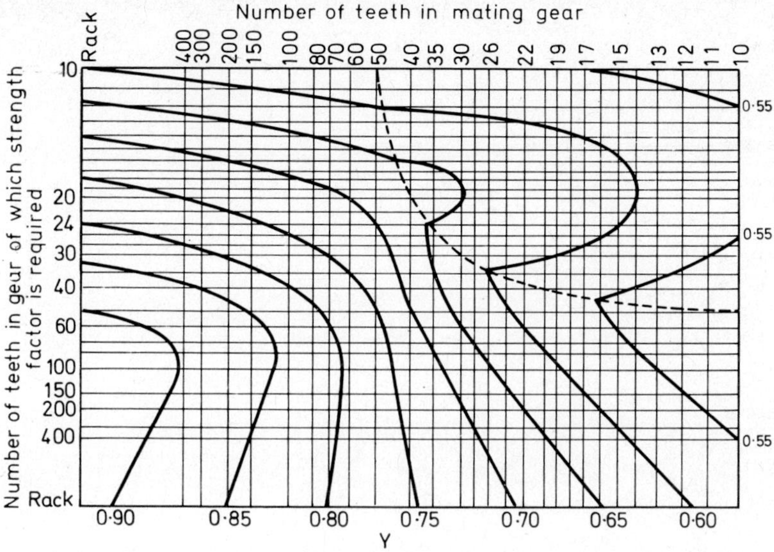

Fig. B. 1. Strength factor (Y) for spur gears and helical gears with 30° helix angle (20° normal pressure angle) (Reproduced from B.S. 436 by kind permission of the British Standards Institution)

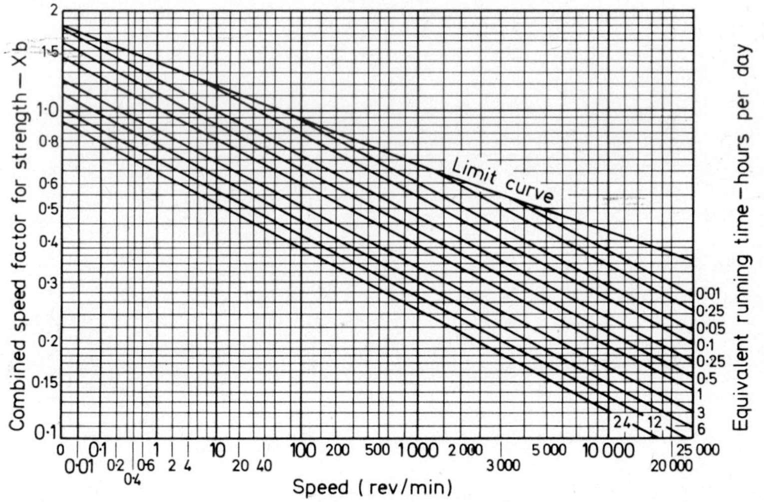

Fig. B. 2. Speed factor for strength (X_b) for spur and helical gears (Reproduced from B.S. 436 by kind permission of the British Standards Institution)

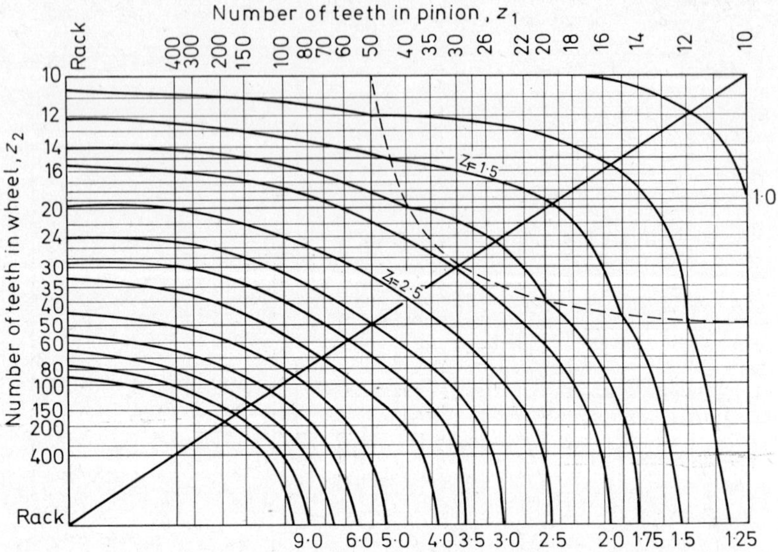

Fig. B. 3. Zone factor (Z) for full depth spur gears (20° Pressure angle) (Reproduced from B.S. 436 by kind permission of the British Standards Institution)

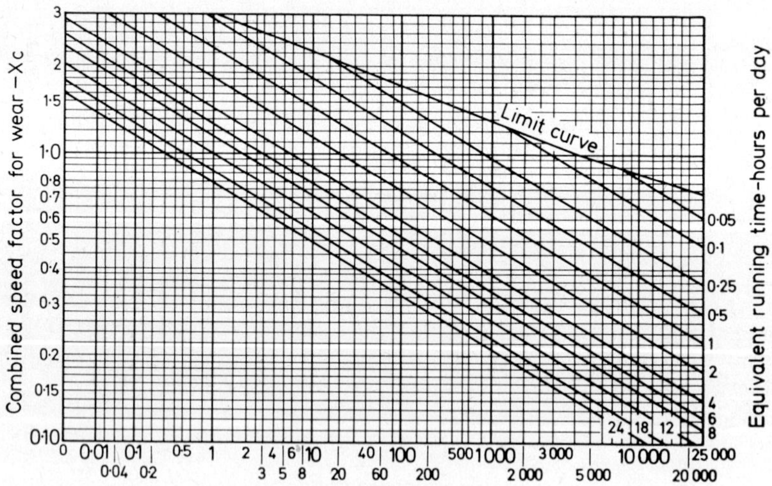

Fig. B. 4. Speed factor for wear (X_c) for spur and helical gears (Reproduced from B.S. 436 by kind permission of the British Standards Institution)

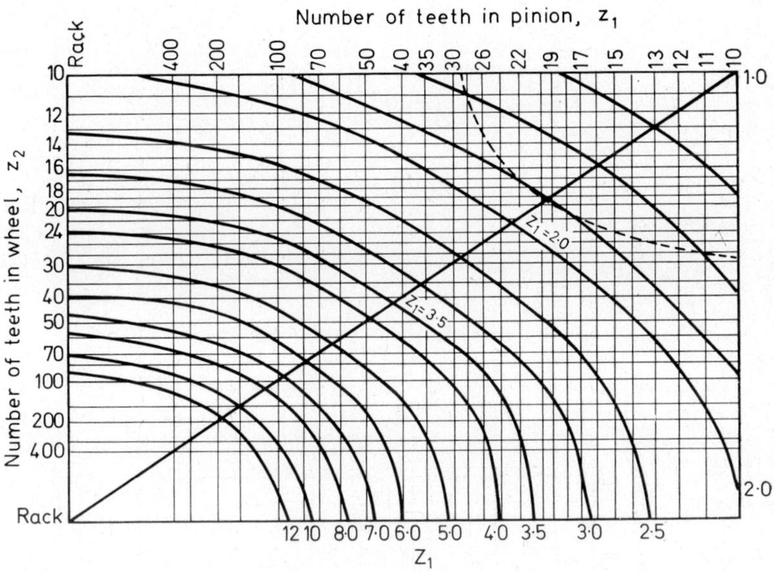

Fig. B. 5. Zone factor (Z_1) for helical gears with 30° helix angle (20° normal pressure angle) (Reproduced from B.S. 436 by kind permission of the British Standards Institution)

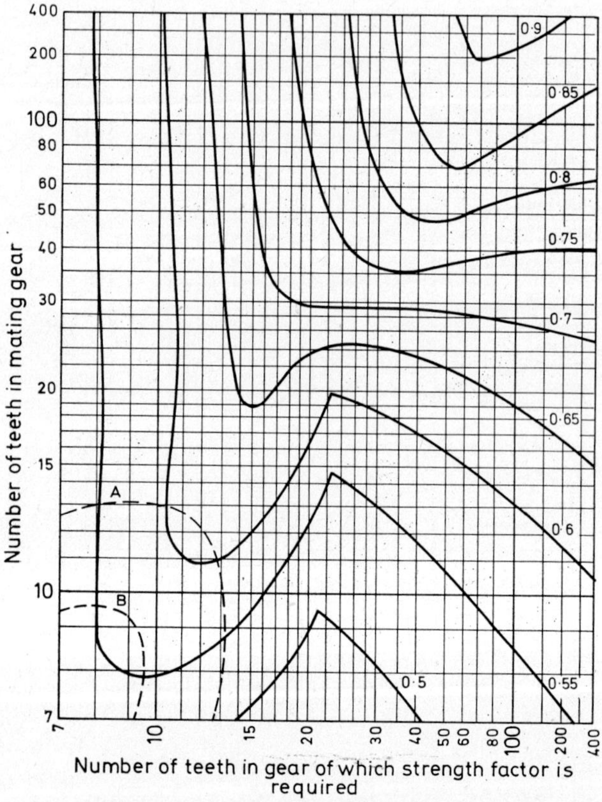

Fig. B. 6. Strength factor (Y) for bevel gears (Reproduced from B.S. 545 by kind permission of the British Standards Institution)

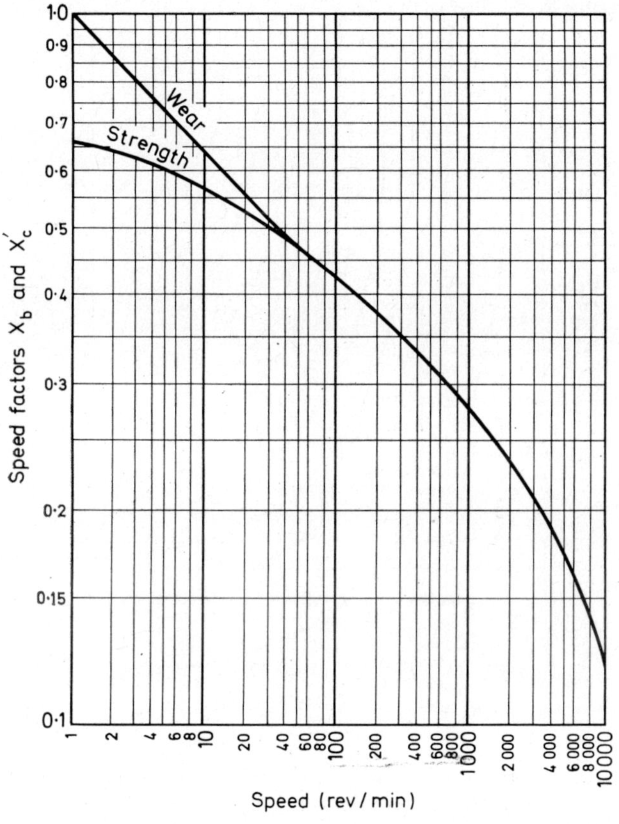

Fig. B. 7. Speed factors for strength (X_b) and wear (X_c) for bevel gears (Reproduced from B.S. 545 by kind permission of the British Standards Institution)

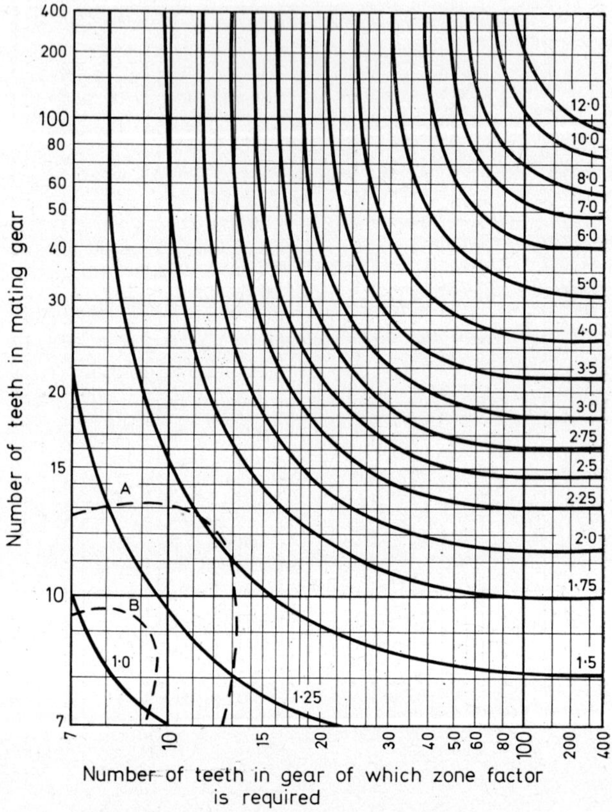

Fig. B. 8. Zone factor (Z) for bevel gears (Reproduced from B.S. 545 by kind permission of the British Standards Institution)

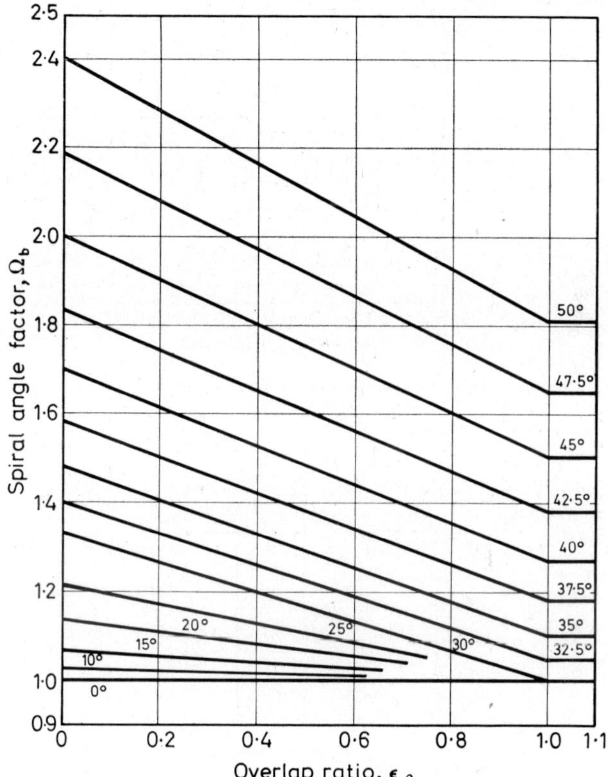

Fig. B. 9. Spiral angle factor for strength (Reproduced from B.S. 545 by kind permission of the British Standards Institution)

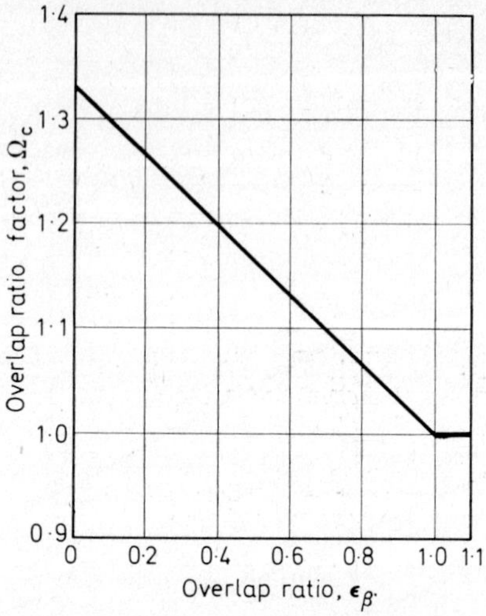

Fig. B. 10. Overlap ratio factor (Reproduced from B.S. 545 by kind permission of the British Standards Institution)

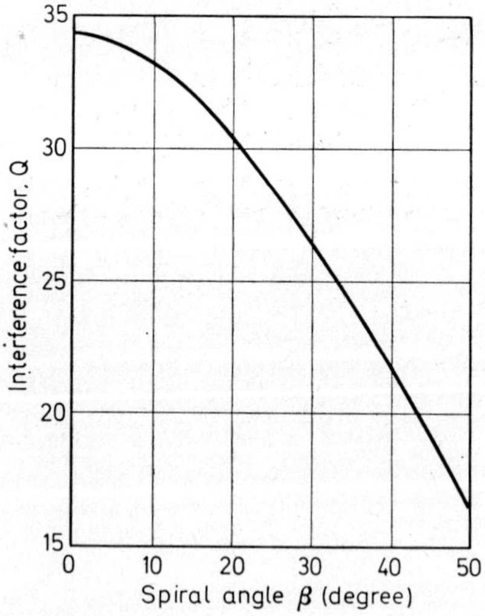

Fig. B. 11. Interference factor (Q) for bevel gears (Reproduced from B.S. 545 by kind permission of the British Standards Institution)

Index

Aluminium-base alloys 142
Amplitude of vibration 16–17, 20
Area-moment method 37

Babbitts metals 141
Ball bearings. *See* under Bearings, rolling contact
Base circle 205
Beam, cantilever 35
  continuous 44
  transverse vibrations 34–44
Bearing characteristic number 105
Bearing reactions, calculation of 85
Bearings 6
  anti-friction 143
  axial thrust 239
  choice of 177
  classification of 100–101
  effect of 32
  journal 101–115
    design examples 131–140
    elements of 102
    heat dissipated from 111–114, 130–131, 133, 135
    heat generated in 113, 133
    methodical procedure for design of 130–131
    performance assessment 119, 136–140
    radial and axial pressure distribution 103
    Raimondi—Boyd design and analysis method 119–130, 136–140
    thermal equilibrium 110–115
  plain, externally pressurised 143
    materials selection 140–143
    maximum bearing pressures 142
  release 303
  rolling contact 101, 143–176
    advantages 177
    axial location of 169–173
    ball 145–148
    dynamic capacity 160
    fillets 172, 176
    life of 159–160
    load carrying capacity 158–169, 175
    lubrication 174
    main types 145
    roller 145, 148–151
    selection 152–158
    shaft and housing fits 171
    shoulder height 173
    standard units 152–158
  size determination 87
  sliding 101, 177
  externally pressurised 177
  throw-out 303
  thrust 100

Index

Bearings *continued*
tilting pad 100
types 6
Belt creep 184
Belt drives 5
shaft loading due to 201–203
Belt failure 183
Belt length 184–185
Belt materials 185
Belt slip 182, 184, 189
Belt speed 184
Belts, flat 181–186
timing. *See* Belts, toothed
toothed 180, 194–196
V- 187–194
advantages over flat belts 189
categories of 191–192
design details 191–194
dimensions of cross-section 189
extra high power 190
initial tension of belt 192
narrow or wedge 190
operating principle 187
standard sizes 190
Bending moment 41, 43, 86
Bending moment diagram 53, 58, 85
Bending strength factor 219
Bending stress 46, 58
Bending stress equation 58
Bending theory 216
Bevel gear factor 251, 252
Bevel gears. *See* Gears
Bibby coupling 81
Bolts in couplings 77–78
Brakes 3
band 306–310
differential 310
force analysis 307–310
reversible or equal acting 308
torque 308
design precautions 314
disk 300–306
force analysis 304–306
plate type 303–304
spot type 303–304
torque capacity 305
energy dissipation 310–313
function 287
reactions at support 299
rim 289–300
force analysis 294–300
self-energising 297–298, 300
shoe type 294
torque capacity 298
types 288–289
Brinell Hardness numbers 220
Buckingham wear equation 219, 226
251, 266, 268

Chain drives 5, 196–203
chain types 197–198
shaft loading due to 201–203
Clutches 3
centrifugal 291
cone 289
design precautions 314
disc 300–306
basic principles 302
force analysis 304–306
torque capacity 305
energy dissipation 310–313
friction 291
function 287
rim 289–300
force analysis 294–300
shoe type 294
types 288–289
Complementary function 15, 16
Contact ratio 234
Copper-base alloys 141
Couplings 2, 52
Bibby 81
bolts in 77–78
cushion type flexible 82
design of 76–82
flange 78
flexible 79–82, 85
flexible pin 79
flexible spring 80
gear 81
rigid 77, 85
selection of 76
Cramer's rule 27
Critical speed 44–48

Damping, critical 15

Damping *continued*
element of mathematical model 11
equivalent viscous 12
heavy 15
light 15
viscous 11
Damping coefficient 11, 24
Damping ratio 16, 18
Deflection 20, 54
equation for 41
permissible 66–67
rule of thumb 66
shear 44
transverse 64
Degrees of freedom 13–14
multiple 21, 25, 28
three 22, 25
Design assumptions 64
Design decisions 1, 64
Diameter factor 262, 263, 269, 355
Diameter quotient 262
Direct stress 59
Disk slope 43, 46–48
Displacements response curves 18
Dissipative energy, equation for 23
Distortion energy theory 59, 61
Dunkerley's method 38–39
Dynamic capacity 160, 166

Eccentricity ratio 105–106, 107, 132
End load 44
Energy dissipation in clutches and brakes 310–313
Equations of motion 14, 21, 23, 28, 35
Equivalent damage concept 59
Equivalent second moment of area 65–66
Equivalent viscous damping 12

Face contact ratio 258
Failure theory 59, 61
Fatigue 49, 59, 72, 183
Fatigue failure 46
Fatigue life 159
Fatigue strength 217
and shrink fits 75–76
Fatigues trength reduction factor 75–76
Flexibility equations 25, 43
Flexibility influence coefficients 25, 37
Flexible couplings. *See* Couplings
Flexible power transmission elements 5, 180–203
Flexible shaft 5
Flow ratio 139
Flywheels 7
Force analysis, band brakes 307–310
disc clutches and brakes 304–306
rim blutches and brakes 294–300
Force transmissibility 18
Form factor, Lewis 216, 224
worm gears 266
Formative gearset 224–225, 229, 232
Fraction of critical damping 16
Free body diagram 40
Frequency, fundamental 37–39
of vibration 13
Frequency equation 26, 43
Fretting corrosion 76
Friction, between moving parts 6
dry or Coulomb 93, 95
generated in journal bearing 103
Friction coefficient 104, 108, 109, 116, 131, 132, 188, 265
Friction energy in clutches and brakes 310–313
Friction equation, Ocvirk's 108
Frictional force 104, 188
Frictional resistance 93
Frictional torque 104
Fundamental frequency 37–39

Gear blanks 207–208
Gear case, design requirements 279
Gear couplings 81
Gear cutters 208
standard 230, 276
Gear forces, helical gears 221
spiral bevel gears 253
spur gears 213
straight bevel gears 246
worm gears 263
Gear teeth, allowable tangential load per unit length of face width 219, 220

Gear *continued*
dynamic load 218
forces on 215
strength of 215
Gears 3–4, 204–286
advantages of 4
bevel 239–260, 348, 349, 350, 352
applications 239
dimensions 241
load carrying capacity 258
maximum face width 258
nomenclature 241
reduction ratio 259–260
spiral 240
spiral, gear forces 253
spiral, overlap ratio 258–259
spiral bevel, strength requirements 255
spiral bevel, wear requirements 256
straight 239
straight, gear forces 246
straight, strength requirements 248
straight, wear requirement 251–252
terminology and symbols 243–245
tooth modifications 256–258
crossed-helical 235–239, 282
application 235
design 237
face width 239
helix angles 237
pitch circle diameter 238
shaft centre distance 238
cycloid tooth form 204
design data sheets 339
design requirements 205
design stress 217
disadvantages 4
failure by wear 219
helical 205–213, 345, 347
aixal thrust in 221–223
double-helical 206, 223
face contact ratio 235
face width 232–233
gear forces 221
handing of helix 222
single 206
standard tooth forms 213
strength forces 221
terminology 209
undercutting 227–232
wear requirements 226
see also Gears, crossed-helical
herringbone 207, 223
housing arrangements 279
hypoid 282
industrial units 282
intermediate 282
internal 282
lubrication 277–279
manufacturing procedures 271–276
material properties 339
materials selection 270
maximum pinion bore 233
modification 228
mounting requirements 276
reduction ratios 235
spur 205–213, 345, 346
face width 232–233
gear forces 213
standard tooth forms 213
straight spur 205
strength requirements 213
undercutting 227–232
wear requirements 213
terminology and symbols 209
tooth forms 204
cycloid 204
involute 204, 209, 213
standard 213
types 4
worm 260–270, 282, 284, 353, 354
application 260
axial pitch 261
dimensions 261
efficiency of 265–266
elements of 260
face width 269–270
form factors 266
gear forces 263
helix angle 261
lead angle 262, 265
lead of thread 262

Gears *continued*
life correction factor 269
load carrying capacity 266
modification of 269
reference diameter 262
rigidity requirements 269–270
strength requirements 266
terminology 261–262
undercutting 269
wear requirements 268
Geometrical constants 60
Gleason Works system 257
Goodman relationship 59, 61
Grease lubrication 174
Gun-metal 142
Gyroscopic couple 47–48

Heat dissipation, clutches and brakes 310–313
journal bearing 111–114, 131, 133, 135
Heat generation in journal bearing 110, 113, 133
Heat transfer coefficient 111, 112, 312 313
Helical gears. *See* Gears
Hertzian stress 160, 219, 237
Holzer method 28–30, 32
Hypoid gears 282

Impact factors 50
Impact loading 49–50
Internia coefficient 24
Influence coefficient 25–27, 37, 41
Interference factor 257–258, 352
Involute 205
see also Gears, tooth forms
Involute functions 229
Involute interference in gears 227–232

Jockey pulley 193–194

Keys (and keyways) 68–74
failure of 70
gibheaded 70
length of 70, 71
parallel 69
standards 68, 316
strength requirements 71–74
stress concentration factors 74
tangential 70
taper 70
types of 69
Woodruff 70
Kinetic energy, equation for 23

Lag, angle of 15
Lagrange equation 22–25
Lasche's equation 112, 131, 133
Lead angle of worm gear 262, 265
Lead-bronze 141
Lewis assumption 215
Lewis equation 215, 216
modified 223–225, 248–249, 255, 266
Lewis form factor 216, 224
Load equation, Ocvirk's 106
Load number 106, 116, 132, 134
Loading diagram 85
Loading ratio 161
tables of 161
Lubricant 6, 94
selection of 133, 138
viscosity. *See* Viscosity
Lubrication 6, 94–96
boundary 95, 101
fluid, mechanism of 96–100
grease 174
hydrodynamic 97–102, 107, 108, 140, 177, 277–278
hydrostatic 143
of gears 277–279
of rolling contact bearings 174
perfect 97
thick film 95, 97, 101
thin film 101

McKee equation 109, 116, 131, 132
Magnification factor 17–18, 68
Mass, element of mathematical model 11

Index

Mass density 117, 119
units 118
Material and form factor 220, 252
Material constants 60
Mathematical model 11, 21, 30–31, 36, 40, 49
Maximum shear strain energy theory 59
Maximum shear stress theory 59, 61
Maxwell's reciprocal theorem 26
Mechanisms 7
synthesis of 8
Mises criterion 59, 61, 89
Modification, of gear teeth 228
of worm fears 269
Modification factors 230
Multi-rotor system 30
Myklestad–Prohl method 40, 48

Natural frequency 12, 20, 67–68
calculation of 28, 37, 40, 49
Newton–Raphson iteration method 27
Newton's law of cooling 111
Newton's law of viscous flow 96, 98, 103
Newton's second law 14, 21, 23–24, 28, 35, 45

Ocvirk's friction equation 108
Ocvirk's load equation 106
Overlap ratio 256, 258–259
Overlap ratio factor 352

Particular integral 15
Period of vibration 12, 13
Petroff's law 105, 108, 109
Phosphor-bronze 142
Pinion addendum coefficient 257
Pitch cones 241–242
Plastic materials 143
Porous metal bearings 143
Potential energy, equation for 23
Power transmisson elements, flexible 5, 180–203
Power transmission methods 1–9

Press-fit fatigue strength 76
Profile contact ratio 225, 234
Prohl method. *See* Myklestad–Prohl method
Pulleys, flat belt 186
jockey 193–194

Radial factor 161
Raimondi–Boyd method for journal bearings 119–130, 136–140
Ratio factor 220, 252
Rayleigh (or Rayleigh–Ritz) method 37–38
Rectilinear and torsional systems, comparison between 32
Reference diameter 262
Reserve factor 54, 61, 71, 90
Resonance 17, 20, 46, 49, 67
Response time 50
Reynolds equation 99
solutions to 105–106
Rigid couplings. *See* Couplings
Rigidity, shaft. *See* Shafts
Roller bearings. *See* under Bearings, rollings contact
Rope drives 5
Rotation factor 161
Rotors, multi-rotor system 30

Seals, bearing 157
Seizure 94
Serrations 74
standards 316
Service factor 202–203
Shaft couplings. *See* Couplings
Shafts, critical section 54
design of 2, 52
example 83–90
methodical procedure for 53–55
requirements 52
diameter for 54, 89
effect of shrink fits on fatigue strength 75–76
flexible 5
loading 53
due to belt and chain drives 201–203

Shafts *continued*
mass effect of 32
material selection 54, 56, 61–63, 87
rigidity 54, 61, 64–67
stepped 31, 33, 34, 64
strength considerations 56–61
stresses in 56
tapered and stepped 34
tapered section 33
vibration characteristics 55, 67–68
see also Keys; Splines
Shear deflection 44
Shear force 41
Shear stress 58
Shock loading. *See* Impact loading
Shrink fits and fatigue strength 75
Slope, equation for 41
Soderberg relationship 61
Sommerfeld equation 137
Sommerfeld number 105, 130, 137, 138
Sommerfeld's solution to Reynolds equation 105
Specific gravity 119
Specific heat capacity 114–115
Speed factor 219, 221, 345, 346, 349, 353, 354
Spiral angle factor 256, 351
Spiral gears. *See* Gears
Splines 74–75
standards 316
Spring suspension system 19–20
Spur gears. *See* Gears
Static deflection curve 37, 38
Steady state vibration 16
Stiffness coefficient 24
Strain energy 11
Strategies 1
Strength factor 219, 226, 345, 348
Stress concentration factor, keyways 74
Sudden loading. *See* Impact loading
Surface strength factor 221
Suspension system, spring 19–20

Teflon 143
Thermal equilibrium of journal bearings 110–115
Thrust factor 161, 167
Torque, frictional 104
Torsion 56
Torsional and rectilinear systems, comparison between 32
Torsional equations 32–33
Torsional shear stress 56–58
Torsional stiffness 31
Torsional vibrations 30–34
Transmissibility 18–19, 20–21
Transmissibility curves 19
Transverse vibrations 34–44
Tribology 94
Twisting moment 56–57, 86
Twisting moment diagrams 85

Undercutting, gears 227–232
worm gears 269

Value analysis 55
Value engineering 55
Velocity factor 217, 218
Vibration 10–51
amplitude of 16–17, 20
damped 11
designing against 49
forced 11, 12
forced-damped 14
free 11
basic equations 12
frequency of 13
natural 12
of shafts. *See* Shafts
period of 12, 13
steady state 16
torsional 30–34
transverse 34–44
see also Natural frequencies
Vibration isolators 18–21, 49
Vibratory forces 49
Vibratory motion 10, 48
Viscosity 95–96, 115–119
absolute or dynamic 115, 117, 132, 133

Viscosity *continued*
determination of 117
equation for 137
function of temperature 119
kinematic 117
units 118
units 115
Viscosity-temperature chart 119

Wear 94–96
Buckingham equation 219, 226, 251, 266, 268
classification of processes 94–95
phenonomenon of 94
Wear requirements, helical gears 226
spiral bevel gears 256
spur gears 213
straight bevel gears 246
worm gears 268
Whirling and whirling speed 44–48
White metals 141
Wire rope 5
Worm gears. *See* Gears

Zone factor 221, 226, 346, 347, 350